The Pyrethroid Insecticides

The Pyrethroid Insecticides

Edited by
John P. Leahey

ICI Plant Protection Division, Jealot's Hill, Berkshire

Taylor & Francis
London and Philadelphia
1985

UK	Taylor & Francis Ltd, 4 John St., London WC1N 2ET
USA	Taylor & Francis Inc., 242 Cherry St., Philadelphia, PA 19106-1906

British Library Cataloguing in Publication Data

The Pyrethroid insecticides.
   1. Pyrethrum (Insecticide) — Environmental aspects
   I. Leahey, John P.
   632′.951   TD 196.P9
ISBN 0–85066–283–4

Library of Congress Cataloging in Publication Data is available

Cover design by S. Robertson
Typeset by Activity Ltd, Salisbury, Wilts
Printed in Great Britain by Redwood Burn Limited,
Trowbridge, Wiltshire.

Contents

Preface

At the time I started to plan this book the Woody Allen film "All you ever wanted to know about sex but never dared to ask" was being shown on British television. The title seemed somewhat ambitious, but it nevertheless encouraged me to view this film. The film gave me a number of ideas, but the only one relevant to this preface was that the title for this book should perhaps be "All you ever wanted to know about pyrethroids but never had the time to look up all the available literature". Again such a title did seem to be a little ambitious considering the wealth of information which has been published on these compounds, especially since the discovery of pyrethroids with potential in agriculture. Nevertheless, I still decided to set myself and my co-authors the task of achieving such a target, at least within the six specific areas we have covered in this book. One frustrating aspect of attempting such a target is that even when one believes that all essential information has been covered and the final typescript has been sent to the publisher, the continuous flow of pyrethroid data into the literature means that the comprehensive survey is already out of date. My postulated title should therefore bear a disclaimer for all information published later than the beginning of 1984.

Having decided on the aspects of the pyrethroids to be covered in this book, my next task was to find authors willing and able to take up the challenge. My stars were obviously in the right configuration at the time I was looking for these authors, since I believe that I did manage to find the right people. It certainly has been a pleasure to work with my co-authors on this book, even though it was far more work than I originally expected.

Once I had finished editing the final chapter and sent it to the publishers I was able to sit back and view the book as a total entity. I had already realized that an enormous amount of effort has been devoted to the study of this group of pesticides, but I had not quite realised the extent and detail which this work has covered. Even more surprising than the amount of work which has been completed was the fact that a high proportion of this work has been published in the open literature. Undoubtedly additional information still remains unpublished; this will mainly be work carried out by the pesticide industry, where competition between companies means that not all data will reach the open literature. Nevertheless, all data, published or unpublished, on hazard evaluation will have been assessed by the world's registration authorities. However, the published data alone makes it clear that the pyrethroids are a class of insecticides of great importance for the protection of man's crops and his health. The large differential toxicity to mammals and insects means that these compounds can be used safely by man to generate very unsafe conditions for insects. Pyrethroids also have very many of the properties required for a minimal deleterious effect on the environment, i.e., ready degradation in soil, virtually zero mobility in soil and rapid metabolism and excretion by animals. No insecticide is perfect, but the pyrethroids do "come close". Such a claim obviously requires justification, and it is the literature summarized and discussed in this book which, I believe, provides this justification. Those interested should find that this book allows them to evaluate this claim and I hope to agree with it.

John Leahey
Bracknell
December 1984

Contributors

Dr J H Davies
Organic Chemistry Division
Shell Biosciences Laboratory
Shell Research Ltd
Sittingbourne, Kent
England

Dr Davies is an organic chemist who has been engaged in the synthesis of chemicals with potential as agricultural pesticides for many years. During most of his career he has specialized in insecticide chemistry and is currently head of the insecticide chemistry group in the Organic Chemistry Division of Shell Research Ltd.

Professor T A Miller
University of California
Riverside, CA
USA

Professor Miller has had a strong interest in insect neurophysiology for many years. He has pioneered a number of techniques for the study of insects and has used these techniques to make major advances in the understanding of the mode of action of pyrethroid and other insecticides. Professor Miller continues to make fundamental contributions in the areas of insect physiology, insecticide mode of action and insect resistance.

Dr V L Salgado
Rohm and Haas Research Laboratories
727 Norriston Road
Spring House
Philadelphia, PA 19477
USA

Dr Salgado is an insect physiologist and biochemist who, both as a PhD student and research associate, has been actively involved in the study of pyrethroids for many years. His researches have produced major contributions to the understanding of the toxic action of these compounds and also to the difference between type 1 and type 2 pyrethroids. Dr Salgado is currently working with the exploratory insecticides group of the Rohm and Haas company.

Dr M H Litchfield
ICI Plant Protection Division
Fernhurst
Haslemere, Surrey
England

Dr Litchfield worked for 20 years at the ICI Central Toxicology Laboratory (CTL), where he led research into the toxicology of pyrethroid insecticides along with many other agricultural and industrial chemicals. He is a former chairman of the British Agrochemical Association Toxicology Committee and is currently chairman of the GIFAP Toxicology Committee. Dr Litchfield recently moved from CTL to join ICI's Plant Protection Division.

Dr I R Hill
ICI Plant Protection Division
Jealott's Hill Research Station
Bracknell, Berkshire
England

Dr Hill is a senior research scientist in the environmental sciences group at Jealott's Hill Research Station, where he has conducted extensive studies over the past nine years on the environmental fate and effects of the pyrethroid insecticides and many other pesticides. He is on the editorial staff of the Journal of Applied Bacteriology and has edited and contributed to a book entitled *Pesticide Microbiology*.

Dr J P Leahey
ICI Plant Protection Division
Jealott's Hill Research Station
Bracknell, Berkshire
England

Dr Leahey has been involved in the study of the metabolism and environmental degradation of agricultural pesticides for many years. He is currently a senior scientist at Jealott's Hill Research Station where he leads a team which has been working on a variety of pyrethroid insecticides for over nine years.

Dr J J Hervé
Societe Procida/Roussel-Uclaf
Centre de Recherches de Biologie Appliquées
Saint-Marcel
13367 Marseilles Cedex 11
France

Dr Hervé began his career as a plant physiologist at the French National Research Institute. After 15 years he moved to Roussel-Uclaf, where he has been heavily involved in the development of pyrethroids as agricultural, veterinary and public health insecticides. Dr Hervé is currently the director of agroveterinary development and registration at Roussel-Uclaf.

Acknowledgement

I should also like to acknowledge the invaluable help of my colleague, Dr M. D. Collins, who contributed much time and effort to help me edit sections of the book where my knowledge and expertise were minimal or even non-existent.

1. The pyrethroids: an historical introduction

J. H. Davies

Introduction

Many excellent reviews have already been written on the subject of the pyrethroids, and the purpose of this survey is not to add gratuitously to their number, but to provide an easily accessible historical account of pyrethroid research over the last 70 years as a background to the more specialised chapters that follow, and to give some account of very recent developments. Inevitably I have made great use of earlier accounts.

I have divided this review into four parts, that follow an approximately chronological order: a brief description of pyrethrum and the elucidation the the structure of its active components (the pyrethrins); a description of the synthesis of synthetic analogues (pyrethroids) between 1924 and the early 1970s, from which a number of synthetic substitutes for the pyrethrins emerged; an account of the work between 1970 and about 1977, which resulted in the introduction of light-stable pyrethroids to agricultural use; and a section on more recent developments. There is nothing fundamental about such a division as each of these activities forms part of a continuum. Historical reviews often suffer from the weakness that they preserve historical continuity at the expense of scientific coherence. For this reason only the first two sections are treated chronologically – the work discussed has been thoroughly reviewed elsewhere – whilst the section on more recent developments is organised according to chemical structure and source.

The picture that I present is inevitably greatly oversimplified. Investigations such as those described in the following pages involve a great deal of apparently fruitless labour, and include periods when understanding appears to decrease rather than increase. A glance at the list of patents given in a recent book on pyrethroid synthesis (Naumann, 1981) will give some idea of the amount of effort that has been expended on this endeavour.

1

These accounts of the progress of pyrethroid synthesis are preceded by an attempt at a working definition of a pyrethroid and a brief explanation of the stereochemical principles involved. The latter is aimed at those to whom chemistry is a peripheral subject, and those who are familiar with the subject will find it superfluous. Before the final section, on recent developments, I have interpolated an outline of the relation between structure and activity in the pyrethroids and a short discussion of the spectrum of activity displayed by pyrethroids in an attempt to illuminate the significance of this recent work.

I have not discussed the synthesis of pyrethroids as a great deal of space would be required to do the subject justice, and the topic has been thoroughly reviewed (Naumann, 1981; Arlt et al., 1981; Tessier, 1982). Photochemistry is relevant to this review only insofar as compounds are light-stable or not, but the topic is fully covered in chapter 6. Metabolic chemistry is mentioned only where it is implicated in the relationship between structure and activity, and this topic also is covered in full in chapter 6.

Patent references

In a review of this kind it is inevitable that many references will be made to the patent literature because publication of information in the open literature relating to compounds of possible commercial importance is often considerably delayed. Patent references serve both to indicate the source of the work and to give an approximate indication of chronology. For this latter purpose the year in which the patent applications were made is included in brackets after the year of publication. For instance, a reference to a Sumitomo patent published in 1970, based on applications made in 1967 and 1968, would appear in the text as "(Sumitomo, 1970[1967/8])".

Nomenclature

Where full names are required I have used IUPAC nomenclature. For the purposes of discussion I have used less cumbersome abbreviations and trivial systems of nomenclature. For instance, the fenvalerate series is treated as a series of 2-substituted phenylacetic acid derivatives, as this emphasises the structural relationships.

Common chemical names are used where possible. These are contrived names that reflect some structural features of the molecule. Before coming into general use they must be approved by the appropriate national body (the British Standards Institute in the UK), and the International Standardisation Organisation. Their function is to provide a convenient alternative to cumbersome chemical names which also avoids the use of proprietary names. They are sometimes uncouth, and occasionally risible, but they serve a useful purpose. The task of obtaining an approved common chemical name is usually undertaken only if development is, at least, contemplated. Possession of such a

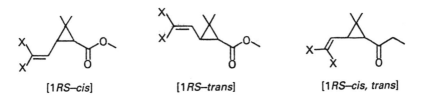

[1*RS–cis*]   [1*RS–trans*]   [1*RS–cis, trans*]

Figure 1.1. Conventions employed for the representation of racemic and partly resolved cyclopropanecarboxylic acid derivatives

name is, of course, no guarantee that a compound has been or will be, commercialised.

Representation of individual isomers by perspective drawings, and of racemic mixtures by two dimensional drawings, is a well understood convention, but representation of intermediate degrees of resolution requires the definition of appropriate conventions. Accordingly *cis-*, *trans-*, and *cis,trans*-cyclopropanecarboxylic acids will be represented as shown in Figure 1.1. Structures are numbered in arabic numerals and named, if possible, with the common chemical names. Where appropriate, the elements of stereochemistry are indicated in brackets in the following order: geometry of the double bond of the 3-substituent in the cyclopropane ring; configuration at the 1- and 3-positions in the cyclopropane ring, or at the 2-position for the phenylacetate series; configuration at the hydroxyl-bearing carbon of the alcohol component. The absence of any indication of stereochemistry denotes a racemic mixture.

I have used modern stereochemical notation throughout, as there seems to be no reason for the reader to labour under the difficulties suffered by the early investigators.

A definition

We could probably do very well without a definition of a pyrethroid, for compounds become pyrethroids in the same way that men become members of some clubs, by a kind of mutual agreement that no-one will object. But it is customary to make such an attempt, even though the task becomes more difficult as time passes.

Pyrethroids have been arrived at by synthesis based on natural pyrethrins as models, generally by systematic variation of parts of the molecule for the purpose of determining the effect on biological activity. A few jumps springing from inspiration or from the availability of materials may have been made, but by and large most pyrethroids may be recognised as having been derived from the constituent esters of pyrethrum, perhaps in several stages, even if a direct comparison between the starting point and the end product shows quite startling differences. It is also true that pyrethroids exhibit a fairly well defined range of biological properties, although compounds at opposite ends of this spectrum will differ markedly.

One reason for this difficulty of definition may be that structure/activity relationships in pyrethroids are based mainly on considerations of shape and stereochemistry rather than on electronic properties, so that they cannot be defined straightforwardly in terms of specific chemical groups. What seem to be hard-and-fast rules in this respect tend to melt away. The only definition that is likely to last therefore is one based on evolution. A compound may be said to be a pyrethroid if its structure can be reasonably derived from that of the natural pyrethrins, and if it exhibits a range of biological properties that overlap to a considerable degree with those of existing members of the group.

Stereochemistry: general observations

This section is intended as a brief guide to the principles and vocabulary of stereochemistry for those who are not familiar with the subject.

The four groups attached to a saturated carbon atom are arranged in space as though they were at the points of a regular tetrahedron (Figure 1.2). It is a geometrical fact that four different objects placed at the points of a regular tetrahedron can be arranged in two different ways. These arrangements are not superimposable; they are mirror images (enantiomers). Interchange of any two groups always produces the alternative arrangement. It follows that any molecule containing a carbon atom to which are attached four different chemical groups (an asymmetric carbon atom) may exist in two forms that are mirror images, differing only in the disposition of the four dissimilar groups about the asymmetric carbon atom. Interchange of any two groups will produce the enantiomer.

If a compound contains two asymmetric carbon atoms, then each can exist in two forms independently, and the total number of arrangements must be four (2^2). If there are n asymmetric carbon atoms, the total number† of possibilities is 2^n.

If a compound contains more than one asymmetric carbon atom, then the various possible forms (isomers) may be related to one another in two general ways. If all the asymmetric centres in one isomer are mirror images of the corresponding centres in the other, then the isomers are mirror images or enantiomers. If some of the asymmetric centres are of the same configuration in the two isomers and others are different, the compounds are diastereoisomers. Enantiomers have the same physical properties, for example, solubility and melting point, and cannot be separated except by using reagents that are

---

† There are exceptions to this generalisation in the case of compounds possessing special symmetry properties but they have no bearing on this discussion.

themselves asymmetric. Diastereoisomers have different, sometimes very different, physical properties and may often be quite readily separated.

Compounds that possess asymmetric carbon atoms made by routes that do not involve asymmetric reagents are obtained as racemic mixtures; that is, mixtures in which each isomer is accompanied by an exactly equal amount of its enantiomer.

In order to distinguish one possible arrangement of dissimilar chemical groups about a saturated carbon atom from the other, a set of rules known as the CIP rules (Cahn et al., 1956) are used. The four groups are arranged in an order of priority based on atomic number, and the carbon atom is viewed from the side opposite to that occupied by the group of lowest priority. In the cases considered here this is often a hydrogen atom. If the three groups (excluding that of lowest priority) appear to be arranged clockwise in order of descending priority, the arrangement is described as *R*; if anticlockwise, then as *S*. This is illustrated in Figure 1.2 in three different ways, using regular tetrahedra, the conventional three-dimensional representation of a saturated carbon atom, and the 'steering wheel' model used by Cahn, Ingold, and Prelog.

Fenvalerate is properly named (*RS*)-α-cyano-3-phenoxybenzyl (*RS*)-2-(4-chlorophenyl)-3-methylbutyrate. The prefix (*RS*) indicates that the fragment that it precedes is present in both *R* and *S* forms. Since both the acid and the alcohol have one asymmetric centre, the compound is a mixture of four isomers. The biologically most active isomer of fenvalerate is (*S*)-cyano-3-phenoxybenzyl (*S*)-2-(4-chlorophenyl)-3-methylbutyrate (Figure 1.3), sometimes known as fenvalerate A alpha. It may be seen that if the asymmetric carbon atoms are viewed from the side opposite to the hydrogen atom, then the groups are arranged anticlockwise in descending order of priority *1-2-3* for both asymmetric carbon atoms, and both arrangements are therefore designated *S*.

This set of rules gives unambiguous descriptions of stereochemistry, but is not always easy to use quickly, and can conceal stereochemical relationships because the description is dependent on an order of priority, which may change when one group is replaced by another. For this reason a simplified version is often adopted to describe cyclopropanecarboxylates which eliminates this problem (Elliott et al., 1974c). An example will make this clear.

Deltamethrin (Figure 1.4), (*S*)-α-cyano-3-phenoxybenzyl (1*R*,3*R*)-3-(2,2-dibromovinyl)-2,2-dimethylcyclopropanecarboxylate, is the single most active isomer of the eight (2^3) possible. The cyclopropane ring is in the plane of the paper and numbered as shown. The configurations shown are 1*R*, 3*R* and alpha *S*. If the bromine atoms are replaced by methyl groups, to give an isomer of chrysanthemic acid, the description of the stereochemistry at C-3 changes to *S* because the priority of the C-3 substituent changes. Confusion is removed by defining the stereochemistry of the 1-position in absolute terms, since this does not change in the series of compounds that interests us, and defining the stereochemistry of the 3-position relative to it, that is, as *cis* or

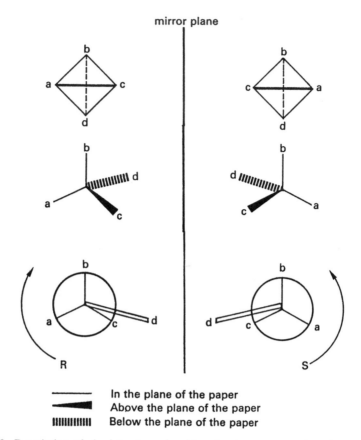

Figure 1.2. Description of absolute stereochemistry about an asymmetric carbon atom. The order of priority is a > b > c > d

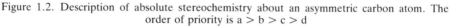

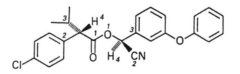

Figure 1.3. Absolute stereochemistry of the active isomer of fenvalerate, (*S*)-α-cyano-3-phenoxybenzyl (*S*)-3-methyl-2-(4-chlorophenyl) butyrate. Priority of substituent groups decreases in the order *1–4*

Deltamethrin acid is therefore (1*R*)-*cis*-3-(2,2-dibromovinyl)-2,2-dimethylcyclopropanecarboxylic acid, and whatever replaces the dibromovinyl group at C-3, the stereochemical description will remain unchanged.

In reading the older literature on pyrethroids and pyrethrins it is necessary to remember that before 1951 it was not possible to determine the absolute

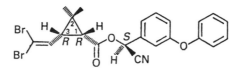

Figure 1.4. Absolute configuration of deltamethrin, (*S*-α-cyano-3-phenoxybenzyl (1*R*,3*R*)-*cis*-3-(2,2-dibromovinyl)-2,2-dimethylcyclopropanecarboxylate

configuration of any organic compound. Up to that date individual enantiomers were distinguished by the direction in which they rotated plane polarised light, a property that could not, and cannot be related to absolute configuration. In some cases the configuration of compounds could be related, by a series of chemical transformations, to either of the two enantiometric forms of glyceraldehyde, to which the two possible absolute configurations were arbitrarily assigned, but these were relative configurations, and had a 50 per cent chance of being correct. Nomenclature, particularly where more than one asymmetric centre was involved, was therefore confusing. The determination of the absolute structure of a derivative of (+)-tartaric acid by X-ray analysis permitted the assignment of absolute configurations thereafter. It so happens that the configurations arbitrarily assigned to the enantiomers of glyceraldehyde were correct and no reversals of previously assigned configurations were necessary.

Pyrethrum

History and commercial development

The terms "pyrethrum" and "insect powder" refer to the dried and powdered flower heads of *Chrysanthemum cinerariaefolium*. This is now the only commercially important species, although others have historical importance. Solvent extraction of the flower heads yields "pyrethrum extract". The active constituents of the extract are collectively referred to as "pyrethrins", which can give rise to confusion as two specific constituents are known as pyrethrin I and pyrethrin II.

The time and place of the discovery of the insecticidal activity of pyrethrum are unknown. It is likely that it was discovered at least twice, in the Caucasus–Iran region of Asia, that region between the Black and Caspian seas, and in Dalmatia, now part of the Adriatic coast of Yugoslavia, where *C. cinerariaefolium* is a native plant. One author states that its use was known to the Chinese in the first century A.D. (Lhoste, 1964).

Insect powder was introduced into a number of European countries from both of the first two regions mentioned above in the second and third quarters of the nineteenth century, although it was in use in Russia well before that period. The plant was introduced to Japan in 1885 (Katsuda, 1982).

There are a number of accounts, stories, and traditions concerning the discovery of the useful properties of pyrethrum flowers and these have been drawn together in an early USDA bulletin (McDonnell et al., 1920 [revised 1926]). This review provides an extensive bibliography, largely of nineteenth century writings, and seems to be the source of most assertions to be found in later reviews concerning the very early history of pyrethrum. In addition to the early history the review covers the importing of insect powder into Europe and the USA and the history of the cultivation of the plant in various countries, including Japan and the USA, up to about 1918. The subsequent history of the commercial development of pyrethrum, especially in Eastern Africa, has been well documented (Gnadinger, 1936, 1945; McLaughlin, 1973; Moore & Levy, 1975) and need not be set out again here. Pyrethrum extract remains an important article of commerce and pyrethrins are still, by modern standards, effective domestic insecticides that possess very low mammalian toxicity. In the context of this introduction the pyrethrins are of interest mainly because of their function as models for the synthetic analogues (pyrethroids), insecticides of even greater efficacy and much wider application than their naturally occurring predecessors.

The chemical structure of the pyrethrins

The elucidation of the structures of the pyrethrins was a process which lasted for a period of about 60 years, from the first correct surmise, that they were esters (Fujitani, 1909), to the final settling of the absolute stereochemistry. This achievement has been exhaustively reviewed (Crombie & Elliott, 1961; Elliott & Janes, 1973; Crombie, 1980). The following brief account will serve to introduce the structures of the active principles and to give some idea of the time scale of the investigation and the number of contributors to the undertaking.

There are six active constituents of pyrethrum extract (Figure 1.5). These are esters of two carboxylic acids, chrysanthemic acid (1) and pyrethric acid (2), and three cyclopentenolones (known collectively as rethrins), pyrethrolone (3), cinerolone (4), and jasmolone (5). In naming these six esters the alcohol component is distinguished by name and the acid by number. Thus esters of chrysanthemic acid are distinguished by a roman I, and esters of pyrethric acid by a roman II. The name given to a particular ester depends upon the alcohol portion. Esters of pyrethrolone, for instance, are known as pyrethrins. Thus the ester of pyrethrolone with chysanthemic acid is described as pyrethrin I (6). Pyrethrum extract typically contains roughly equal quantities of chrysanthemic and pyrethric acid esters. Pyrethrolone esters (pyrethrins I (6) and II (7)) account for 73%, cinerolone esters (cinerins I (8)) and II (9)) for 19% and jasmolone esters (jasmolins I (10) and II (11)) 8% of the total (Head, 1973).

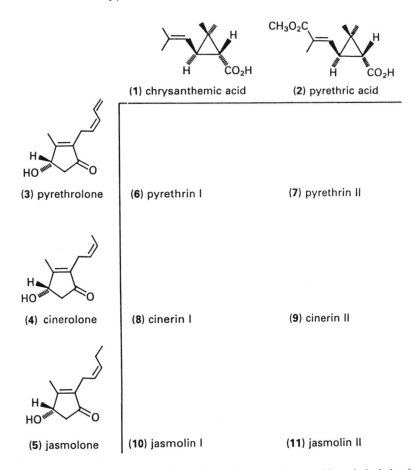

(1) chrysanthemic acid   **(2) pyrethric acid**

(3) pyrethrolone	**(6) pyrethrin I**	**(7) pyrethrin II**
(4) cinerolone	**(8) cinerin I**	**(9) cinerin II**
(5) jasmolone	**(10) jasmolin I**	**(11) jasmolin II**

Figure 1.5. Structures and absolute configurations of the constituent acids and alcohols of the natural pyrethrins and the names of the six esters known collectively as pyrethrins

Staudinger & Ruzicka (1924a) in their well known investigation concluded that the active principles of pyrethrum were derived from two carboxylic acids, and assigned correct structures to them as single (dextrorotatory) isomers of *trans*-2,2-dimethyl-3-(2-methylprop-1-enyl)cyclopropanecarboxylic acid and *trans*-2,2-dimethyl-3-(2-methoxycarbonylprop-1-enyl)cyclopropane-carb-oxylic acid. The absolute configuration could not be determined at that time but was later deduced by means of a series of chemical transformations leading to products of known stereochemistry (Crombie & Harper, 1954). The geometry of the side chain acid was settled as *trans* by synthesis (Crombie et al., 1957).

The original investigators isolated what they thought was a single ketoalco-hol, to which they gave the name pyrethrolone, and assigned to it the structure of 5-hydroxy-3-methylcyclopentanone bearing a doubly unsaturated side chain in the 2-position. It required much further work to recognise the presence of another, endocyclic, 2,3-double bond (LaForge & Haller, 1936), to assign the hydroxyl group correctly to the 4-position in the rethrolones (LaForge & Soloway, 1947a, b) and to confirm this conclusion by synthesis (Soloway & LaForge, 1947a, b). Ultra-violet spectroscopic data indicated that the side chain was a 2′,4′-diene (Gillam & West, 1944; West, 1946), but the correctness of this assignment was at first obscured by the results of oxidation experiments, designed to address the same problem, which were complicated by the presence of the then-unknown cinerolone.

The geometry at the 2′-position of the side chain was shown to be *cis* by synthesis (Crombie et al., 1951; 1956). The absolute configuration at C-4 was correctly assigned in 1959 (Katsuda et al., 1958, 1959), although this work has been criticised (Elliott & Janes, 1973). Unambiguous proof of the *S*-configuration at C-4 of the rethrolones was provided in the early 1970s by means of optical rotatory dispersion (Miyano & Dorn, 1973) and X-ray data (Begley et al., 1972, 1974).

The discovery of the existence of cinerolone (**4**) (LaForge & Barthel, 1945) as a component of the rethrolone fraction of pyrethrum extract removed an element of confusion from attempts at structure determination. Another source of confusion, the ease of racemisation at C-4 of the cyclopentenolone ring, is discussed in the earliest of the reviews quoted (Crombie & Elliott, 1961). The hydroxyl group in cinerolone was at first placed in the 5-position, but the assignment was changed in due course to the 4-position for the same reasons already given for the case of pyrethrolone. The geometry of the side chain was again established by synthesis (Crombie & Harper, (1950).

The isolation of jasmolins I (**10**) and II (**11**) had to await the advent of chromatographic techniques, and, once isolated, the structures were fairly readily identified by spectroscopic comparisons with relatives of known structure (Godin et al., 1964; Beevor et al., 1965).

Synthetic analogues of the pyrethrins: pyrethroids

Synthetic substitutes for pyrethrin: synthesis 1924–70

Although synthesis of analogues of the natural pyrethrins began almost as soon as the nature of the active materials was perceived (Staudinger & Ruzicka, 1924b), the first work to result in commercial success was published 25 years later. At the time the structures of pyrethrins and cinerins were known, although the absolute configurations were not. The key to this first success was the development of a general synthesis of cyclopentenolones bearing an unsaturated substituent in the 2-position. This meant that the natural rethrolones and their synthetic analogues were readily available as racemic mixtures. A contemporary improvement in the route to chrysanthemic acid

(Campbell & Harper, 1945) led to the synthesis of the first pyrethrin analogue (pyrethroid) to find some practical application in insect control. This compound, named allethrin (**12**), was the ester of racemic 2-allyl-3-methyl-cyclopent-2-ene-4-ol-1-one (allethrolone) with racemic *cis/trans* chrysanthemic acid (Schechter et al., 1949). The ester of allethrolone with the naturally derived (1*R*)-*trans* acid (Schechter et al., 1949) is now known as bioallethrin (**13**), and the ester of the (1*R*)-*trans* acid with (*S*)-allethrolone as (*S*)-bioallethrin (**14**) (Gersdorff & Mitlin, 1953). The prefix "bio" is generally used to indicate esters of the natural, dextrorotatory (1*R*)-*trans*-chrysanthemic acid.

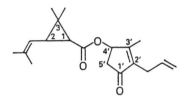

(**12**) allethrin
(**13**) bioallethrin [1*R–trans*, 4′*RS*]
(**14**) (*S*)-bioallethrin [1*R-trans*, 4′*S*]

Although these latter two esters were separated or synthesised shortly after the synthesis of allethrin (LaForge et al., 1954, 1956) they were not commercialised until much later, when sufficiently sophisticated chemistry had been developed (Naumann, 1981). An account of the early commercialisation and manufacture of allethrin has been published (Sanders & Taff, 1954). Allethrin proved to be a satisfactory substitute for natural pyrethrins in some respects as it had similar properties (Barthel, 1961; Elliott, 1954).

Most of the work carried out during this period was devoted to variation of the alcohol portion of the ester, so that this group consists mainly of esters of chrysanthemic acid, resolved to a greater or lesser degree. Staudinger and Ruzicka had made the piperonyl ester of chrysanthemic acid in the course of their original investigation of the structure of pyrethrum constituents, and had observed slight insecticidal activity (Staudinger & Ruzicka, 1924b). This observation was later followed by a more extensive investigation of benzyl esters of chrysanthemic acid by Barthel and his colleagues (Barthel, 1961), which revealed the connection between biological activity and the substitution pattern in the benzene ring. Dimethrin (the 3,4-dimethylbenzyl ester of racemic chrysanthemic acid) was one of the more active compounds made. Synthesis of a group of allybenzyl esters demonstrated that unsaturated substituents increase potency in this series as they do in the rethronyl esters (Elliott et al., 1965). A development that owes something to the empirical approach occurred at about this time with the discovery that chrysanthemate esters of *N*-hydroxymethylphthalimide and its 3,4,5,6-tetrahydro derivative had very high knockdown activity on houseflies (Kato et al., 1964; Sumitomo, 1965[1963]). The latter compound, tetramethrin (**15**), still finds substantial use as a constituent of aerosols.

Elliott and Janes and their colleagues at Rothamstead continued their systematic approach to structural variation and replaced the benzyl group by 2- and 3-furylmethyl groups (Elliott et al., 1974a). Of these alcohols the 5-benzyl-3-furylmethyl proved to give by far the most active derivatives. Resmethrin (**16**), NRDC 104, 5-benzylfurylmethyl (1*RS*)-*cis*,*trans*-chrysanthe-mate, and bioresmethrin (**17**), NRDC 107, the ester of the (1*R*)-*trans* acid (Elliott et al., 1967; National Research Development Corporation, 1969[1965/6]) possessed activity of the same order as, or greater than the most active of the natural esters, pyrethrin I (**6**), and were considerably more active than allethrin (**12**) and pyrethrum extract. They also had lower mammalian toxicity than the natural product. Both have found considerable practical application, as non-residual sprays for glasshouse use and against nuisance flies, and as domestic insecticides.

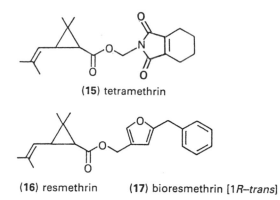

(15) tetramethrin

(16) resmethrin   **(17)** bioresmethrin [1*R–trans*]

A number of other compounds described during this period appear to have been considered as candidates for commercial development as domestic insecticides. These include prothrin (**18**) (Dainippon Jochugiku, 1968[1966]; Katsuda et al., 1969), proparthrin (**19**) (Yoshitomi, 1969[1967/8]), butethrin (**20**) (Taisho, 1970[1969]; Sota et al., 1971) and the allethronyl ester of 2,2,3,3-tetramethylcyclopropanecarboxylic acid, terallethrin (Matsui & Kitahara, 1967).

Some time after the discovery of the benzylfuryl chrysanthemates a group of Japanese workers (Sumitomo 1969[1968]; Fujimoto et al., 1973) made the chrysanthemic acid ester of 3-phenoxybenzyl alcohol, phenothrin (**21**), S-2539. As the form enriched in the 1*R*-isomers (d-phenothrin) this compound is in use as a domestic insecticide. Hindsight informs us that this work represented a considerable step towards the achievement of light stability

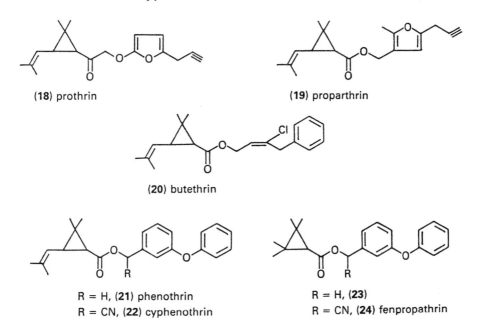

(**18**) prothrin   (**19**) proparthrin

(**20**) butethrin

R = H, (**21**) phenothrin   R = H, (**23**)

R = CN, (**22**) cyphenothrin   R = CN, (**24**) fenpropathrin

in pyrethroids. It is a matter of archival interest only that the same patent contains a description of what was probably the first light-stable pyrethroid made, 3-phenoxybenzyl 2,2,3,3-tetramethylcyclopropanecarboxylate (**23**). It has, however, never achieved commercial status.

As well as 3-phenoxybenzyl alcohol the related 3-benzylbenzyl and 3-benzoybenzyl alcohols were investigated (Elliott, 1971). Esters of these alcohols are rather less active.

The result of the work carried out over this period was to provide a number of effective synthetic substitutes for the natural pyrethrins; many of these are still very evident in commercial use. Among the most visible are bioallethrin (**13**), allethrin containing a high proportion of 1R-isomers ("pynamin-forte"), d-phenothrin (**21**), and tetramethrin (**15**).

Introduction of light-stable pyrethroids: synthesis 1968–74

Following the discovery of the activity of 3-phenoxylbenzyl esters, Sumitomo chemists investigated the effect of substitution at the benzylic carbon atom (alpha substitution). They found that most substituents sharply reduced the activity, but that two exceptions were the ethynyl group, which brought about a modest reduction in activity, and the cyano group, which markedly enhanced the activity (Matsuo et al., 1976). A 1971 patent application (Sumitomo, 1973[1971]) describes a number of α-cyano-3-phenoxybenzyl cyclopropane-carboxylates, including the chrysanthemate, cyphenothrin (**22**), and the

2,2,3,3-tetramethylcyclopropanecarboxylate, fenpropathrin (**24**). The latter compound was the first of those light-stable pyrethroids now used for agricultural purposes to be synthesised, although it was not commercialised until 1980, rather later than some more recently synthesised compounds. The former, as d-cyphenothrin, the form enriched in 1*R*-isomers, is still under investigation as a domestic insecticide (Matsunaga et al., 1983). 2,2,3,3-Tetramethylcyclopropanecarboxylic acid is of interest as the first known acid component of light-stable pyrethroid, although there were to be few further developments in this area as most other tetraalkylcyclopropanecarboxylic acid esters are relatively inactive (Matsui & Kitahara, 1967).

The dichlorovinyl analogue of chrysanthemic acid was first synthesised in or around 1957 (Farkas et al., 1959) in both *cis* and *trans* forms. The authors made the allethrolone ester of the *trans* (**25**) but not the *cis* acid, but screening on houseflies revealed only that the new compound was of similar activity to allethrin. The investigation was not pursued further at that time. Some years later, examination of this acid in combination with the recently developed synthetic pyrethroid alcohols (Elliott et al., 1975b) revealed that the 5-benzyl-3-furylmethyl ester of (1*R*)-*trans*-3-(2,2-dichlorovinyl)-2,2-dimethylcyclopropanecarboxylate (**26**) was twice as active on houseflies and mustard beetles as bioresmethrin (**17**) (Elliott et al., 1973a). Further investigation led to the synthesis of the 3-phenoxybenzyl ester of the racemic acid (Elliott et al., 1973b, 1973c). This compound, NRDC 143 (**27**), was given the common name permethrin (National Research Development Corporation, 1975[1972/3]). The new compound was shown to be very much more stable to light than any of the chrysanthemates or benzylfuryl esters previously synthesised, to have activity of the same order as bioresmethrin, or greater, and to possess low mammalian toxicity.

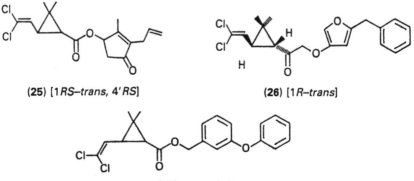

(**25**) [1*RS–trans*, 4′*RS*]   (**26**) [1*R–trans*]

(**27**) permethrin

This work was then systematically extended to include derivatives of the dibromo- and difluorovinylcyclopropanecarboxylic acids with 3-phenoxybenzyl alcohol and α-cyano-3-phenoxybenzyl alcohol. Examination of the relationship between stereochemistry and activity in this series (Burt et al., Elliott et al., 1974c, 1975a) showed that, as is the case with the chrysanthemate esters, it is only derivatives of acids having the 1R-configuration that possess substantial biological activity. Unlike chrysanthemates the (1R)-*cis*-derivatives usually show greater activity than the (1R)-*trans*-derivatives.

Esters of single isomers of these acids with 3-phenoxybenzyl alcohol are, of course, single isomers themselves, since the alcohol contains no asymmetric centre. Esters of single isomers of these acids with the unresolved alpha-cyano alcohol consist of two diastereoisomers, because a second centre of asymmetry has been introduced. One of these pairs of diastereoisomers, (RS)-α-cyano-3-phenoxybenzyl (1R)-*cis*-3-(2,2-dibromovinyl)-2,2-dimethylcyclopropane-carboxylate, NRDC 156, deposited crystals from a solution, which were shown to consist of the form having the S-configuration at the benzyl carbon atom (Elliott et al., 1974b; Owen, 1974). This single isomer, deltamethrin (**28**), NRDC 161, (NRDC, 1974[1973/4]) was also shown to be the more biologically active of the two comprising NRDC 156, and is therefore the most active single isomer of the eight possible. At the time of its discovery it was the most active insecticide known. It has since been commercialised and its biological properties, uses and synthesis have recently formed the subject of a monograph (Roussel-Uclaf, 1982).

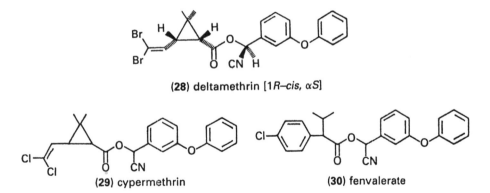

(**28**) deltamethrin [1R–*cis*, αS]

(**29**) cypermethrin   (**30**) fenvalerate

One further compound emerged from this series of investigations conducted by the group at Rothamstead, which, like permethrin (NRDC 143) attained considerable importance as an agricultural insecticide. This is cypermethrin (**29**), NRDC 149, (RS)-α-cyano-3-phenoxybenzyl (1RS)-*cis*, *trans*-3-(2,2-dichlorovinyl)2,2-dimethylcyclopropanecarboxylate. The active isomers are described (Elliott et al., 1974c; 1975a), but the racemic compound makes no appearance in the open literature until later (Elliott, 1977), although the compound appears in a patent published some time earlier (NRDC, 1975[1972/3]), and was launched as a commercial product in 1977.

In addition to the 3-dihalovinylcyclopropanecarboxylic acids the same workers investigated a number of other cyclopropanecarboxylic acids bearing ethylenic (Elliott et al., 1976a), and non-ethylenic substituents (Elliott et al., 1976b) in the 3-position. Unsaturated substituents conferred high activity in

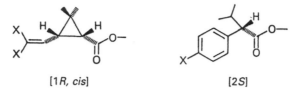

[1R, cis]   [2S]

Figure 1.6. The 1R-cis-configuration of cyclopropanecarboxylates compared with the 2S-configuration of phenylacetates

general, and saturated substituents low activity. The 3-alkoxyiminomethyl-cyclopropanecarboxylates (52) have been the subject of more recent work, which is discussed in a later section.

One of the more suprising developments in pyrethroid synthesis was the discovery of insecticidal activity in a group of phenylacetic acid esters (Ohno et al., 1974) which led to the synthesis of fenvalerate (Sumitomo, 1976[1972/3]), S5602, (RS)-α-cyano-3-phenoxybenzyl (RS)-2-methyl-3-(4-chlorophenyl) butyrate (30) (Ohno et al., 1976). One of the four possible stereoisomers of fenvalerate, described as A alpha (Yoshioka, 1978), is responsible for most of the insecticidal activity, and the stereochemical requirements for maximum activity reinforce the assumption that fenvalerate should be regarded as an analogue of the insecticidal cyclopropanecarboxy-lates. Only esters of (S)-2-methyl-3-(4-chlorophenyl)butyric acid have insecticidal activity and this configuration is easily reconciled with the 1R-configuration of active cyclopropanecarboxylates (Figure 1.6). As is the case with cyclopropanecarboxylates, esters of α-cyano-3-phenoxybenzyl alcohol with the (S)-configuration have much greater insecticidal activity than those of the enantiomeric alcohol (Miyakado et al., 1975; Aketa et al., 1978; Nakayama et al., 1979). Fenvalerate is the fourth of the first group of light-stable pyrethroids to be described from 1973 onwards, and was the first to be introduced into commerce in 1976.

Not all effort during this period was devoted to the synthesis of compounds that were to find application in agriculture. Chemists at Roussel-Uclaf investigated a number of cycloalkylidenecyclopropanecarboxylates and demonstrated that the 5-benzyl-3-furylmethyl ester of (1R)-trans-3-cyclo-pentylidenmethylcyclopropanecarboxylic acid, bioethanoresmethrin (31), was more active on houseflies than the analogous chrysanthemate (Velluz et al., 1969). This was followed by the synthesis of (1R)-cis-2,2-dimethyl-3-(tetra-hydro-2-oxo-3-thienyl)cyclopropanecarboxylate, (32), RU 15525 (Roussel-Uclaf. 1972[1970]), probably the most active housefly knockdown agent known (Lhoste & Rauch, 1976).

The development that permitted the use of pyrethroids in agriculture was the replacement of light-unstable centres in both the acid and alcohol portions of the pyrethrins by light-stable groups of otherwise similar steric and chemical properties (Elliott et al., 1978). Chrysanthemic acid, natural and synthetic

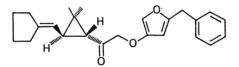

(31) bioethanoresmethrin [1*R–trans*]

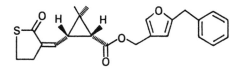

(32) RU 15525 [1*R–cis*]

rethrolones and furylmethyl alcohols are all too unstable in the presence of light and air to show sufficient persistence for agricultural purposes, whilst dihalovinylcyclopropanecarboxylic acids, tetraalkylcyclopropanecarboxylic acids, phenylacetic acids, and their esters with benzyl alcohols show sufficient stability in light and air to make them effective crop insecticides.

Whilst increased persistence permitted their use, the factors that led to their undoubted success were the possession of very intense biological activity compared with other classes of insecticide (Elliott, 1979). This compensated for synthesis chemistry rather more sophisticated, and more expensive, than that usually associated with pesticides, and a spectrum of activity, especially on lepidopterous larvae, which made them suitable for use in the larger and more commercially attractive markets, particularly cotton.

Before continuing with more recent work it is appropriate to discuss the main conclusions that had been arrived at concerning the relationship between structure and biological activity at this stage, and to consider to what extent the practical use to which the insecticidal activity of pyrethroids may be put is determined or affected by physical properties. The significance of more recent developments may thus be made clearer.

Structure–activity relationships and their limitations

Structure–activity relationships in pyrethroids have been reviewed in great detail (Elliott, 1977; Elliott & Janes, 1978; 1979). These reviews are based on a great volume of published work, only a portion of which has been referred to in this text, and numerous useful correlations have been made, some of which are firmer than others, with the result that a mass of detail has been reduced to a manageable body of information. What follows is a selection of the more important conclusions that have been drawn.

There is a small group of structural features that pyrethroids invariably require if they are to possess high insecticidal activity irrespective of the detailed structure of the rest of the molecule, or the nature of the test species employed. Departure from these requirements will always produce a sharp reduction in insecticidal activity. The effects of structural changes that fall outside this group are less easy to predict, because they do depend upon the identity of the test species, and are influenced by the parts of the molecule not directly affected by the change. The two groups are dealt with separately, beginning with the first, which forms the basis of a number of satisfactory generalisations. Pyrethroids are known to act at many different sites, and different compounds can produce different symptoms. Nevertheless, for the purposes of this discussion it is assumed that there are certain basic similarities among the sites of action.

Until recently all known active pyrethroids were esters of carboxylic acids. Attempts to replace the ester with another linkage had been unsuccessful (Berteau & Casida, 1969). There is now a single group of active compounds that falls outside this category. These compounds are ethers of alkyl aryl ketoximes (**53**) and will be discussed later. This exception probably demonstrates that the near-universal requirement for an ester linkage has to do with steric rather than chemical requirements.

Active esters of 3-substituted cyclopropanecarboxylic acids all have the 1R-configuration. This was observed rather early in the history of synthetic pyrethroids for derivatives of chrysanthemic acid (Gersdorff & Mitlin, 1953; Barthel, 1961), and was later also found to be the case for the dihalovinylcarboxylates (Burt et al., 1974). Active isomers in the phenylacetate series have the corresponding S-configuration at C-2 (see Figure 1.6; Miyakado et al., 1975).

Another requirement for activity is a *gem*-dimethyl substituent at C-2 of the cyclopropane ring (Barlow et al., 1971), which may be replaced by a spiroalkyl group (Davis & Searle, 1977). The methyl group *cis* to the carboxyl group seems to exert the greater effect, but the *trans* methyl group also contributes to potency (Sugiyama et al., 1975). Similarly only those phenylacetates that have a corresponding substituent in the 2-position show biological activity. Thus 2-methyl phenylacetic acid esters are inactive, whilst the ethyl, isopropyl and t-butyl analogues all have insecticidal activity (Ohno et al., 1974).

The configuration at the 3-position of the cyclopropane ring is not critical. Both *cis*- and *trans*-substituted compounds show activity. The only constraint appears to be one of shape. Bulky saturated groups are not tolerated (Barthel, 1961). Staudinger & Ruzicka (1924b) first noticed that reduction of the isobutenyl group of pyrethrins abolished the activity. Substituents may be quite extensive in two dimensions, that is, they may be large and flat. 2,2,3,3-Tetra-methylcarboxylic acid might be considered an exception to this generalisation. It is possible that this is connected with the small size of the substituents.

It has been suggested that the facts listed above are best explained by assuming the only essential requirements for biological activity to be a *gem*-dimethyl group held in a certain relationship to a suitably esterified carboxyl group, the effect of the 3-substituent depending only on whether it obstructs this arrangement or not (Barlow et al., 1971). The evidence presented broadly supports this interpretation in that it explains the activity of both enantiomers of suitable esters of 2,2-dimethylcyclopropanecarboxylic acid. The 3-substituent presumably has some function concerning susceptibility to metabolic detoxification and ease of access to the site of action, as compounds lacking the 3-substituent are of relatively low activity.

Only derivatives of alkyl or aralkyl alcohols show activity. The carbon atom to which the hydroxy group is attached must be sp^3 hybridised (tetrahedral). Phenyl esters do not possess insecticidal activity.

The question of stereochemistry at the benzylic carbon atom (alpha-carbon atom) is discussed more fully below, but one observation seems to be generally applicable. Substitution of the alpha-carbon atom of 3-phenoxybenzyl alcohol with a cyano group increases the activity of the esters to most species. Where this increased activity is seen, it is invariably associated with only one of the two enantiomeric configurations, the *S*-configuration (Elliott et al., 1974b; Aketa et al., 1978).

These requirements are not species specific, so whatever species of insect had been used to evaluate pyrethrins and pyrethroids the results would probably have been the same. The consistency of the observations suggests that these features are related to intrinsic activity, that is the ability of the molecule to produce a physiological change at the site or sites of action, and must reflect the overall shape and asymmetry of the site of action.

The effects of structural features and modifications other than those described above have proved to be inconsistent and unpredictable, except in a limited way, perhaps towards one species, or for a limited group of compounds. Whilst such limited correlations are very useful in a practical sense, they do not increase our understanding unless the factors which are responsible for the overall effect can be disentangled. Fortunately considerable progress has been made in this direction.

A given modification of structure may, or may not, produce a change in intrinsic activity, but even if such a change is produced it may be obscured by changes in routes and rate of metabolism, and by changes in penetration and distribution consequent upon changes in physical properties. All of these variables will affect the ease of access of a compound to its site of action and thus its effective toxicity. However, the total effect need not be the same for different insect species, nor need a given modification applied to a series of related compounds produce the same modification of biological activity in each case. The following discussion illustrates these points by reference to substitution at C-3 of the cyclopropane ring of cyclopropanecarboxylates, to the effect of the alcohol portion of the pyrethroids on the spectrum of activity, and to the effect of substitution at the alpha-carbon of pyrethroids. Reference is also made to the phenomenon of antagonism, which can complicate the assessment of the biological activity of partly resolved compounds.

It has already been remarked that stereochemistry at the 3-position of cyclopropanecarboxylates is not crucial, although it is generally true, with minor exceptions, that in the chrysanthemate series the *trans*-isomers are more active (Barthel, 1961), while the *cis*-series is more active if the isobutenyl methyl groups are replaced by halogen (Elliott et al., 1975a). It is also apparent, especially when more recent work is considered, that considerable variation is possible in the nature of the substituent at the 3-position of the cyclopropane ring without loss of activity. Evidence in support of the view that the configuration at C-3, and the nature of the substituent, affects properties such as susceptibility to metabolic detoxification and ease of access to the site of action, rather than exerting a major influence on intrinsic activity, comes from a number of recently published observations.

Synergists, compounds that suppress oxidative or hydrolytic degradation of the toxic material in the insect, are commonly employed as constituents of pyrethrin and pyrethroid formulations in the practical control of domestic insects, and are often used in experimental work designed to measure intrinsic toxicity. It is reported (Casida, 1983) that the synergised toxicities of (1*R*)-*trans*-chrysanthemates, (1*R*)-*cis*-3-(2,2-dibromovinyl)-2,2-dimethyl-cyclopropanecarboxylates and 2,2,3,3-tetramethylcyclopropanecarboxylates differ by a factor of less than 2. This seems to support the contention that, within limits, the nature of the 3-substituent has little effect on intrinsic activity. That the configuration at C-3 has little effect on intrinsic activity is suggested by the observation that *cis*- and *trans*-isomers of permethrin (**27**) and phenothrin (**21**) are equally effective in causing repetitive firing in the cercal sensory nerve of the American cockroach, although *cis*-permethrin has much the greatest whole insect toxicity.

An approach based on the observation of pharmokinetic processes in the American cockroach (Soderlund, 1983) has produced evidence that the configuration at C-3 has relatively little effect on the intrinsic activity, but a large effect on availability at the site of action. The just toxic dose for 3-phenoxybenzyl (1*R*)-*cis*-3-(2,2-dibromovinyl)-2,2-dimethylcyclopropane-carboxylate is about one-third of that of the (1*R*)-*trans*-isomer. Nevertheless the steady-state concentration of the (1*R*)-*cis*-isomer in the ventral nerve cord is twice that of the (1*R*)-*trans*-isomer after the topical application of equitoxic doses. If concentration in the nerve cord is an indication of intrinsic activity, then the *trans*-isomer is twice as active as the *cis*-isomer, at least on this insect. The difference in effective toxicity in favour of the *cis*-isomer is therefore not caused by a difference in intrinsic toxicity, but by the greater ease of access to the active site possessed by the *cis*-isomer. The efficiency of transport to the site of action, measured as a ratio of steady state concentration to the applied dose necessary to produce it was about eight-fold greater for the *cis*-isomer than for the *trans*. This greater availability of the *cis*-isomer can be ascribed to the greater ease with which the *trans*-isomer is cleaved by insect esterases (Soderlund et al., 1983). However, the possibility also exists that other processes, such as the rate of penetration of the cuticle, play some part.

Variation of the structure of the alcohol portion of pyrethroids again produces an overall effect dependent upon changes brought about in intrinsic activity and in routes and rates of penetration, distribution and metabolism. Structure–activity relationships in pyrethroid alcohols have been extensively reviewed (Elliott & Janes, 1978), but the following generalisations constitute a serviceable summary.

The majority of alcohols that yield active esters are substituted benzyl alcohols, or the closely related pyridylmethyl and furylmethyl alcohols. To these must be added N-hydroxymethyl-3,4,5,6-tetrahydrophthalimide, fur-furyl alcohols, the rethrolones and a group of acyclic unsaturated secondary alcohols (Matsuo, T. et al. 1980), that may be considered to be rethrolone analogues, whose properties fit them to the domestic rather than the agricultural market. These secondary alcohols are of some interest from the stereochemistry viewpoint because they have provoked a study of stereo-chemistry at the α-carbon atom of secondary pyrethroid alcohols in general (Matsuo, N. et al., 1983).

Esters of the rethrolones (excepting pyrethrin I, which is active on a number of species), furfuryl alcohols, N-hydroxymethyl-3,4,5,6-tetrahydrophthali-mide, and the unsaturated acyclic alcohols tend to show high knockdown activity on flies, and sometimes on other flying insects, and low toxicity to other species. Because they are readily metabolised their biological activity may be greatly increased by synergists. Esters of 5-benzyl-3-furylmethyl alcohol possess high toxicity to many species, whilst benzyl esters, particularly 3-phenoxybenzyl and 2-methyl-3-phenylbenzyl esters (the latter is discussed in the final section), show very high toxicity to many species, especially to lepidopterous larvae, but rather poor knockdown activity. Because of their much reduced susceptibility to metabolic breakdown, compared to the pyrethrins, the last group is synergised to a much smaller degree.

The substitution of the α-carbon atom of pyrethroid alcohols has an unpredictable effect on the biological activity of the esters, which may be increased, decreased, or scarcely affected (Elliott, et al., 1983a), depending upon the structure of the alcohol, and the nature of the substituent. In the case of most practical importance to date, substituted 3-phenoxybenzyl alcohols, a cyano substituent generally increases activity, although not necessarily in all species (Brempong-Yeboah et al., 1982). The increased activity is invariably associated with the (S)-α-cyano-phenoxybenzyl esters and not with the enantiomeric configuration. Similarly, the activity of the rather less active esters of α-ethynyl-3-phenoxybenzyl alcohol is associated principally with the enantiomer (**33**) possessing the same configuration as the active cyano-deriva-tive (described in this case as the *R*-configuration because of the sequence rules). This preferred stereochemistry cannot be readily reconciled with the preferred stereochemisty of the rethrolones (**34**) nor with that of a series of unsaturated acyclic secondary alcohols (**35**) which may be regarded as open-chain analogues of the rethrolones (Matuso, T. et al., 1980; Matsuo, N.

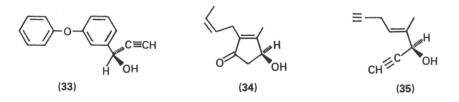

(33)   (34)   (35)

et al., 1983). The evidence for the preferred stereochemistry in all these cases is quite firm. On the face of it this appears to be an uncomfortable observation and one difficult to accommodate. It is doubtful, however, whether these apparent discrepancies merit detailed discussion at this stage for the following reasons. In the course of this discussion the assumption has been made that all pyrethroids act, not at the same site, but at sites that have at least some properties in common. This simplifies the discussion and permits a few useful and generally applicable empirical rules to be stated. In reality the situation is rather more complex. In particular esters of α-cyanophenoxybenzyl alcohol show effects on both insect and mammalian nervous systems that are not seen with the pyrethrins or the earlier synthetic compounds (Lawrence & Casida, 1983). Under these circumstances, while it is reasonable to attach some importance to observations that seem to be generally applicable – that is to say, the empirical rules referred to – the significance of discrepancies is uncertain at our present state of knowledge of the effect of these active molecules at the molecular level. Speculation is therefore likely to be unprofitable.

It is evident from this discussion that the effect of changes in substitution or geometry at C-3 of the cyclopropane ring (or in the aromatic substitution pattern in the phenylacetate series), or of a change in the nature of the alcohol component, will depend upon the structure of the whole molecule. Other factors that will determine the effect of a given structural feature on activity are the property used as a measure of biological activity, and the test species employed. Knockdown activity, which is a very desirable property in domestic insecticides, follows a pattern of structure–activity relationship quite different from that observed for lethality (Barlow et al., 1971), although the basic stereochemical rules still hold good. It is likely that this property is related to physical properties (Briggs et al., 1974, 1976). As is the case with other classes of insecticide, activity on one species of insect cannot always be used as a guide to activity on another. This is well illustrated by a short series of alkyl aryl ketoxime ethers, the activity of which to two species of lepidoptera moves in precisely opposite directions (Table 1.1). Such selectivity, which was observed early in the history of pyrethroids (Elliott et al., 1950), may render a compound useless for practical purposes, or it may confer an advantage if, for instance, selectivity is shown between a pest and its parasite (Elliott et al., 1983b).

A possible pitfall in the assessment of biological activity of pyrethroids is the occurrence of antagonism between stereoismomers. Fortunately this was not a problem with the first light-stable compounds. Whereas fenvalerate A alpha,

Table 1.1. Biological activity of substituted cyclopropyl phenyl ketone oxime 3-phenoxybenzyl ethers against two species of lepidoptera

General structure (53) E-isomers		Toxicity index[a] against	
R	X	S.l.	H.z.
cyclopropyl	F	380	56
cyclopropyl	Br	210	200
cyclopopyl	Cl	160	280

[a] Toxicity index = 100 (LC50 for parathion)/LC50 for test compound.
S.l. = *Spodoptera littoralis* (Boisd.); H.z. = *Heliothis zea* (Boddie).

Table 1.2. Antagonism between diastereoisomers. Biological activity of pentachlorbenzyl esters (36) and (37)

Compound	Stereochemistry of acid	Toxicity index[a] against				
		M.d.	S.l.	A.f.	T.u.	H.z.
(36)	1RS-cis	9	170	96	4	670
(37)	1R-cis	78	530	370	210	2800

[a] Toxicity index as defined in Table 1.
S.l. and H.z. are defined in Table 1.
M.d. = *Musca domestica* (L.); A.f. = *Aphis fabae* Scopoli; T.u. = *Tetranchus urticae* (Koch).

(*S*)–α-cyano-3-phenoxybenzyl   (*S*)-3-methyl-2-(4′-chlorophenyl)butyrate, the isomer of fenvalerate responsible for the activity of the compound, is about four times as active as the racemate (Yoshioka, 1978), such simple relationships do not always exist between the activity of individual isomers and their mixtures. Table 1.2 shows the results of laboratory screens with pentachlorobenzyl   (1*RS*)-*cis*-3-(2,2-dichlorovinyl)-2,2-dimethylcyclopropanecarboxylate (36) and the fully resolved (1*R*)-*cis*-isomer (37). The increase in

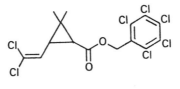

(36) [1*RS–cis*]

(37) [1*R–cis*]

activity on complete resolution varies from three-fold on the caterpillar *Spodoptera littoralis* (Egyptian cotton leaf worm) to at least fifty-fold on the mite *Tetranychus urticae*. This result implies that the insecticidally active isomer is capable of exerting a blocking effect (possibly, but not necessarily, at the site of action), and should not be regarded as biologically inert. This is the only really convincing example of this phenomenon that I am aware of, but it is necessary to consider the possibility of other cases coming to light.

Spectrum of activity: physical properties

The spectrum of activity and the practical applications of the pyrethroids are subjects dealt with fully in other chapters, but a few general observations are put forward here, mainly with the objective of illuminating the significance of more recent work.

The pyrethrins and earlier synthetic compounds, all of which were esters of chrysanthemic acid, were evaluated principally for their knockdown and lethal effects on flying insects. Although it was known that these compounds were active on a number of other species including caterpillars (Butt & Keller, 1964; Cantu & Wolfenbarger, 1970; Elliott et al. 1950) – indeed pyrethrum seems to have been evaluated against cabbage and cotton worms as early as 1880 (McDonnell et al., 1920[revised 1926]) – this aroused little interest because of their high cost and photo-instability.

Up to 1970 or thereabouts, investigators had spent 20 years investigating the biological properties of esters of chrysanthemic acid. The discovery of highly active synthetic alcohols, including 5-benzyl-3-furylmethyl alcohol, not itself the progenitor of light-stable compounds, permitted the more effective evaluation of variation in the structure of the acid component of the molecule. This in turn permitted and encouraged a more imaginative approach, and many types of structure were screened, mainly in Britain and Japan. Light-stability, an exceedingly difficult property to predict, was a by-product of this greater variation.

The consequences of light-stability were rapidly recognised, and permethrin, cypermethrin, deltamethrin, and fenvalerate were subjected to extensive screening. Fortunately they turned out to be as active on a wide variety of economically important pests as they had been on the rather small number of insects screened up to that point. They proved to be broad-spectrum contact insecticides with activity on many lepidopterous larvae that was phenomenal by the standards of the day. This made them suitable for use in cotton, one of the few single crops which might provide a financial reward capable of justifying the cost of introducing a new compound. Had this not been so, it is most unlikely that four new compounds would have been introduced in the space of two years. The compounds have of course found a variety of applications, and their performance in various areas has been extensively reviewed Breese, 1977; Breese & Highwood, 1977; Mowlam et al., 1977; Elliott et al., 1978; Ruscoe, 1977, 1979, 1980; Piedallu et al., 1982).

In spite of their exceedingly high contact activity, these four compounds have their limitations. One is to do with the spectrum of activity; the others are imposed by physical properties. Although they have useful activity on ticks, the activity on phytophagous mites, pests of cotton, top fruit and other crops, is low in comparison with activity on other arthropods. The light-stable synthetic pyrethroids are lipophilic compounds of high molecular weight and consequent low volatility. They are thus ineffective in soil, as they can move neither in the vapour phase nor in the water phase, and they do not exhibit systemic activity. The manner in which the physical properties of pyrethroids are reflected in their behaviour has recently been reviewed in some detail (Briggs et al., 1983).

Pyrethroid synthesis 1975–83

The exposure of the pyrethroids to extensive screening on economically significant pests has permitted some discrimination with regard to acaricidal activity. One result of this has been the quite recent development of fenpropathrin (**24**), a compound synthesised in the early 1970s, for use as an insecticide and miticide on top fruit (Nicholls & Peecock, 1981).

Recent work on synthetic pyrethroids may be looked at from two viewpoints. Firstly, what ingenious structural variations have been devised to produce novel products? Secondly, to what extent do the recent introductions overcome the defects of the older compounds? The first point is best dealt with before the newer work has been considered.

Most recent variations in the acid component of pyrethroids are based on 3-vinylcyclopropanecarboxylic acids, although some successful work has been reported on variants of the acyclic phenylacetate series. Spirocyclopropanecarboxylic acid esters, which are the only progeny of tetramethylcyclopropanecarboxylic acid also feature, but rather faintly.

The only alcohols to yield active light-stable pyrethroids are still substituted benzyl alcohols, with the exception of the closely related 4-phenoxy-2-pyridylmethyl alcohol. A number of differently substituted benzyl alcohols have been described, the most intersting of which are the phenylbenzyl alcohols, which appear to yield very active esters. It is rather interesting to see that work is still devoted to analogues of the pyrethrins. Prallethrin, an analogue of allethrin, in which the allyl group is replaced by a propargyl group, has recently been reported as an experimental insecticide for indoor use (Matsunaga et al., 1983). Some work has also been reported on analogues and isomers of pyrethric acid, possibly a rather under-exploited area (Martel, 1983)

A group of workers at American Cyanamid have investigated esters of a series of spiro-substituted cyclopropanecarboxylic acids with 3-phenoxybenzyl alcohol and its α-cyano derivative (Brown & Addor, 1979; American

Cyanamid, 1976[1975]). (*RS*)-α-cyano-3-phenoxybenzyl (*RS*)-3,3-dimethyl-spiro[cyclopropane-1,1'-indene]-2-carboxylate, cypothrin (**38**), has achieved commercial status in animal health as a tickicide. It is not in use in other areas. The structure is an interesting demonstration of the variation that may be accommodated in substitution at the 3-position without losing all activity. A series of simple spiroalkane cyclopropanecarboxylic acid esters have also been shown to possess substantial activity, especially knockdown (Davis & Searle, 1977).

(38) cypothrin

A series of phenylacetic acid esters, essentially variations on the fenvalerate model, emerged a few years later from the same source (American Cyanamid, 1980[1976/7]). One of these, α-cyano-3-phenoxybenzyl (*S*)-2-(4-difluoro-methoxyphenyl)-3-methylbutyrate, flucythrinate (**39**), a broad spectrum insecticide reported also to be effective in suppressing phytophagous mites (Whitney & Wettstein, 1979; Wettstein, 1981), is at present in commercial use.

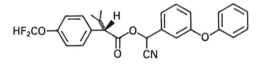

(39) flucythrinate [2*S*, α*RS*]

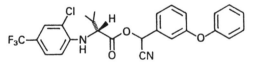

(40) fluvalinate [2*R*, α*RS*]

One of the more imaginative variations has come from a group at Zoecon, who have reported the activity of esters of a series of *N*-phenyl amino acids (Zoecon, 1978[1977/8]), which may be regarded as derived from the phenylacetate series by insertion of nitrogen between C-2 and the phenyl ring. The idea was said to have been arrived at by consideration of DDT, which may be modified in this way without loss of activity. The substitution pattern required for maximum activity is not the same as that observed in the phenylacetate series, but the preferred stereochemistry at C-2 is the same, and fluvalinate (**40**), (*RS*)-α-cyano-3-phenoxylbenzyl (*R*)-2-(2-chloro-4-trif-luoromethylanilino)-3-methylbutyrate, has a rather similar spectrum of activity (Henrick et al., 1980). It is reported that fluvalinate provides adequate control of phytophagous mites. One difference from the phenylacetate series may be noted. Introduction of a racemic α-cyano group gives at best a two-fold increase of activity on *Heliothis zea*, and may result in decreased activity. The compound seems to have been introduced in the first place as a racemic mixture, but the present product is a pair of diastereoisomers derived from the *R*-acid.

A relatively minor change, introduction of a 4-fluoro atom into the benzyl group of α-cyano-3-phenoxybenzyl alcohol, gave Bayer cyfluthrin (**41**), the ester of this alcohol with 3-(2,2-dichlorovinyl)-2,2-dimethylcyclopropane-carboxylate (Bayer, 1978[1977]). This compound, which has a level and spectrum of activity resembling those of cypermethrin (Hamman & Fuchs, 1981), was introduced as a cotton insecticide in 1981.

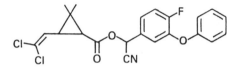

(41) cyfluthrin

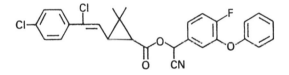

(42) flumethrin [*EZ*, 1*RS*–*trans*, α*RS*]

The ester of the same alcohol with *trans*-3-[2-chloro-2-(4-chloro-phenyl)]-2,2-dimethylcyclopropanecarboxylic acid (**42**), flumethrin (Bayer, 1981[1979]), has very high activity on cattle ticks (Hopkins & Woodley, 1982) and is in use as a tickicide. The rather large 3-substituent is another example of the toleration of extensive substituents in the 3-position.

Bromination of the double bonds of both deltamethrin, to yield tralomethrin (**43**) (Roussel-Uclaf, 1978[1976/7]), and cypermethrin (Ciba-Geigy, 1978[1977/8]) has been described. As bromination introduces a new centre of asymmetry, tralomethrin consists of two isomers, while bromination of cypermethrin produces sixteen (**44**). The proposed common name tralocythrin (Roussel-Uclaf) is usually applied to the pair of isomers derived from cypermethrin which possess the same stereochemistry as tralomethrin. The bromination product of cypermethrin appears to have a level and spectrum of

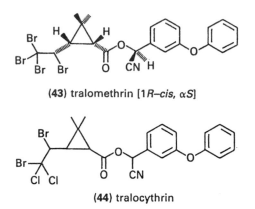

(43) tralomethrin [1*R–cis*, α*S*]

(44) tralocythrin

activity similar to those of the parent compound, and the configuration of the 1-carbon of the tetrahaloethyl group does not appear to be important (Ackermann et al., 1980).

The replacement of the isobutenyl group of chrysanthemic acid by the dichlorovinyl group led to insecticidal cyclopropanecarboxylates of greater metabolic- and photo-stability. An investigation reported in 1979 (Bentley et al., 1980) demonstrated that replacement of one or both of the chlorine atoms by trifluoromethyl groups gave α-cyano-3-phenoxybenzyl esters of substantial insecticidal activity. The most active compound to emerge from this work, α-cyano-3-phenoxybenzyl   (*RS*)-*cis*-3-[(Z)-2-chloro-3,3,3-trifluoroprop-1-enyl]-2,2-dimethylcyclopropanecarboxylate **(45)**, cyhalothrin, (Imperial Chemical Industries, 1979[1977/8]) has been introduced as an ectoparasiticide, although it possesses a broad spectrum of activity.

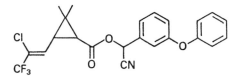

(45) cyhalothrin [*Z*, 1*RS–cis*, α*RS*]

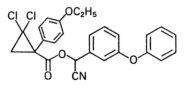

(46) cycloprothrin

If there is any doubt about the status, as pyrethroids, of any of the compounds mentioned in these pages, it probably attaches to those that form the next topic for discussion. The perception of a relationship between the structural requirements for activity in the pyrethroids and in DDT led Holan and his colleagues at the CSIRO laboratories in Melbourne to the synthesis of a rather unusual group of cyclopropane derivatives (Holan et al., 1978; CSIRO, 1977[1975]). One of these (**46**) now has the name cycloprothrin (Wakita et al., 1982) and is undergoing trials against crop pests in Japan and pests of livestock in Australia.

A number of variations on the phenoxybenzyl alcohol theme have been described. Dow's fenpyrithrin (**47**), Dowco 417, α-cyano-6-phenoxy-2-pyridylmethyl 3-(2-2-dichlorovinyl)-2,2-dimethylcyclopropanecarboxylate, an analogue of cypermethrin in which the benzyl group is replaced by 2-pyridylmethyl, is a broad-spectrum insecticide (Dow, 1979[1977]; Malhotra et al., 1982). A positional isomer of this compound, the corresponding 3-(2-pyridyloxy)benzyl ester (Ciba-Geigy, 1978[1977/8]) does not appear to have merited much attention.

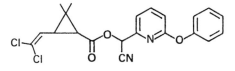

(47) fenpyrithrin

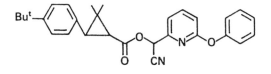

(48) NC1–85193 [1*RS–trans*, α*RS*]

Esters of α-cyano-6-1-pyridylmethyl alcohol have also been reported to have good miticidal activity. The ester with *trans*-3-(4-t-butylphenyl)-2,2-dimethylcyclopropanecarboxylic acid (**48**) is specifically mentioned (Ozawa et al., 1982).

The most interesting development to have appeared in the recent literature, from the point of view of structural innovation and possibly also biological activity, is a series of phenylbenzyl esters of various cyclopropanecarboxylic acids. Simple unsubstituted phenylbenzyl esters have ben reported to possess quite interesting activity (Plummer & Pincus, 1981; Plummer, 1983), which is not enhanced by substitution of the benzyl carbon with a cyano group (Elliott et al., 1983a). Recent reports also describe esters of 3-phenyl-2-methylbenzyl (**49**) and 3-phenyl-2,6-dimethylbenzyl (**50**) alcohols, claimed to possess good insecticidal (FMC, 1982[1979]) and especially miticidal (FMC, 1982[1980]) activity (Engel et al., 1983b; Plummer et al., 1983; VanSaun et

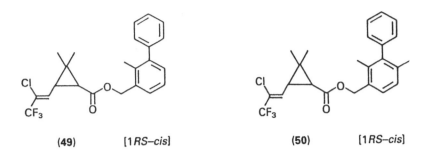

(49)   [1*RS–cis*]   **(50)**   [1*RS–cis*]

al., 1983). Even the most insecticidally active pyrethroids have not shown substantial activity on phytophagous mites hitherto, and we may expect to hear more of this group of compounds. The geometry of the side-chain of the acid component of the group (ICI's cyhalothrin is an ester of the same acid) has recently been settled as *Z* by X-ray crystallography (Engel et al., 1983a).

Another recently synthesised compound is an ester of *N*-methyl-3-(4-chloro-anilino)benzyl alcohol (Ishimitsu et al., 1982; Minamite et al., 1982). α-Cyano-*N*-methyl-3-(4-chloroanilino)benzyl 2-(4-chlorophenyl)-3-methyl-butyrate (**51**) is reported to be active on the green rice leafhopper and to possess low fish toxicity (Nippon Soda, 1981[1979/80]).

Quite a lot of effort has been devoted in a number of laboratories to the investigation of esters of simple benzyl alcohols bearing alkyl and halogen substituents (Bayer, 1978[1976/7]), which have proved to possess quite interesting miticidal and insecticidal activity (see Table 1.2, p. 23). The lower molecular weight of simple benzyl alcohols is naturally accompanied by greater volatility, which raises some hope that they might exhibit activity in the soil. Two groups have looked at this possibility (Shell, 1983[1981]; ICI, 1983[1981]).

A group of compounds originally described in 1974 (NRDC, 1974[1972/3]), the 3-alkoxyiminomethylcyclopropanecarboxylates (**52**), has been of sufficient interest to be the subject of more recent work (Shell, 1981[1978/9]; 1980[1978]).

One exception, already mentioned earlier, seems to have been found to the long-standing rule that all pyrethroids must be carboxylic acid esters. Whether or not this exception will prove to have any consequences, or whether it will remain a curiosity is not possible to predict. A group of oxime ethers (**53**) has been shown to exhibit insecticidal activity rather similar in nature and intensity to that shown by the phenylacetate series (Bull et al., 1980). The similarity between the groups is further underlined by the observation that the same structural changes in the two series produce parallel changes in biological activity. Only those of the *E*-series show biological activity. This work has had no commercial outcome but it has added to the difficulty of finding a satisfactory definition of a pyrethroid.

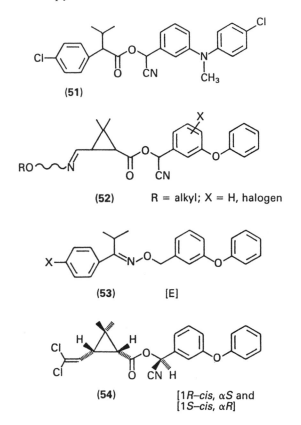

(51)

(52) R = alkyl; X = H, halogen

(53) [E]

(54) [1R–cis, αS and
[1S–cis, αR]

One new product that has been recently introduced does not involve a new compound. This is a modification of cypermethrin consisting of two isomers only, the single most active isomer, (S)-α-cyano-3-phenoxybenzyl (1R)-cis-3-(2,2-dichlorovinyl)-2,2-dimethylcyclopropanecarboxylate (54), and its inactive enantiomer, in equal proportions (FMC, 1981 [1979]; Shell 1981[1979]; Fisher et al., 1983). Separation of a racemic compound from a pyrethroid mixture has been reported before in the case of allethrin (Schechter et al., 1951), but on that occasion the material did not contain one of the active isomers.

Conclusions

From this account it can be seen that most recent work has resulted mainly in the synthesis of a number of compounds having biological properties very similar to those of the first four pyrethroids to find use as agricultural insecticides. There are some exceptions, and one of the deficiencies of the original compounds that will possibly be overcome by some of the newer compounds is the relatively low activity on phytophagous mites. The possibility of soil activity has at least been investigated, but so far without success. The absence of systemic activity is probably something that will have to be lived with. The lipophilicity which is the cause may well be intimately connected with the activity of the group, and talk of a systemic pyrethroid brings to mind the old joke which ends, "If I was going to Dublin, I wouldn't start from here".

The success of the newer pyrethroids will depend upon their ability to match or surpass the performance of the older compounds, or to offer a broader, or different spectrum of biological activity. With each new introduction the criteria for commercial success became more stringent, but many groups of workers, in industry and in other centres of research, continue to work actively in this area. It would be a rash observer who declared that no further improvement was possible.

Acknowledgements

I would like to thank my colleagues, Dr. G. Boocock and Dr. J. A. Rigby for considerable help with searching of the literature, particularly the patent literature. I am grateful to Lawrence J. King, Professor Emeritus at the State University College of Arts and Science, Geneseo, NY, for drawing to my attention the review of early pyrethrum literature (McDonnell et al., 1920). My thanks are due to colleagues in the Biological Evaluation Division of Sittingbourne Research Centre for permission to use original biological data.

References

Ackerman, P., Bourgeois, F. & Drabek, J., 1980, The optical isomers of α-cyano-3-phenoxybenzyl 3-(1,2-dibromo-2,2-dichloroethyl)-2,2-dimethylcyclopropanecarboxylate and their insecticidal activities. *Pestic. Sci.*, **11**, 169–79.
Aketa, K., Ohno, N., Itaya, N., Nakayama, I. & Yoshioka, H., 1978, Synthesis of diastereoisomers of the recent pyrethroids fenvalerate and cypermethrin (NRDC 149) from (−)-3-phenoxy-mandelic acid and determination of their absolute configurations. *Agric. Biol. Chem.*, **42**, 895–6.
American Cyanamid, 1976[1975], Insecticidal and acaricidal pyrethroid compounds, US Patent 3 966 959.
American Cyanamid, 1980[1976/7], *m*-Phenoxybenzyl and α-cyano-*m*-phenoxybenzyl esters of 2-haloalkyl (oxy-, thio-, sulfinyl-, or sulfonyl) phenylalkanoic acids, US Patent 4 199 595.
Arlt, D., Jautelat, M. & Lantzsch, R., 1981, Synthesis of pyrethroid acids. *Angew. Chem. int. edn.*, **20**, 703–22.
Barlow, F., Elliott, M., Farnham, A. W., Hadaway, A. B., Janes, N. F., Needham, P. H. & Wickham, J. C., 1971, Insecticidal activity of pyrethrins and related compounds. IV. Essential features for insecticidal activity in chrysanthemates and related cyclopropane esters. *Pestic. Sci.*, **2**, 115–8.
Barthel, W. F., 1961, Synthetic pyrethroids. In *Advances in Pest Control Research*, edited by R. L. Metcalf, Vol. 4 (New York: Interscience Publishers, Inc), pp. 33–74.

Bayer, 1978[1977], Substituted phenoxybenzyl carboxylates. Preparation and use as insecticides and acaricides, Ger. Offen. 2 709 264.

Bayer, 1978[1976/7], New halobenzyl cyclopropanecarboxylates, preparation and use as insecticides and acaricides, Belgian Patent 862 109.

Bayer, 1981[1979], (±)-*Trans*-3-(*E*,*Z*-2-chloro-2-(4-chlorophenyl)-vinyl)-2,2-dimethylcyclopropanecarboxylic acid (±)-α-cyano-3-phenoxy-4-fluorobenzyl ester, the individual *E* and *Z* isomers and their use as ectoparasiticides, Ger. Offen. 2 936 864.

Beevor, P. S., Godin, P. J. & Snarey, M., 1965, Jasmolin I, cinerin I and a new method for isolating research quantities of the pyrethroids. *Chem. & Ind.*, 1342–3.

Begley, M. J., Crombie, L., Simmonds, D. J. & Whiting, D. A., 1972, Absolute configuration of the pyrethrins. Configuration and structure of (+)-allethronyl (+)-*trans*-chrysanthemate   6-bromo-2,4-dinitrophenylhydrazone   by   X-ray methods. *Chem. Commun.*, **23**, 1276–7.

Begley, M. J., Crombie, L., Simmonds, D. J. & Whiting, D. A., 1974, X-Ray analysis of synthetic (4*S*)-2-(prop-2-enyl)rethron-4-yl (1*R*, 3*R*) chrysanthemate 6-bromo-2,4-dinitrophenylhydrazone, and chiroptical correlation with the six natural pyrethrin esters. *J. Chem. Soc., Perkin Trans. I*, 1230–5.

Bentley, P. D., Cheetham, R., Huff, R. K., Pascoe, R. & Sayle, J. D., 1980, Fluorinated analogues of chrysanthemic acid. *Pestic. Sci.*, **11**, 156-64.

Berteau, P. E. & Casida, J. E., 1969, Synthesis and insecticidal activity of some pyrethroid-like compounds including ones lacking cyclopropane or ester groupings. *J. Agric. Food Chem.*, **17**, 931–8.

Breese, M. H., 1977, The potential for pyrethroids as agricultural, veterinary, and industrial insecticides. *Pestic. Sci.*, **8**, 264–9.

Breese, M. H. & Highwood, D. P., 1977, Cypermethrin, a new synthetic pyrethroid insecticide. *Proceedings of the 9th British Insecticide and Fungicide Conference (Brighton)*, Vol. 2, (Croydon, UK: British Crop Protection Council), pp. 641–8.

Brempong-Yeboah, C. Y., Saito, T., Miyata, T. & Tsubaki, Y., 1982, Topical toxicity of some pyrethroids. *J. Pestic. Sci.*, **7**, 47–51.

Briggs, G. G., Elliott, M., Farnham, A. W. & Janes, N. F., 1974, Structural aspects of the knockdown of pyrethroids. *Pestic. Sci.*, **5**, 643–9.

Briggs, G. G., Elliott, M., Farnham, A. W. Janes, N. F., Needham, P. H., Pulman, D. A. & Young, S., 1976, Relation of polarity with activity in pyrethroids. *Pestic. Sci*, **7**, 236–40.

Briggs, G. G., Elliott, M. & Janes, N. F., 1983, Present status and future prospects for synthetic pyrethroids. In *Pesticide Chemistry. Human Welfare and the Environment*, Proceedings of the 5th International Congress of Pesticide Chemistry, Kyoto, Japan, 1982, edited by J. Miyamoto & P. C. Kearney, Vol, 2, (Oxford & New York: Pergamon Press), pp. 157–64.

Brown, D. G. & Addor, R. W., 1979, Benzospiro pyrethroids. In *Advances in Pesticide Science (Zurich, 1978)*, edited by H. Geissbuhler, Vol. 2, (Oxford: Pergamon Press), pp. 190–5.

Bull, M. J., Davies, J. H., Searle, R. J. G. & Henry, A. C., 1980, Alkyl aryl ketone oxime *O*-ethers: a novel group of pyrethroids. *Pestic. Sci.*, **11**, 249–56.

Burt, P. E., Elliott, M., Farnham, A. W., Janes, N. F., Needham, P. H. & Pulman, D. A., 1974, The pyrethrins and related compounds. XIX. Geometrical and optical isomers of 2,2-dimethyl-3-(2,2-dichlorovinyl)-cyclopropanecarboxylic acid and insecticidal esters with 5-benzyl-3-furylmethyl and 3-phenoxybenzyl alcohols. *Pestic. Sci.*, **5**, 791–9.

Butt, B. A. & Keller, J. C., 1964, Materials evaluated as insecticides and acaricides at Brownsville, Tex., September 1955 to June 1961. *USDA Agriculture Handbook*, No. 236, pp. 22–24 and 73.

Cahn, R. S., Ingold, C. & Prelog, V., 1956, The specification of asymmetric configuration in organic chemistry. *Experientia*, **12**, 81–124.

Campbell, I. G. M. & Harper, S. H., 1945, Experiments on the synthesis of the pyrethrins. Part I. Synthesis of chrysanthemum monocarboxylic acid. *J. Chem. Soc.*, 283–6.

Cantu, E. & Wolfenbarger, D. A., 1970, Toxicity of three pyrethroids to several insect pests of cotton. *J. Econ. Entomol.*, **63**, 1373–4.

Casida, J. E., 1983, Novel aspects of metabolism of pyrethroids. In *Pesticide Chemistry. Human Welfare and the Enviroment*, Proceedings of the 5th International Congress of Pesticide Chemistry, Kyoto (Japan), 1982, edited by J. Miyamoto & P. C. Kearney, Vol. 2 (Oxford & New York: Pergamon Press), pp. 187–92.

Ciba-Geigy, 1978[1977/8], Cyclopropanecarboxylic acid esters, a method of preparation and their use in pest control, Ger. Offen. 2 805 226.

Ciba-Geigy, 1979[1977], Cyclopropanecarboxylic acid esters, their salts with inorganic acids and their use in pest control, Ger. Offen. 2 829 329.

Commonwealth Scientific and Industrial Research Organisation, 1977[1975], Cyclopropane compounds, preparation and insecticides containg them, Ger. Offen. 2 653 189.

Crombie, L. & Harper, S. H., 1950, Experiments on the synthesis of pyrethrins. Part IV. Synthesis of cinerone, cinerolone and cinerin I. *J. Chem. Soc.*, 1152–60.

Crombie, L., Harper, S. H. & Thompson, D., 1951, Experiments on the synthesis of the pyrethrins. Part VII. Synthesis of *trans*-pyrethrone, *trans*-pyrethrolone and pyrethrin I. *J. Chem. Soc.*, 2906–15.

Crombie, L. & Harper, S. H., 1954, The Chrysanthemum carboxylic acids. Part VI. The configurations of the chrysanthemic acids *J. Chem. Soc.*, 470.

Crombie, L., Harper, S. H. & Newman, F. C., 1956, Experiments on the synthesis of the pyrethroids. Part XI. Synthesis of *cis*-pyrethrolone and pyrethrin I: introduction of the *cis*-penta-2:4-dienyl system by selective hydrogenation. *J. Chem. Soc.*, 3963–71.

Crombie, L., Harper, S. H. & Sleep, K. C., 1957, Experiments on the synthesis of the pyrethrins. Part XIII. Total synthesis of (±)-*cis*- and *trans*-chrysanthemumdicarboxylic acid, (±)-*cis*- and *trans*-pyrethric acid and rethrins II. *J. Chem. Soc.*, 2743–54.

Crombie, L. & Elliott, M., 1961, Chemistry of the natural pyrethrins. *Fortschr. Chem. Org. Naturstoffe*, **19**, 120–64.

Crombie, L., 1980, Chemistry and biosynthesis of natural pyrethrins. *Pestic. Sci.*, **11**, 102–18.

Dianippon Jochugiku, 1968[1966], Substituted furfuryl chrysanthemate insecticides, Fr. Demande 1 550 606.

Davis, R. H. & Searle, R. J. G., 1977, Pyrethroid insecticides derived from some spiroalkane cyclopropanecarboxylic acids. In *Synthetic Pyrethroids, ACS Symposium Series No. 42*, edited by M. Elliott (Washington, DC: American Chemical Society), pp. 37–44.

Dow Chemical Co., 1979[1977], Substituted pyridine methyl esters of cyclopropanecarboxylic acids and their use as insecticides, US Patent 4 163 787.

Elliott, M., Needham, P. H. & Potter, C., 1950, The insecticidal activity of substances related to the pyrethrins I. *Ann. Appl. Biol.*, **37**, 490–507.

Elliott, M., 1954, Allethrin. *J. Sci. Food Agric.*, **5**, 505–14.

Elliott, M., Janes, N. F., Jeffs, K. A., Needham, P. H. & Sawicki, R. M., 1965, New pyrethrin-like esters with high insecticidal activity. *Nature, Lond.*, **207**, 938–40.

Elliott, M., Farnham, A. W., Janes, N. F., Needham, P. H. & Pearson, B. C., 1967, 5-Benzyl-3-furylmethyl chrysanthemate: a new potent insecticide. *Nature, Lond.*, **213**, 493–4.

Elliott, M., 1971, The relationship between the structure and activity of pyrethroids. *Bull. Wld. Hlth. Org.*, **44**, 315–24.

Elliott, M. & Janes, N. F., 1973, Chemistry of the natural pyrethrins. In *Pyrethrum*, edited by J. E. Casida, Chap. 4 (New York & London: Academic Press), pp. 56–100.

Elliott, M., Farnham, A. W., Janes, N. F., Needham, P. H. & Pulman, D. A., 1973a, Potent pyrethroid insecticides from modified cyclopropane acids, *Nature, Lond.*, **244**, 456.

Elliott, M., Farnham, A. W., Janes, N. F., Needham, P. H. Pulman, D. A. & Stevenson, J. H., 1973b, NRDC 143, a more stable pyrethroid. *Proceedings of the 7th British Insecticide and Fungicide Conference (Brighton)*, Vol. 2, pp. 721–8.

Elliott, M., Farnham, A. W., Janes, N. F., Needham, P. H., Pulman, D. A. & Stevenson, J. H., 1973c. A photostable pyrethroid, *Nature, Lond.*, **246**, 169–70.

Elliott, M., Farnham, A. W., Janes, N. F. & Needham, P. H., 1974a, Insecticidal activity of the pyrethrins and related compounds. VI. Methyl-, alkenyl-, and benzylfurfuryl and 3-furylmethyl chrysanthemates. *Pestic. Sci.*, **5**, 491–6.

Elliott, M., Farnham, A. W., Janes, N. F., Needham, P. H. & Pulman, D. A., 1974b, Synthetic insecticide with a new order of activity. *Nature, Lond.*, **248**, 710–11.

Elliott, M., Janes, N. F. & Pulman, D. A., 1974c, The pyrethrins and related compounds. Part XVIII. Insecticidal 2,2-dimethylcyclopropanecarboxylates with new unsaturated 3-substituents. *J. Chem. Soc., Perkin Trans.* **1**, 2470-4.

Elliott, M., Farnham, A. W., Janes, N. F., Needham, P.H. & Pulman, D. A., 1975a, Insecticidal activity of the pyrethrins and related compounds VII. Insecticidal dihalovinyl analogues of *cis* and *trans* chrysanthemates. *Pestic. Sci.*, **6**, 537–42.

Elliott, M., Janes, N. F. & Graham-Bryce, I. J., 1975b, Retrospect on the discovery of a new insecticide. *Proceedings of the 8th British Insecticide and Fungicide Conference (Brighton)*, Vol. 2, pp. 373–9.

Elliott, M., Farnham, A. W., Needham, P. H. & Pulman, D. A., 1976a, Insecticidal activity of the pyrethrins and related compounds. X. 5-Benzyl-3-furylmethyl 2,2-dimethylcyclopropanecarboxylates with ethylenic substituents at position 3 on the cyclopropane ring. *Pestic. Sci.*, **7**, 499–502.

Elliott, M., Farnham, A. W., Needham, P. II. & Pulman, D. A., 1976b, Insecticidal activity of the pyrethrins and related compounds. IX. 5-Benzyl-3-furylmethyl 2,2-dimethylcyclopropanecarboxylates with non-ethylenic substituents at position 3 on the cyclopropane ring. *Pestic. Sci.*, **7**, 492–8.

Elliott, M., 1977, Synthetic pyrethroids. In *Synthetic Pyrethroids, ACS Symposium Series No. 42*, edited by M. Elliott, Chap. 1 (Washington D.C: American Chemical Society).

Elliott, M. & Janes, N. F., 1978, Synthetic pyrethroids–a new class of insecticide. *Chem. Soc. Revs.*, **7**, 473–505.

Elliott, M., Janes, N. F. & Potter, C., 1978, The future of pyrethroids in insect control. *Ann. Rev. Entomol.*, **23**, 443–69.

Elliott, M., 1979, Progress in the design of insecticides. *Chem. & Ind.*, 757–68.

Elliott, M. & Janes, N. F., 1979, Recent structure–activity correlations in synthetic pyrethroids. In *Advances in Pesticide Science*, symposia papers presented at the 4th International Congress of Pesticide Chemistry, Zurich (Switzerland), 1978, edited by H. Geissbuhler, Part 2, (Oxford & New York: Pergamon Press), pp. 166–73.

Elliott, M., Janes, N. F., Khambay, B. P. S. & Pulman, D. A., 1983a, Insecticidal activity of the pyrethrins and related compounds. XIII. Comparison of the effects of alpha-substituents in different esters. *Pestic. Sci.*, **14**, 182–90.

Elliott, M., Janes, N. F., Stevenson, J. H. & Waters, J. H. H, 1983b, Selectivity of pyrethroid insecticides between *Ephestia kuhniella*, and its parasite *Venturia canescens*. *Pestic. Sci.*, **14**, 423–6.

Engel, J. F., McPhail, A. T. & Miller, R. W., 1983a, Synthesis and X-ray crystal structure of a new potent pyrethroid acid, (±)-*cis*-3-[(*Z*)-2-chloro-3,3,3-trifluoroprop-1-enyl]-2,2-dimethylcyclopropanecarboxylic acid. *J. Chem. Soc., Perkin Trans. 1*, 1737–40.

Engel, J. F., Plummer, E. L., Stewart, R. R., VanSaun, W. A., Montgomery, R. E., Cruickshank, P. A., Harnish, W. N., Nethery, A. & Crosby, G., 1983b, Synthesis and biological activity of a new group of broad-spectrum pyrethroid insecticides. In *Pesticide Chemistry. Human Welfare and the Environment*, Proceedings of the 5th International Congress of Pesticide Chemistry (IUPAC), Kyoto (Japan), 1982, edited by J. Miyamoto & P.C. Kearney, Vol. 1, (Oxford & New York: Pergamon Press), pp. 101–6.

Farkas, J., Kourim, P. & Sorm, F., 1959, Relation between chemical structure and insecticidal activity in pyrethroid compounds. I. An analogue of chrysanthemic acid containing chlorine in the side chain. *Coll. Czech. Chem. Communs.*, **24**, 2230–6.

Fisher, J. P., Debray, P. H. & Robinson, J., 1983, WL85871 – a new multipurpose insecticide. In *Plant Protection for Human Welfare Proceedings of the 10th International Congress on Plant Protection (Brighton)*, Vol. I (Croydon; UK: BCPC Publications), pp. 452–9.

FMC, 1981[1979], Process for preparation of a crystalline insecticidal pyrethroid enantiomer pair, US Patent 4 261 921.

FMC, 1982[1979], Insecticidal perhaloalkylvinyl cyclopropanecarboxylates, US Patent 4 332 815.

FMC, 1982[1980], Biphenylmethyl perhaloalkylvinyl cyclopropanecarboxylates for control of acarids, UK Patent 2 085 005.

Fujimoto, K., Itaya, N., Okuno, Y., Kadota, T. & Yamaguchi, T., 1973, A new insecticidal pyrethroid ester. *Agric. Biol. Chem.*, **37**, 2681–2.

Fujitani, J., 1909, Chemistry and pharmacology of insect powder. *Arch. Exp. Path. Pharmak.*, **61**, 47–75.

Gersdorff, W. A. & Mitlin, N., 1953, Effect of molecular configuration on relative toxicity to house flies as demonstrated with the four *trans* isomers of allethrin. *J. Econ. Ent.*, **46**, 999–1003.

Gillam, A. E. & West, T. F., 1944, Absorption spectra and the structure of pyrethrins I and II. Part II. *J. Chem. Soc.*, 49–51.

Godin, P. J., Sleeman, R. J., Snarey, M. & Thain, E. M., 1964, Jasmolin II: a new constituent of pyrethrum extract. *Chem. & Ind.*, 371–2.

Gnadinger, C. B., 1936, *Pyrethrum Flowers*, 2nd edn (Minneapolis: McLaughlin Gormley King Co.), pp. 1–4.

Gnadinger, C. B., 1945, *Pyrethrum Flowers*, suppl. 2nd edn, (Minneapolis: McLaughlin Gormley King Co.).

Hamman, I. & Fuchs, R., 1981, Baythroid, a new insecticide. *Pflanzenschutz-Nachrichten Bayer*, **34**, 121–51.

Head, S. W., 1973, Composition of pyrethrum extract and analysis of pyrethrins. In *Pyrethrum*, edited by J. E. Casida, Chap. 3 (New York & London: Academic Press).

Henrick, C. K. et al., 1980, 2-Anilino-3-methylbutyrates and 2-(isoindolin-2-yl)-3-methylbutyrates, two novel groups of synthetic esters not containing a cyclopropane ring. *Pestic. Sci.*, **11**, 224–41.

Holan, G. B., O'Keefe, D. F., Walser, R. & Virgona, C. T., 1978, Structural and biological link between pyrethroids and DDT in new insecticides. *Nature, Lond.*, **272**, 734–6.

Hopkins, T. J. & Woodley, I. R., 1982, The efficacy of flumethrin (Bayticol) against susceptible and organophosphte resistant strains of the cattle tick *Boophilus microplus* in Australia. *Vet. Med. Rev.*, 130–9.

Imperial Chemical Industries Ltd, 1979[1977/8], Halogenated cyclopropanecarboxylic acid esters, UK Patent 2 000 764.

Imperial Chemical Industries Ltd, 1983[1981], Methods and compositions for combatting soil pests, European Patent 73 566.

Ishimitsu, K., Kasahara, I., Yamada, T., Soma, S. & Kamimura, H., 1982, An attempt at the introduction of nitrogen into the molecule of pyrethroid and the selective toxicity. *Abstracts of the 5th International Congress of Pesticide Chemistry (IUPAC), Kyoto*, Paper Ia-4.

Kato, T., Ueda, K. & Fujimoto, K., 1964, New insecticidally active chrysanthemates, *Agric. Biol. Chem*, **28**, 914–5.

Katsuda, Y., Chikamoto, T. & Inoue, Y., 1958, The absolute configuration of naturally derived pyrethrolone. *Bull. Agric. Chem. Soc. Japan*, **22**, 427–8.

Katsuda, Y., Chikamoto, T. & Inoue, Y., 1959, The absolute configuration of (+)-pyrethrolone and (+)-cinerolone. *Bull. Agric. Chem. Soc. Japan*, **23**, 174–8.

Katsuda, Y., Chikamoto, T., Osami, H., Kunishige, T. & Susii, Y., 1969, Novel insecticidal chrysanthemic esters. *Agric. Biol. Chem.*, **33**, 1361–2.

Katsuda, Y., 1982, Pyrethroids research and development centennial in Japan, *J. Pestic. Sci.*, **7**, 317–27.

LaForge, F. B. & Haller, H. L., 1936, Constituents of pyrethrum flowers. VI. The structure of pyrethrolone. *J. Amer. Chem. Soc.*, **58**, 1777–80.

LaForge, F. B. & Barthel, W. F., 1945, Constituents of pyrethrum flowers. XIX. The structure of cinerolone. *J. Org. Chem.*, **10**, 222–7.

LaForge, F. B. & Soloway, S. B., 1947a, Constituents of pyrethrum flowers. XXI. Revision of the structure of dihydrocinerolone. *J. Amer. Chem. Soc.*, **69**, 2932–5.

LaForge, F. B. & Soloway, S. B., 1947b, The structure of dihydrocinerolone. *J. Amer. Chem. Soc.*, **69**, 186.

LaForge, F. B., Green, N. & Schechter, M. S., 1954, Allethrin. Resolution of *dl*-allethrolone and synthesis of the four optical isomers of *trans*-allethrin, *J. Org. Chem.*, **19**, 457–62.

LaForge, F. B., Green, N. & Schechter, M. S., 1956, Allethrin. Synthesis of four isomers of *cis*-allethrin, *J. Org. Chem.*, **21**, 455–6.

Lawrence, L. J. & Casida, J. E., 1983, Stereospecific action of pyrethroid insecticides on the *gamma*-aminobutyric acid receptor-ionophore complex. *Science*, **221**, 1399–1401.

Lhoste, J., 1964, Les pyrethrines. *Phytoma, defense des cultures*, No. 161, 21–5.

Lhoste, J. & Rauch, F., 1976, RU 15525, a new pyrethroid with a very strong knockdown effect. *Pestic. Sci.*, **7**, 247–50.

Malhotra, S. K., Van Heertum, J. C., Larson, L. L. & Ricks, M. J., 1982, Dowco 417: a potent synthetic pyrethroid insecticide. *J. Agric. Food Chem.*, **29**, 1287–9.

Martel, J. J., 1983, Chirality as a major factor in the selection of highly active compounds in pyrethrinoid series. In *Pesticide Chemistry: Human Welfare and the Environment*, Proceedings of the 5th International Congress of Pesticide Chemistry, Kyoto (Japan), 1982, edited by J. Miyamoto & P. C. Kearney, Vol. 2 (Oxford & New York: Pergamon Press), pp. 165–70.

Matsui, M. & Kitahara, T., 1967, Studies on chrysanthemic acid. Part XVIII. A new biologically active acid component related to chrysanthemic acid, *Agric. Biol. Chem.*, **31**, 1143–50.

Matsunaga, T., Yoshida, K., Shinjo, G., Tsuda, S., Okuno, T. & Yoshioka, H., 1983, New pyrethroid insecticides for indoor applications. In *Pesticide Chemistry. Human Welfare and the Environment*, Proceedings of the 5th International Congress on Pesticide Chemistry, Kyoto (Japan), 1982, edited by J. Miyamoto & P. C. Kearney, Vol. 2 (Oxford & New York: Pergamon Press), pp. 231–8.

Matsuo, T., Itaya, N., Mizutani, T., Ohno, N., Fujimoto, K., Okuno, Y. & Yoshioka, H., 1976, 3-Phenoxy-alpha-cyanobenzyl esters, the most potent synthetic pyrethroids. *Agric. Biol. Chem.*, **40**, 247–9.

Matsuo, T., Nishioka, T., Hirano, M., Suzuki, Y., Tsushima, K., Itaya, N. & Yoshioka, Y., 1980, Recent topics in the chemistry of synthetic pyrethroids containing certain secondary alcohol moieties. *Pestic Sci.*, **11**, 202–8.

Matsuo, N., Yano, T., Yoshioka, H., Kuwahara, S., Sugai, T. & Mori, K., 1983, The absolute structure and activity relationships of major secondary alcohols of pyrethroid insecticides. In *Pesticide Chemistry. Human Welfare and the Environment*, Proceedings of the 5th International Congress of Pesticide Chemistry, Kyoto (Japan), 1982, edited by J. Miyamoto & P. C. Kearney, Vol. 1 (Oxford & New York: Pergamon Press), pp. 279–84.

McDonnell, C. C., Roark, R. C., LaForge, F. B. & Keenan, G. L., 1920 (revised 1926), Insect Powder. *United States Department of Agriculture, Department Bulletin No. 824*, 94 pp.

McLaughlin, G. A., 1973, History of pyrethrum. In *Pyrethrum*, edited by J.E. Casida, Chap. 1. (New York and London: Academic Press), pp. 3–15.

Minamite, Y., Tsuji, Y., Hirobe, H., Ohgami, H. & Katsuda, Y., 1982, Anilinobenzyl alcohol esters as novel pyrethroid insecticides. *J. Pestic. Sci.*, **7**, 349–55.

Miyakado, M., Ohno, N., Okuno, Y., Hirano, M., Fujimoto, K. & Yoshioka, H., 1975, Optical resolution and determination of absolute configurations of alpha-isopropyl-4-substituted phenylacetic acids and insecticidal activities of their 5-benzyl-3-furylmethyl esters. *Agric. Biol. Chem.*, **39**, 267–72.

Miyano, M. & Dorn, C. R., 1973, Prostaglandins. VI. Correlation of the absolute configuration of pyrethrolone with that of the prostaglandins. *J. Amer. Chem. Soc.*, **95**, 2664–9.

Moore, J. B. & Levy, L. W., 1975, Commercial sources of pyrethrum. In *Pyrethrum Flowers*, 3rd edn, edited by R.H. Nelson (Minneapolis: McLaughlin Gormley King Co.), pp. 1–9.

Mowlam, M. D., Highwood, D. P., Dowson, R.J. & Hattori, J., 1977, Field evaluation of fenvalerate, a new synthetic pyrethroid insecticide. *Proceedings of the 9th British Insecticide and Fungicide Conference (Brighton)*, Vol. 2, (Croydon, UK: British Crop Protection Council), pp. 649–56.

Nakayama, I., Ohno, N., Aketa, K. Suzuki, Y., Kato, T. & Yoshioka, H., 1979, Chemistry, absolute structures and biological aspect of the most active isomers of fenvalerate and other recent pyrethroids. In *Advances in Pesticide Science*, symposia papers presented at the 4th International Congress of Pesticide Chemistry, Zurich (Switzerland), 1978, edited by H. Geissbuhler, Part 2 (Oxford & New York: Pergamon Press), pp. 174–81.

National Research Development Corporation, 1969[1965/6], Cyclopropanecarboxylic acid derivatives and their use as insecticides, UK Patent 1 168 797.

National Research Development Corporation, 1974[1972/3], Synthetic insecticides of the pyrethrin type, Fr. Demande 2 211 454.

National Research Development Corporation, 1974[1973/4], Insecticidal stereoisometric esters of 2,2-dimethyl-3-(2,2-dihalovinyl)cyclopropanecarboxylic acids, Belgian Patent 818 811.

National Research Development Corporation, 1975[1972/3], 3-Substituted -2,2-dimethyl-cyclopropanecarboxylic acid esters, their preparation and their use in pesticide compositions. UK Patent 1 413 491.

Naumann, K., 1981, Chemie der Synthetischen Pyrethroid-Insektizide. In *Chemie der Pflanzenschutz und Schadlingsbekampfungsmittel*, edited by R. Wegler, Vol. 7 (Berlin, Heidelberg & New York: Springer-Verlag).

Nicholls, R. & Peecock, P. R., 1981, Control of the fruit tree red spider mite with the synthetic pyrethroid fenpropathrin. *Proceedings of the 11th British Insecticide and Fungicide Conference (Brighton)*, Vol. 1, (Croydon, UK: British Crop Protection Council), pp. 89–95.

Nippon Soda K K, 1981[1979/80], Benzyl alcohol derivatives of isovaleric acid-acaricides and insecticides especially effective against green rice cicadas, Ger. Offen. 3 033 358.

Ohno, N., Fujimoto, K., Okuno, Y., Mizutani, T., Hirano, M., Itaya, N., Honda, T. & Yoshioka, H., 1974, A new class of pyrethroidal insecticides; alpha-substituted phenylacetic acid esters. *Agric. Biol. Chem.*, **38**, 881–3.

Ohno, N., Fujimoto, K., Okuno, Y., Mizutani, T., Hirano, M., Itaya, N., Honda, T. & Yoshioka, H., 1976, 2-Arylalkanoates, a new group of synthetic pyrethroid esters not containing cyclopropanecarboxylates. *Pestic. Sci.*, **7**, 241–6.

Owen, J. D., 1974, Absolute configuration of NRDC 161. *Chem. Commun.*, 859.

Ozawa, K., Ishii, S., Hirata, K. & Hirose, M., 1982, Phenyl cyclopropane carboxylic acid esters as new pyrethroid acaricides and insecticides. *Abstracts of the 5th International Congress of Pesticide Chemistry, Kyoto (Japan)*, Paper Ia-7.

Piedallu, C., Roa, L., Colas, R., Carle. P.-R., Delabarre, M., Escuret, P. & Scheid, J.-P., 1982, Applications of Deltamethrin. In *Deltamethrin*, Chap. 7 (Roussel-Uclaf).

Plummer, E. L. & Pincus, D. S., 1981, Pyrethroid insecticides derived from [1,1-biphenyl]-3-methanol. *J. Agric. Food Chem.*, **29**, 1118–22.

Plummer, E. L., 1983, Pyrethroid insecticides derived from [1,1'-biphenyl]-3-methanol. 2. Heteroaromatic analogues, *J. Agric. Food Chem.*, **31**, 718–21.

Plummer, E. L., Cardis, A. B., Martinez, A. J., VanSaun, W. A., Palmere, R. M., Pincus, D. S. & Stewart, R. R., 1983, Pyrethroid insecticides derived from substituted biphenyl-3-ylmethanols. *Pestic Sci.*, **14**, 560–70.

Roussel/Uclaf, 1972[1970], New derivatives of cyclopropanecarboxylic acids and a method of preparation, Fr. Addn. 2 097 244.

Roussel-Uclaf, 1978[1976/7], New esters of cyclopropanecarboxylic acids with a polyhalogenated substituent. Preparation and pesticidal compositions, Ger. Offen. 2 742 546.

Roussel-Uclaf, 1982, *Deltamethrin*. This book is not for private sale, but copies have been deposited with major libraries. See *Chemistry in Britain*, Jan. 1983, p. 24.

Ruscoe, C. N. E., 1977, The new NRDC pyrethroids as agricultural insecticides. *Pestic. Sci.*, **8**, 236–42.

Ruscoe, C. N. E., 1979, The impact of photostable pyrethroids as agricultural insecticides. *Proceedings of the 10th British Insecticide and Fungicide Conference (Brighton)*, Vol 3, (Croydon UK: British Crop Protection Council), pp. 803–14.

Ruscoe, C. N. E., 1980, Pyrethroids as cotton insecticides. *Outlook Agric.*, **10**, 167–75.

Sanders, H. J. & Taff, A. W., 1954, Allethrin. *Ind. Eng. Chem.*, **46**, 414–26.

Schechter, M. S., Green, N. & LaForge, F. B., 1949, Constituents of pyrethrum flowers. XXIII. Cinerolone and the synthesis of related cyclopentlenolones. *J. Amer. Chem. Soc.*, **71**, 3165–73.

Schechter, M. S., LaForge, F. B., Zimmerli, A. & Thomas, J. M., 1951, Crystalline allethrin isomer. *J. Amer. Chem. Soc.*, **73**, 3541–2.

Shell Oil, 1980[1978], Oxyimino-substituted (1R, cis)-cyclopropanecarboxylate pesticides, US Patent 4 282 249.

Shell Oil, 1981[1978], Oxyimino-substituted (1R, trans)-cyclopropanecarboxylate pesticides, US Patent 4 211 792.

Shell Int. Res. Mij. BV., 1981[1979], Cyclopropane carboxylic acid ester derivatives, UK Patent 2 064 528.

Shell Int. Res. Mij. BV., 1983[1981], Protection of crops against soil pests employing an alkylbenzyl cyclopropanecarboxylate, European Patent 74 132.

Soderlund, D. M., 1983, Pharmokinetic approaches to the activity of pyrethroids in insects. In *Pesticide Chemistry. Human Welfare and the Environment*, Proceedings of the 5th International Congress of Pesticide Chemistry, Kyoto (Japan), 1982, edited by J. Miyamato & P. C. Kearney, Vol 3 (Oxford & New York: Pergamon Press), pp. 69–74.

Soderlund, D. M., Sanborn, J. R. & Lee, P. W., 1983, Metabolism of pyrethrins and pyrethroids in insects. In *Progress in Pesticide Biochemistry and Toxicology*, edited by D. H. Hutson & T. R. Roberts, Vol 3, Chap 8, (Chichester: John Wiley & Sons) pp. 401–35.

Soloway, S. B. & LaForge, F. B., 1947, The synthesis of dihydrocinerolone. *J. Amer. Chem. Soc.*, **69**, 979–80.

Sota, K., Amano, T., Aida, M., Noda, K., Hyashi, A. & Tanaka, I., 1971, New synthetic pyrethroids. 4-Phenyl-2-buten-1-yl and 4-aryl-2-butyn-1-yl chrysanthemates. *Agric. Biol. Chem.*, **35**, 968–70.

Staudinger, H. & Ruzicka, L., 1924a, Insektentotende Stoffe, Parts 1–6. *Helv. Chim. Acta*, **7**, 177–259 and 377–90.

Staudinger, H. & Ruzicka, L., 1924b, 'Insektentotenstoffe', Parts 7–10. *Helv. Chim. Acta*, **7**, 390–458.

Sugiyama, T., Kobayashi, A. & Yamashita, K., 1975, Configurational relation between substituents on cyclopropane ring of pyrethroids and their insecticidal toxicity. *Agric. Biol. Chem.*, **39**, 1483–8.

Sumitomo Chemical Co., 1965[1963]. Chrysanthemum monocarboxylates, Japanese Patent 658 535, *Chem. Abstr.*, **63**, 5689e.

Sumitomo Chemical Co., 1969[1968], Cyclopropanecarboxylic acid esters. Insecticides, Ger. Offen. 1 926 433.

Sumitomo Chemical Co., 1973[1971], Alpha-cyanobenzyl esters of cyclopropanecarboxylic acid compounds. Preparation and use as insecticides and acaricides, Ger. Offen. 2 231 312.

Sumitomo Chemical Co., 1976[1972/3], Substituted acetates and pesticidal compositions containing them, UK Patent 1 439 615.

Taisho Pharmaceutical Co., 1970[1969], Insecticidal 4-phenyl-2-butenyl and 4-phenyl-2-butynyl 2,2-dimethylcyclopropanecarboxylates, Ger. Offen. 2 000 636.

Tessier, J., 1982, Stereoisomers-specific synthesis of the most potent isomers. In *Deltamethrin*, Chap. 2 (Roussel-Uclaf), pp. 38–44.

VanSaun, W. A., Roush, D. M. & Gingrich, H. L., 1983, A regiospecific synthesis route to the 3-substituted benzyl alcohol portion of some highly active broad-spectrum pyrethroid insecticides. *Abstracts of the 185th ACS National Meeting, Seattle, Division of Pesticide Chemistry*, Paper 86.

Velluz, L., Martel, J. & Nomine, G., 1969, Synthese d'analogues de l'acide *trans*-chrysanthemique. *C. R. Acad. Sci., Paris, ser. C*, **268**, 2199–2203.

Wakita, S., Kurokawa, T., Tejima, I., Kato, S., Hirose, K. & Holan, G. B., 1982, DDT–pyrethroid derivatives and their insecticidal activities. *Abstracts of the 5th International Congress of Pesticide Chemistry, Kyoto (Japan)*, Paper IIa–19.

West, T. F., The Structure of Pyrethrolone and Related Compounds, Part V. *J. Chem. Soc.*, 463–6.

Wettstein, K., 1981, AC 222705 pyrethroidal insecticide for use in top fruit and hops. *Proceedings of the 11th British Insecticide and Fungicide Conference. (Brighton)*, Vol. 2, (Croydon, UK: British Crop Protection Council), pp. 563–71.

Whitney, K. W. & Wettstein, K., 1979, A new pyrethroid insecticide: performance against crop pests. *Proceedings of the 10th British Insecticide and Fungicide Conference (Brighton)*, Vol. 2, pp. 387–94.

Yoshioka, H., 1978, Development of fenvalerate, a new and unique synthetic pyrethroid containing the phenylisovaleric acid moiety. *Rev. Plant Prot. Res.*, **11**, 39–52.

Yoshitomi Pharmaceutical Industries, 1969[1967/8], Pesticidal cyclopropane-carboxylic esters, Fr. Demande 1 578 385.

Zoecon, 1978[1977/8], Esters and thiolesters of aminoacids, a method of preparation and their use as insecticides, Belgian Patent 865 114.

2. The mode of action of pyrethroids on insects

T. A. Miller and V. L. Salgado

The insect nervous system

Insects are segmented animals. The central nervous system (CNS) is distinctly segmented with bilateral symmetry. Each of the left and right halves of the CNS of any segment receives sensory axons from receptors on the ipsilateral side of that segment and sends efferent (motor) axons back to the muscles, organs and tissues on the same side. There are some exceptions to this rule where motor axons situated on one side of ganglion might send axons to muscles on the contralateral or opposite side of the same segment, but overall the same-sided symmetry is generally maintained.

All sensory and motor functions are monitored, controlled and integrated by a large population of interneurons. The interneurons may reside wholly in one-half of one ganglion (intraganglionic) or communicate via the nervous connectives with other ganglia (interganglionic). The interganglionic interneurons best known to toxicologists are the giant axons of the ventral nerve cords of the American cockroach, *Periplaneta americana*, which transfer cercal sensory information to higher centres from the terminal abdominal ganglion. Here, as in other cases, sensory axons synapse in the first available segmental ganglion and all information is transfered via synaptic connections to interneurons, and sometimes directly to motor neurons.

Since motor neurons are usually electrically slaved with no inherent ability to produce patterns of electrical activity, and since many motor output patterns can be produced in the absence of sensory input, the interneurons are the key elements in the nervous system for the control of behaviour. It is the interneurons that appear to form and direct the motor output (Pearson, 1981).

Another part of the insect nervous system has been described in some detail in recent years and is made up of neurons that have been called the

43

neurosecretory neurons. The neurosecretory system does not have a well conceived name yet because so little is known of its physiology. Possibly the best known elements in this system are the corpora cardiaca, a pair of neurohemal organs located just behind and innervated by the brain. The corpora cardiaca release neurohormones to perform various functions. For example, the neuropeptide, adipokinetic hormone, is released to mobilise lipid reserves in the locust, in response to various bodily energy needs (Mordue, 1982). Other elements of the neurosecretory system appear to send axons to muscles to modulate tension (O'Shea & Evans, 1979), or to perform direct synaptic control of a tissue, such as the light organ of the firefly, *Photuris versicolor* (Christensen & Carlson, 1981). The characterisation of numerous and regularly placed neurosecretory elements in the median nervous system has led to their being described as the perisympathetic system, although their function is not well known (Raabe et al., 1974).

All manifestations of nerve poisoning are mediated by the motor neurons regardless of the site of action. To determine the site of action of insecticides one must first determine if the motor neuron is being poisoned itself. If not, one searches elsewhere for the cause of an effect. However, if insecticides are acting in some way on the motor neurons, then the study of mode of action is complicated. One must determine the role of the motor neuron in poisoning, and its contribution to an ultimate physiological lesion. Such is the case for pyrethroid insecticides. Indeed, the pyrethroids appear to be acting at virtually every part of the insect nervous system: on sensory neurons (Roeder & Weiant, 1946) on interneurons (Narahashi, 1971) on motor neurons (Yeager & Munson, 1945), and on neurosecretory neurons (Singh & Orchard, 1982).

The sodium channel is probably common to most insect nerve cells. Control of sodium permeability is so vital to nerve function that any long-term disruption here would have drastic consequences for the insect. The mode of action of DDT and pyrethroids on the sodium channel has been studied by only a few groups, principally those of Narahashi (1982) using the squid and crayfish giant axons, and van den Bercken whose work has largely confirmed that of Narahashi. Recently Vijverberg et al. (1982) and Narahashi (1982) concluded that all pyrethroids have a common site of action, the sodium channel. Although the proof for this conclusion is missing, because it involves eliminating all other possible sites of action, it represents a convenient starting point for summarising the mode of action of pyrethroids on target organisms.

Pyrethroid action on the nervous system

The connection between DDT and pyrethroids

Evidence suggests that DDT and pyrethroids act at the same or a very similar site in the insect nervous system, despite the fact that the only obvious underlying chemical similarity is a high lipid solubility and similar partitioning properties; certainly the structures of these groups, the pyrethroids and DDT are dissimilar on first inspection.

DDT and pyrethroids share a number of similar properties. Although some exceptions have been found (Sparks et al., 1982), DDT and pyrethroids are the only insecticides that possess a clear negative temperature coefficient of action on the nervous system (Vinson & Kearns, 1952; Blum & Kearns, 1956). In fact, in examining putative modes of action, it is vital that the actions be examined at different temperatures.

The most compelling evidence for a similar action comes from genetics. Busvine (1951) documented cross-resistance to pyrethrum in an Italian strain of housefly that was resistant to knockdown by DDT. Later termed knockdown resistance (*kdr*), this factor was shown to be recessive, and to be expressed through site or nerve insensitivity (Farnham, 1977). With some important exceptions (Sawicki, 1978) this gene selectively confers resistance to the entire category of both DDT and pyrethroid compounds. The *kdr* resisistance, like temperature, has remained one of the most powerful research tools in the study of the mode of action of pyrethroids and DDT.

Certain pyrethroids, such as barthrin (3,4-methylenedioxy-6-chloro-benzyl-(1*R*, *trans*)-chrysanthemate), produce symptoms of poisoning in motor axons of adult houseflies that are identical to those produced by DDT and DDT analogues and distinguishable from those produced by "pyrethroids" (Adams & Miller, 1980). Both DDT and pyrethroid compounds act on the peripheral nervous system as well as on the central nervous system. Other major groups of insecticides lack any action on peripheral axons.

Both DDT and pyrethroid chemicals produce similar lesions in the motor nerve terminals of a variety of insect species (Salgado et al., 1983a,b), underscoring a similar mode of action that is independent of the structure. The nerve terminal lesions have been examined in about two dozen species of Diptera, Lepidoptera and Coleoptera so far and in every case the nerve terminal sensitivity is lower in strains with a suspected *kdr* resistance mechanism (S. N. Irving, 1983, personal communication and T. A. Miller et al., 1983).

Structure–activity studies are the subject of another chapter in this volume. However, the link between DDT and pyrethroids was emphasised by the work of George Holan and his colleagues who made structural hybrid molecules that resemble both separate groups of compounds as shown below with toxicity data on adult housflies (from Holan et al., 1978). The fact that the hybrid molecules are active is interesting, but does not prove any biological link between the two groups of compound. The DDT analogue and permethrin could be acting on two different receptors, and the hybrid may be acting on either or both of these receptors.

Chang & Plapp (1983) studied the specific binding of radiolabelled DDT and pyrethroids to a membrane fraction prepared from housefly heads. They found

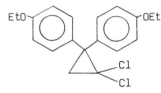

DDT Analog, 6.5 µg/g (2.0)

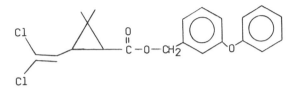

Permethrin, LD$_{50}$ = 2.5 µg/g
(synergized with sesamex = 0.45)

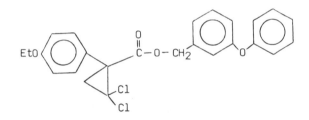

Hybrid, 9.5 µg/g (0.9)

that excess permethrin and cypermethrin inhibit [^{14}C]DDT binding and, conversely, that DDT inhibits [^{14}C]permethrin binding. However, the inhibition was not competitive, suggesting that DDT and pyrethroids might act at two distinct receptor sites that are linked allosterically.

In sum, DDT and pyrethroids share many characteristics in their actions as insecticides, and probably act on the same biomolecule but at different receptor sites. For this reason the mode of action of DDT will be considered similar to, if not identical with, that of pyrethroids for the purpose of the present review.

The concept of primary lesion

When considering the insecticidal properties of chemicals, we often talk of the "receptor". The receptor is even more dramatically emphasised in structure–activity studies where very slight changes lead to dramatic differences in toxicity. Indeed, the tendency to think of a toxic molecule fitting a membrane or enzyme surface in three dimensions is hard to overcome, especially when all one has to go on is toxicity data and the structure of the toxic molecule itself.

The majority of traditional insect toxicology effort has been chemically or biochemically oriented, and very few studies have examined the insect itself or the sequence of events in the poisoning process. As a result of the emphasis on chemical structure, the time factor in poisoning is often overlooked. This is due to the complexity of the poisoning steps and the perception that everything occurring after the "primary lesion" is either too complex for analysis (cf. Richards & Cutkomp, 1945) or due to advanced degenerative changes (Moriarty, 1969). For example, cholinesterase inhibition is toxic when inhibition lasts for a specific length of time, but the reasons for this are obscure because we have not followed the events in the insect started by cholinesterase inhibition, the "primary lesion".

Likewise, there is a time factor in pyrethroid poisoning. It is not a simple matter of causing convulsions or symptoms of poisoning; a given pyrethroid must be allowed to act over a given length of time. What happens in the first few hours following treatment often bears no resemblance to the ultimate fate of the insect being examined.

Cause of death from pyrethroid poisoning

There has been a body of opinion that the secondary actions of insecticide poisoning include loss of ability to retain water in the insect body. The arguments for this view were recently reviewed by Gerolt (1983) and previously by Moriarty (1969) among others. Unfortunately, the consequences of insecticide action on insects very much depend on the chemical, the insect and the conditions.

The consequences of pyrethroid action are manifold, but one recurring theme, autotoxication, persists throughout the literature (Blum & Kearns, 1956; Sternburg, 1963; Moriarty, 1969; Narahashi, 1971). It is said that after a certain point in poisoning with any insecticide, or mechanical or electrical stress, a "toxin" appears in the haemolymph. The key observation is that haemolymph extracted from prostrate DDT-poisoned coackroaches and injected into houseflies produces typical symptoms of DDT poisoning, but the amount of DDT present is insufficient to cause the observed poisoning symptoms. Furthermore, haemolymph from prostrate cockroaches, when assayed on ventral nerve cords, caused high-frequency impulse trains within a few minutes, and these increased in frequency until all activity stopped (cf. Moriarty, 1969).

There has been an assumption or implication in the autointoxication theory that the cessation of activity in the cockroach ventral nerve cord correlates with the prostrate state of poisoning, which has been described as a relaxed paralysis with constant tremors (Sternburg & Hewitt, 1962). However, this basic assumption may be incorrect and the ventral nerve cord assay somewhat misleading. As shown by Gammon (1978b) at 16·5 °C, a cockroach, prostrate from topical application of 5·25 µg of DDT, exhibited after-discharges in the ventral nerve cord in response to cercal nerve stimulation up to 171 hours after treatment. Clearly the cercal-giant fibre synaptic pathway functioned long after poisoning had led to "prostration".

At 32 °C, treatment with 27·6 µg of DDT was followed by the development of increased and prolonged afferent discharge in the ventral nerve cord four to six hours after treatment, but the cockroach exhibited no poisoning symptoms. From 5 to 30 hours following treatment, spontaneous trains were recorded from the ventral nerve cord and the cockroach was ataxic. Tremors were not seen consistently at 32 °C. At about 30 hours after treatment, cockroaches became prostrate; spontaneous activity declined and nervous activity was blocked between 24 and 80 hours after treatment (Gammon, 1978b).

At 25 °C, treatment with 20 µg of DDT produced continuous tremors after 6·5 hours. Prostration at one day post-treatment was accompanied by increasing nerve after-discharges in the ventral nerve cord reaching a maximum at 37 hours and eventually blocking about 50 hours after treatment.

Clearly the spontaneous and neurally evoked responses of the ventral nerve cord are highly temperature-dependent in DDT poisoning. As pointed out by Hutzel (1942), separation of the abdominal ventral nerve cord from the thoracic ganglia at the abdomen–thorax juncture does not change the period of initial excitation in contact-poisoned German cockroaches. Therefore it is rather difficult to determine the cause of ataxia by observing the activity of the ventral nerve cord and not the motor neurons supplying the legs.

When the action of allethrin was examined in a parallel manner, again there were qualitative differences. At colder temperature (15 °C), allethrin was more active on the sensory afferents. Nerve blockage did not occur until long after paralysis and could even be a secondary consequence of allethrin poisoning (Gammon, 1978a).

The American cockroach, *Periplaneta americana*, has an unusual and unique reaction to stress that has been documented many times (Sternburg, 1963). If agitated or shocked, or even simply restrained (Figure 2.1) the cockroaches may show no symptoms (as one of those depicted in Figure 2.1A) or they may show a greatly increased oxygen consumption and gradually develop paralysis of the legs, usually metathoracic legs first, until eventually they become prostrate (Figure 2.1A, arrow) and die. The increase in oxygen consumption due to either DDT poisoning (Figure 2.1B, arrow) or restraint is said to be a "generalised stress syndrome" superimposed over and perhaps obscuring the response of the cockroach to poisoning (Heslop & Ray, 1959) complicating mode-of-action studies. More recent studies show that not only does oxygen consumption increase in pyrethroid-poisoned cockroaches (as would be expected of convulsant poisons), but diuresis activity and reduction in collectable haemolymph occur as well (Soderlund, 1979).

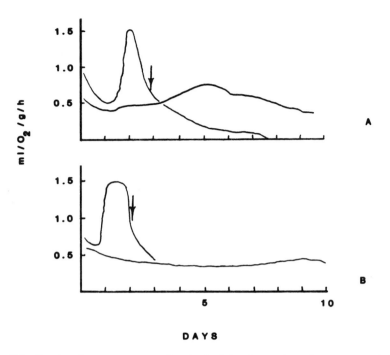

Figure 2.1. Graphs of oxygen consumption by *Periplaneta americana* L. over several days. In (A), two cockroaches were restrained, one of which became paralysed. In (B), oxygen consumption of two unrestrained cockroaches was measured, one of which was treated with DDT, the other a control. The cockroaches showing increased oxygen consumption became prostrate at the arrow. Redrawn from Heslop & Ray (1959)

Gerolt (1983) argues for a single cataclysmic event causing death from insecticides: "poisoning" of the integument leading to loss of water and respiratory functions. The scheme was presented without evidence as an attempt to reconcile many observations reported on treated insects, and refutes (sometimes poorly) evidence to the contrary.

Some clue as to the consequences of DDT poisoning comes from the antidotes used to reverse symptoms of poisoning. Non-anaesthetic doses of phenobarbital controlled tremors and convulsions from DDT poisoning. A host of other materials: urethan, barbital, pentobarbital, $MgSO_4$, calcium and sodium dilantin could reverse symptoms but only at high doses and offered only partial protection from DDT poisoning (Hrdina et al., 1975). The protective action of narcosis on insects and mammals was noted from the beginning (Läuger, 1946; Merrill et al., 1946). Narcosis maintained for 2–3 days with phenylacetylbarbiturate afforded rats a small measure of survival. An oral dose of 0·5 gm/kg DDT killed 12 out of 12 non-narcotised rats, but 2 of 9 rats kept under continued narcosis survived the same dose of DDT.

Desoxylic acid caused a considerable decrease in toxicity of DDT to flies, and injection of calcium ions into the haemolymph of flies retarded death (Läuger, 1946). Although treatment of houseflies with cholesterol and DDT reduced the symptoms, the houseflies became prostrate in the same amount of time as without cholesterol.

The injection of the venom from *Microbracon hebetor* (Say) into wax moth, *Galleria mellonella*, larvae masked the development of DDT symptoms (Beard, 1958). The *Microbracon* venom is thought to block skeletal muscle activity, but does not prevent vegetative functions. DDT is toxic to *Galleria* larvae at 74 µg/gm injected in oil, but relatively non-toxic when applied topically. When pretreated with *Microbracon* venom, *Galleria* larvae failed to develop poisoning symptoms when subsequently injected with lethal doses of DDT. In fact, larvae paralysed with venom alone were indistinguishable from the same paralysed larvae then injected with DDT. Thus DDT injection did not hasten death with the skeletal muscles blocked.

Gammon (1978c) found that allethrin poisoning was somewhat alleviated by pretreatment of American cockroaches with sublethal injections of tetrodo-toxin (TTX). This he interpreted as evidence of a primary action on the peripheral nervous system since the protective action of TTX would presumably be limited to areas outside of the blood–brain barrier which prevents TTX from penetrating past the perineurium. Curiously, the protective action of TTX does not extend to DDT, whose central actions are thought to require fairly high doses.

Symptoms of poisoning and type I versus type II actions

Neurotoxic insectides usually produce ataxia, tremors and convulsions in varying degrees in treated insects. DDT produces quite distinct tremors and pyrethroids produce a variety of symptoms which have been grouped into two main catetories designated I and II, plus a third category that is neither entirely I nor II (Clements & May, 1977; Gammon et al., 1981; Glickman & Casida, 1982; Ray, 1982; Staatz et al., 1982; Scott & Matsumura, 1983).

These classes were constructed from overt behavioural symptoms or based on initial actions of the pyrethroids on nervous activity of both insects and mammals. Most pyrethroid analogues are classified this way, but it isn't clear what these classifications mean in terms of mode of action, although some are inclined to differentiate between primarily central actions by type II pyrethroids (largely α-cyanophenoxybenzyl insecticides, regardless of acid structure) and peripheral actions by other pyrethroids (non α-cyano substituted, and most other pyrethroids such as allethrin and bioresmethrin) which are termed type I compounds (Gammon et al., 1981; Ray and Cremer, 1979; Staatz et al., 1982). However, there are always exceptions to these categories (Gammon et al., 1981; Scott & Matsumura, 1983) giving the impression of a continuum of activity.

Wu et al. (1975) also noticed categories of symptoms produced by DDT analogues on crayfish giant axons. They described excitatory, blocking and dualist actions. The blocking action suppressed the action potential without affecting the resting potential, and the categories had no sharp distinct boundaries. However, the concentrations used in this study were high (0.1mM) and the poisoning symptoms were not examined at different temperatures; thus, it was impossible to tell which, if any, of the symptoms were related to toxicity.

DDT and pyrethroid compounds might be categorised on the basis of their poisoning behaviour on motor neurons at different temperatures. Here, also, two distinct categories were reported, correlated also with the pattern of repetitive discharges in the flight motor nerves during poisoning (Adams & Miller, 1980).

The distinction between peripheral versus central action is not a trivial matter. Very little is known about the details of disruption of the insect nervous system in relation to the poisoning process. Specifically, the nerve centres controlling locomotion have been studied little or not at all with respect to their role in the poisoning process. Actions of pyrethroids on the central or peripheral nervous system are discussed on pp. 64–68.

Burst discharge

The intriguing suggestion was made some time ago (M. Elliott, 1978, personal communication) that α-cyanophenoxybenzyl (or type II) pyrethroids might be displaying the raw toxicity of the pyrethroid molecule free from side effects. The chief observation of note here is that type I pyrethroids produce tremors, convulsions and hyperactivity in insects, while type II pyrethroids, though usually more toxic, are conspicuously less spectacular in producing symptoms. Studies on burst discharges produced in nerves by pyrethroids tend to support Dr Elliott's impression.

Clements & May (1977) cut the metathoracic crural nerve (main nerve supplying the jumping leg) of the locust, *Locusta migratoria*, and recorded ascending nervous activity, that is, orthodromic activity in the sensory nerves, and any antidromic activity in the motor axons travelling from the periphery towards the metathoracic ganglion. They recorded the minimum dose of various pyrethroids necessary to produce one of four responses: continuous firing of the crural nerve (type 1), repetitive after-discharges with muscle contractions (type 2), sustained contractions (type 3), and block of neurally-evoked contractions within 20 minutes (type 4). Laying aside for a moment the possible effects of lower temperatures on these same responses, one may learn a great deal from the results presented.

Compound numbers I and VII (Table 2.1) showed antidromic bursts in the motor nerves within 3 minutes of perfusion of the femur with 10 nM concentrations. These compounds also showed a greatly increased force of

Table 2.1. A summary of the pharmacological actions of pyrethroids perfused through the metathoracic leg of the locust, *Locusta migratoria*. Taken from Clements & May (1977)

| | R_1 | R_2 | R_3 | \multicolumn{4}{c}{Lowest concentration (μM) to cause action of type:[a]} | 1 h KD50 | 5 day LD50 |
				1	2	3	4	\multicolumn{2}{c}{(μg/g)}
I	CH_3	CH_3	H	—	0·01	—	> 100	40·0	6·5
II	CH_3	CH_3	CN	0·1	—	0·01	0·1	1·4	0·87
III	CH_3	CH_3	CH_2CN	—	1·0	—	> 100		
IV	CH_3	CH_3	CH_3	—	0·01	—	10		
V	CH_3	CH_3	C CH	1·0	0·01	—	100		
VI	CH_3	CH_3	$CH_2CH=CH_2$	—	0·1	—	100		
VII	Cl	Cl	H	—	0·001	—	10	150	130
VIII	Cl	Cl	CN	0·1	—	1·0	1·0	99	100

IX	—	—	H	1·0	0·1	—	100	34·0	97
X	—	—	CN	0·1	—	0·1	10	2·5	5·2
XI	pyrethrin I			0·01	—	0·01	1·0	9·8	11·3
XII	pyrethrin II			0·1	—	0·1	1·0	10·4	44·0
XIII	S-bioallethrin			0·1	0·01	0·01	10·0	3·0	11·2
XIV	tetramethrin			0·01	0·01	1·0	1·0	20·0	37·0

[a] Type 1 action – continuous firing of nerve; type 2 – repetitive after-discharges with muscle contractions; type 3 – sustained contraction; type 4 – block of neurally-evoked contraction within 20 min. A bar indicates no response seen up to concentration used for type 4 response.

contraction in response to single shocks of the crural nerve, evidently due to a summated excitatory postsynaptic potential produced by a burst of nervous impulses in the motor axon (burst discharge). Careful examination of Table 2.1 shows that high knockdown activity (marked KD) was not correlated with ability to cause bursts in motor axons (cf. compounds II, XI, XII, XIII, XIV). Compound I, in particular, showed relatively poor knockdown ability, yet produced motor burst discharges at 10^{-8} M. Since this compound was relatively toxic, one might assume that it either penetrated very slowly, or that knockdown is not related to actions on motor units. The latter choice is supported by the overall evidence.

Further comparisons from Table 2.1 show that continuous orthodromic (presumed sensory) discharge was associated with knockdown properties. All compounds reported as having good knockdown produced the discharges without muscle contractions at or below 100 nM. An exception to this was compound VIII which produced discharges at 100 nM but had little or no knockdown and relatively low toxicity. Miller & Adams (1977) also found that knockdown was correlated with ability to cause repetitive firing in sensory nerves.

The data in Table 2.1 also show that a greater concentration of most pyrethroids is required to produce block of motor functions (type 4) than to produce either sensory or motor axon discharges. Possibly the properties of block are more important at lower temperatures; however, in the results of Clements & May (1977), block appears only at concentrations 10 to 1000 times higher than that necessary to produce other axonal actions.

Narahashi (1971) showed that burst discharges induced by allethrin in giant axons of the cockroach at 33 °C reverted to single potentials at 26 °C. He noted that about 20 nM allethrin induced repetitive discharges at 30 °C, but 1·1 μM was necessary to produce the discharge in axons at 15 °C. Repetitive discharge was found to be positively correlated with temperature between 33 and 15 °C. He further identified nerve blocking action as negatively correlated with temperature. The same results were reported for housefly motor axons (Adams & Miller, 1980) and squid giant interneurons (Starkus & Narahashi, 1978).

The restriction of burst discharges to a relatively narrow temperature range as reported several years ago by Narahashi (1963) does not appear to have been fully appreciated, judging from the number of assays employing burst discharges in the interim. In effect, one must discount repetitive discharges and similar signs of hyperactivity in searching for the underlying mode of action of pyrethroids and DDT.

Trains of nerve impulses in sensory axons of the cockroach leg increased in frequency at lower temperatures after injection of 30 nM DDT and the trains (or bursts) were reduced at high temperatures (Narahashi, 1971). Unfortunately, studies of the effects of temperature on poisoning of sensory nerves have seldom used single identified axons nor a full range of temperatures from 1 to 35 °C. Thus, if the entire sensory activity of a cercal nerve is recorded, for example, it is difficult to determine if individual axons burst only over a discrete range of temperatures.

Although Eaton & Sternburg (1964) found a positive coefficient of temperature for sensory axons in the cockroach leg and a negative temperature coefficient for the ventral nerve cord, Narahashi (1971) contends that the response of sensory axons to DDT is dose dependent; that as temperature is lowered the threshold concentration of DDT to produce trains becomes lower.

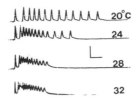

Figure 2.2. The effect of temperature on repetitive discharge in flight motor units measured in an intact fly. The intact house fly was treated with an LD50 (0·2 μg) of bioneopynamin and electrodes were implanted after symptoms developed about 10 min later. Discharges were evoked in dorsolongitudinal motor neuron 5 and recorded from muscle 5. The top trace was taken just above the transition temperature where single potentials became repetitive. As the temperature increases, the frequency of discharge increases while the duration is shortened. Calibration: 50 mV, 20 ms. Taken from Adams & Miller (1979)

This point concerning the response of axons to DDT or pyrethroids at different temperatures is vital to understanding the early symptoms of poisoning.

The number of spikes in a train increased at lower temperatures in the allethrin-treated lateral-line organs of the clawed-frog, *Xenopus laevis* (van den Bercken et al., 1973). However, this is a bit misleading because the frequency of spikes *decreased* at lower temperatures. A preparation employing evoked nervous impulses in the two dorsal-most dorsolongitudinal flight muscles of the adult housefly, *Musca domestica*, poisoned with bioneopynamin, shows a similar result (Figure 2.2). Here the burst duration increased at lower temperatures and the number of impulses remained about the same, but below 20 °C a single stimulus evoked only single potentials (Adams & Miller, 1979). Thus the severity of burst discharge symptoms produced in this motor unit were felt only above 20 °C, and the results of van den Bercken et al. (1973) were showing similar responses.

When examined over a much broader range of temperatures, poisoning by *trans*-barthrin produced burst discharges between 33 and 12 °C (Figure 2.3) in a motor axon preparation identical to that used to obtain the results in Figure 2.2 (Adams & Miller, 1980).

A number of DDT and pyrethroid analogues were examined for their ability to produce evoked burst discharges in the flight motor nerve DLM 5 at various temperatures (Table 2.2). In all cases the burst discharges decreased at lower temperatures to single responses but in a uniform manner. All "pyrethroid" compounds producing bursts reverted to single potentials near 19 °C and all DDT-like compounds reverted to single potentials at about 10 °C. Oddly, the "pyrethroid" *trans*-barthrin fell into the DDT group.

If these effects on motor and interneurons hold true for sensory neurons – the evidence on sensory neurons is a bit cloudy because outside of the lateral line work on frog, single identified sensory structures have not been studied in response to a range of DDT and pyrethroid compounds – then burst discharge cannot be involved in the mode of action of pyrethroid or DDT insecticides. Stated another way: a pyrethroid insecticide does not have to produce burst

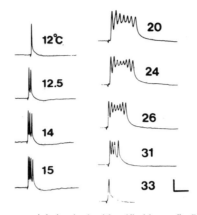

Figure 2.3. Evoked muscle potentials in single, identified housefly flight motor axons at various temperatures during poisoning by *trans*-barthrin. *Trans*-barthrin backfiring persisted only between 12·5 and 33 °C. Burst discharge duration increased as temperature was raised above 12 °C. Bursts were longest in response to single stimuli near 20 °C, but above 33 °C the repetitive activity again reverted to single responses. These trends were fully reversible and responses depended somewhat on frequency of stimulation. Calibration: 12·5–15 °C, 25 mV, 0·1 s; 20–33 °C, 25 mV, 20 ms. From Adams & Miller (1980)

Table 2.2. Transition temperature (temperature at which single stimuli evoked single responses in flight motor axons instead of trains of nerve impulses) of various pyrethroid and DDT analogues acting on adult female houseflies. From Adams & Miller (1980)

Compound	Backfiring[a]	Dose[b] female (μg/fly)	Transition temperature (n)	"Uncoupling"[c]
Tetramethrin	+	0·2 (0·2)	19 ± 2·2 (17)	+
S-bioallethrin	+	0·06 (0·1)	18 ± 2·9 (6)	+
Permethrin	+	0·05 (0·05)	18 (1)	+
Cismethrin	+	0·1 (0·05)	ND	+
Decamethrin	−	0·05 (0·003)	—	+
Kadethrin	−	0·3 (0·07)	—	+
Pydrin	−	0·2 (0·04)	—	+
Trans-barthrin[d]	+	1·0 (1·0)	10.4 ± 1·5 (5)	−
DDT	+	1·0 (0·17)	13·5 ± 3·4 (5)	−
GH-74[e]	+	0·3 (0·3)	11·1 ± 0·6 (4)	−
CRC 12000[e]	+	0·2 (0·2)	ND	−

[a] Indicates antidromic conduction of motor nerve impulses during convulsive bursts.
[b] Values in parentheses = LD50 (μg/female).
[c] Indicates whether central coordination of flight motor axons was thought to reflect an action on the CNS.
[d] Note that *trans*-barthrin, a pyrethroid analogue, shows actions more characteristic of DDT than pyrethroids.
[e] DDT analogues
ND, not experimentally determined.

discharges to be toxic to insects. The worth of burst discharge as an indicator of inherent activity is put into perspective most dramatically by the large number of insecticidal α-cyano-substituted pyrethroids that do not produce burst discharge (Clements & May, 1977; Adams & Miller, 1980). One can conclude from this that either α-cyano pyrethroids are acting on a different site (subscribed to by Casida et al., 1983), or that the pyrethroids as a whole are causing some other more subtle poisoning symptom that is hidden by the more dramatic bursts of nerve activity seen with DDT and the type I pyrethroids.

It has been noted for some time that pyrethroids and DDT cause first excitation (burst discharge) and then block of nerve activity. The block has been shown on numerous occasions to occur more rapidly at lower temperatures or higher concentrations (Roeder & Weiant, 1951; Wang et al., 1972; Adams & Miller, 1980). Thus, block of nerve activity is correlated the correct way with toxicity to pyrethroids and DDT.

Actions of pyrethroids on the neuromuscular junction

The DDT analogue EDO (2,2-*bis*(*p*-ethoxyphenyl)-3,3-dimethyloxetane) had no effect on muscle electrical properties when perfused on to the claw-opener muscle of the crayfish, *Procambarus clarkii*, at a concentration of 10 nM. EDO did, however, potentiate end-plate potentials by causing multiple spiking in the nerve (Farley et al., 1979). Concentrations of EDO higher than micromolar caused spontaneous activity in the motor nerve and increased minature end-plate potentials (mEPPs). In the presence of tetrodotoxin, EDO did not produce nerve excitation or mEPPs. This was interpreted as meaning that the nerve excitation caused mEPPs (Farley et al., 1979). On hind-sight the production of mEPPs probably had nothing to do with nerve discharges, but rather was due to simple nerve terminal depolarisation.

Wouters & van den Bercken (1978) found a similar action of 1 μM allethrin on frog neuromuscular transmission, but the repetitive discharges were found to occur only at the nerve terminal itself and not in the main part of the axon. Presynaptic repetitive firing was also found to occur in the presynaptic terminals of the Xth sympathetic ganglion of the frog.

Wouters & van den Bercken (1978) made the useful observation that the sustained muscle concentration reported by Clements & May (1977) in the absence of repetitive hyperactivity in the pyrethroid-poisoned crural nerve of locust also suggested a local discharge of motor nerve terminals. They also noted that Clements & May reported that all insecticidal pyrethroids eventually blocked muscle contraction altogether, but *not* conduction in the motor axon, pointing specifically to a lesion in the terminal itself. One is tempted to conclude from this that pyrethroids are not strictly axonal poisons.

These observations by Wouters & van den Bercken about the lability of the motor nerve terminal are similar to those of Adams & Miller (1979) about the flight motor nerve terminal of adult houseflies. The specific site of initiation of bursts of impulses in response of pyrethroids and DDT is the nerve terminal, and there is little or no action on the axon itself.

The motor nerve terminals of dorsolongitudinal flight muscles of Diptera are notoriously difficult to locate because of the invaginations of the muscle cell, and miniature postsynaptic potentials (mEPSPs) are not normally recorded there. The body wall muscles of the larval housefly, on the other hand, are an ideal neuromuscular system in which to study synaptic transmission in insects. Osborne and his colleagues at the University of Birmingham characterised this system in a number of remarkable papers. Later, Irving found evidence that the neurotransmitters at the two synapses were different, each mimicked by glutamate and aspartate, respectively (Irving & Miller, 1980). When applying pyrethroids to the larval dipteran nerve–muscle system, Irving first noted a gradual increase in the rate of production of mEPSPs. Further study led to the conclusion that the presynaptic terminal is an important site of action of pyrethroids and DDT (Salgado et al., 1983a,b).

The increase of mEPSP rate was produced by all types of pyrethroids, regardless of α-cyano phenoxybenzyl substitution, and by DDT. Sometimes the compounds also produce nerve discharges, sometimes not. The compounds produced these effects at clearly physiological concentrations at the nanomolar level, but sometimes took a considerable time to show the effects, especially at low concentrations. It was thought that here at last was a common lesion site shared by all members of the pyrethroid–DDT group.

The increase of mEPSP rate by pyrethroid treatment was reversibly blocked by calcium-free solutions or by tetrodotoxin, a specific blocker of sodium channels. This information, along with several more sophisticated experiments, led to the conclusion that these compounds were acting on the sodium channels to depolarise the presynaptic terminals. The magnitude of the depolarisation was estimated to be about 30 mV.

The ability of pyrethroids to depolarise nerve membranes has been noted in passing for years in the midst of all the biophysical work on sodium channels and pyrethroid action. Most recently, Nishimura et al. (1983) noted that pyrethroids and DDT analogues depolarised crayfish giant interneurons. They reported that DDT caused a maximum of only 5% depolarisation, even at massive doses (millimolar), and that pyrethroids caused greater depolarisation. As a note of caution here, it might not be as important how much a nerve terminal, for example, is depolarised as how long the depolarisation persists. Five percent depolarisation might take longer to produce a lesion, but might in the end by as toxic as a compound producing a larger depolarisation when metabolism is taken into consideration. In addition, as yet we have no information on whether sodium channels on different parts of the neuron are exact copies or enjoy a sort of regional specificity and therefore are specific in their responses.

Salgado et al. (1983a) found that increase in mEPSP rate produced by pyrethroids or DDT eventually led to block of neuromuscular transmission. Initially, the block could be reversed by hyperpolarising the nerve, but eventually the mEPSP rate declined almost to zero and the block became irreversible. This later block was thought to represent an irreversible lesion and could be duplicated by a lengthy perfusion of a nerve–muscle preparation by elevated potassium concentrations. Effects on neuromuscular transmission during poisoning were studied in *Culex* larvae kept in dishes treated with permethrin. The severity of neuromuscular block reflected the poisoned condition of the larvae in all cases (Salgado et al., 1983b).

Additional evidence was obtained to demonstrate the importance of the presynaptic lesion in poisoning. Among the compounds examined, depolarising potency was correlated with toxicity. The ability of deltamethrin to increase mEPSP rate and block synaptic transmission had a negative temperature dependence, and *kdr* insects were resistant to the presynaptic actions of pyrethroids in proportion to their resistance (Salgado et al., 1983b).

The motor terminals are in the periphery, but notes by Wouters & van den Bercken (1978) and Dresden (1948) suggest that central presynaptic terminals are also involved in poisoning. Unfortunately, it is not as simple to record from central synapses with the possible exception of the cercal-giant synapse in cockroach, and even here, mEPSPs are normally not recorded and the role of the sixth abdominal ganglion of the cockroach in poisoning is dubious – so this synapse is valuable as a model site of action only. It might be far easier to follow the progress of poisoning by monitoring the function of reflex response pathways that are known and identified in the insect. For this reason it is particularly unfortunate that the approach taken by Dresden (1948) was not followed up and still today remains to be examined in further detail.

Studies on the mode of action of pyrethroids always have been and always will be complicated because the potential sites that can be affected are scattered throughout the nervous system. Any slight change in the partitioning properties of a molecule might change the total number of sites affected and change the toxicity.

Sodium-channel poisons can be toxic entirely by virtue of an action on motor nerve terminals, that is, exclusively in the periphery. Alternatively a few or many lesion sites in the CNS can be recruited to produce toxic effects depending on the compound. This should make structure–activity correlations very difficult to find.

Temperature and pyrethroid action

Temperature and the mode of action of pyrethroids have been inseparably connected since Vinson & Kearns (1952) concluded that DDT was more toxic at colder temperatures by virtue of an intrinsic susceptibility of some physiological system rather than penetration or metabolism. The same conclusion was reached by Blum & Kearns (1956) concerning the action of pyrethrum on the American cockroach.

Table 2.3. Toxicity of DDT and certain 3-phenoxybenzyl pyrethroids against third instar larvae of tobacco budworm, *Heliothis virescens* at 37·8, 26·7 and 15·6 °C. Taken from Sparks et al. (1983)

Compound	Temp (°C)	LD50 (μg/g)	LD50 ratio 37·8 : 15·6 °C
Phenothrin	37·8	4·64	
	26·7	2·51	−24·2
	15·6	0·19	
Permethrin	37·8	1·944	
	26·7	1·440	−9·00
	15·6	0·216	
Cypermethrin	37·8	0·511	
	26·7	0·241	−1·81
	15·6	0·283	
Tralomethrin[a]	37·8	0·056	
	26·7	0·061	+1·55
	15·6	0·087	
Deltamethrin	37·8	0·016	
	26·7	0·044	+5·50
	15·6	0·088	
Fenvalerate	37·8	0·224	
	26·7	0·396	+2·29
	15·6	0·512	
Flucythrinate (Pay-off)	37·8	0·231	
	26·7	0·254	−2·01
	15·6	0·115	
DDT	37·8	62·36	
	26·7	31·49	−15·35
	15·6	4·06	

[a] Tralomethrin is thought to be insecticidal by breaking down rapidly to deltamethrin (Irving & Fraser, 1984).

The temperature dependence of toxicity of some modern pyrethroids shows interesting variations in the tobacco budworm. *Heliothis virescens* (Sparks et al., 1982, 1983; Table 2.3). Note first that a strictly negative temperature coefficient of toxicity of pyrethroids breaks down. Some 3-phenoxybenzyl pyrethroids lacking α-cyano substitution achieve high negative temperature coefficients (permethrin, phenothrin) similar to DDT. However, on *Heliothis virescens*, α-cyano-substituted 3-phenoxybenzyl pyrethroids are weakly or moderately positively correlated with toxicity (tralomethrin, deltamethrin and fenvalerate) or are slightly negatively correlated (flucythrinate and cypermethrin). The temperature dependence is not predictable and each pest species may show a different pattern. The boll weevil, *Anthonomus grandis grandis*, for example, does show a negative temperature dependence for α-cyano compounds, being 13 times more susceptible to deltamethrin and 7·4 times more susceptible to fenvalerate at 15·6 °C as opposed to 37·8 °C.

Considerations of the toxicity of pyrethroids at various temperature should be a vital part of any pest control practice. In the fall of 1981, widespread failure of permethrin to control tobacco budworm on cotton in the Imperial Valley, California, occurred during a 7 to 10-day period when the temperature at night stayed abnormally high, over 100 °F (about 40 °C). One glance at Table 2.3 shows that the toxicity of permethrin to *Heliothis* larvae drops off appreciably at higher temperatures. This negative temperature coefficient was thought to explain the failure of control. When the night-time temperatures returned to normal cooler values, permethrin treatments again were seen to control budworm.

Pyrethroids and DDT sometimes act very quickly to produce symptoms and cause a loss of coordination, often accompanied by spasms or tremors. When compounds act especially quickly they are said to have good knockdown. Because of the negative temperature coefficient of toxicity, the poisoning symptoms produced by a moderate dose may be reversed by raising the temperature. The ability to reverse poisoning symptoms persists until a certain minimum amount of time has passed. These relationships were studied with DDT and pyrethrum (Lindquist et al., 1945; Table 2.4).

Table 2.4. Recovery and kill of flies exposed at 65 °F in a box treated with DDT, transferred to a chamber at 70 °F after knockdown, and after various intervals removed to another chamber at 100 °F. 100 flies in each test. From Lindquist et al. (1945)

Time flies held at 70 °F (h)	Recovery after 1 hour at 100 °F (%)	Kill after 24 hours at 100 °F (%)
0	97	50
1	29	87
2	17	94
3	14	96
4	7	98
5	6	97
6	2	98
24	0	100

A residue of DDT that knocked down houseflies in 63 minutes at room temperature, 65 °F, caused 50% mortality only if the flies were immediately removed to 100 °F (37·8 °C). However, if the houseflies so treated were kept at 70 °F (21·2 °C) for 1 hour, then transferred to 100 °F, an 87% mortality was measured after 24 hours (Lindquist et al., 1945). Although the test regime used in this study was a bit unwieldy (hold on DDT residue until knockdown at 65 °F, then move to untreated chamber at 70 °F, then move to 100 °F at various times), the point illustrated is fairly clear.

Note that 97% of treated houseflies "rescued" by removal to 100 °F recovered and subsequently about half survived (Table 2.4). Leaving a similar group of houseflies at 70 °F for 1 hour longer drastically reduced the number showing recovery at 100°F (29%) and caused higher mortality (87%). When houseflies were left at 70 °F for 4 hours, only 7% recovered after an hour at 100 °F and mortality was 98% after 24 hours at 100 °F. A similar observation was made for the action of DDT on cockroach (Vinson & Kearns, 1952).

Except for a few individuals, the fate of the dose was largely determined over the first seven hours after the initial contact with the DDT residue. Similar data may be obtained from any compound with a negative temperature coefficient, but specific responses depend greatly on the insect and compound. Clearly the initial responses in pyrethroid or DDT poisoning give no clue to the ultimate fate of the insect.

Nishimura et al. (1982) found that adult houseflies were often knocked down for only a few hours with non-synergised doses of allethrin, phenothrin or substituted benzyl (1R)-*trans*-chrysanthemates, and subsequently completely recovered. Used with the synergists piperonyl butoxide plus NiA 16388 (O-propyl-O-propargyl benzenephosphonate), the same dose was uniformly toxic with no recovery from knockdown. This is another way of demonstrating the Lindquist result that pyrethroids or DDT must be present for a definite amount of time before poisoning is not reversible. Simply overtly examining an insect that is knocked down gives no clue as to the fate of a particular unsynergised treatment. The insect can recover from a considerable insult to its nervous sytem. Ford (1979) found no correlation between speed of knockdown and relative toxicity.

The effect of temperature on the nervous activity of individual poisoned axons has been studied in several cases. Measuring the time to onset of trains of nervous impulses in the crural nerve of cockroach following injection of a suspension of 1 p.p.m. DDT, Roeder & Weiant (1951) found little variation in the time to onset of trains between 32 and 12 °C (Figure 2.4). However, at 10 °C, the latency to recorded trains more than doubled from about 6 minutes to nearly 16 minutes.

Some years later, when measuring the actions of pyrethroids and DDT on single identified motor axons of the housefly, M. E. Adams noted that in flight motor axons of housefly repetitive discharges occurred only over a narrow temperture range down to a "critical" value near 10 °C. Examination of other compounds, such as barthrin, showed a similar result (Figure 2.3). All compounds examined fell into three categories (Table 2.2). Either burst discharges reverted to single potentials at 10 °C (DDT and analogues) or reversion occurred near 19 °C (pyrethroids causing burst discharge). A third category was reserved for pyrethroids that did not produce burst discharges and reversion from burst discharge to single potentials by pyrethroids and DDT was said to occur at the "transition temperature" for housefly motor axons, and compounds were said to have characteristic transition tempera-tures. Observations of Narahashi on internerurons showed similar results. Allethrin caused repetitive discharges only above 26 °C in ventral nerve cord

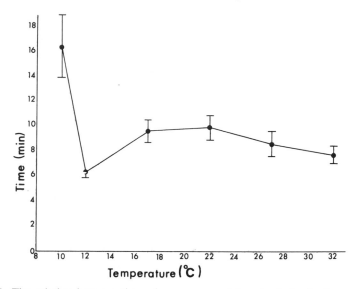

Figure 2.4. The relation between time of appearance of impulse trains in the crural nerve following a single injection of suspension containing 1·0 p.p.m. DDT and the temperature of the preparation. A number of observations were made at each temperature selected, and the standard error of each mean is indicated. Redrawn from Roeder & Weiant (1951)

interneurons of cockroach and only at certain tempertures in giant interneurons of squid (Narahashi, 1971).

Concerning these "transition" temperatures that are characteristic of some pyrethroids, it is curious that Romey et al. (1980) measured a clear break in Arrhenius plots of resting membrane conductance of axons between 17 and 20 °C. The action of veratridine on the sodium channel indicated a change in its properties below 20 °C, and suggested that fluidity of membrane lipids in the vicinity of the channel might influence the degree to which the opening of the channel is modulatable. Since the binding of tetrodotoxin to the sodium channel was similar over the range from 1 to 30 °C in purified crab axonal membrane, interest was focussed on the lipid environment adjacent to the sodium channel rather than the sodium channel itself as the site of veratridine and, by implication, pyrethroid action.

Knockdown

Pyrethroids that act very quickly are said to possess good knockdown. There are two opposing views as to the cause of knockdown, conveniently represented by the work of Burt & Goodchild (1974) and Clements & May (1977). Burt contended that all poisoning symptoms (knockdown and advanced intoxication) resulted from actions on the CNS of insects. Clements & May (1977), on the other hand, subscribed to the interpretation of Page et al. (1949) that knockdown is so fast that there is not time for penetration of a compound into the CNS, and they concluded that knockdown is due to excessive sensory hyperactivity in the peripheral nervous system. We do not have sufficient information as yet to decide clearly which of these cases is correct, partly because one can never completely discount central actions; however, if this is like other matters having to do with pyrethroids, it is probably impossible to generalise, although the extreme interpretation taken by Paul Burt is probably true only for a limited number of compounds. The majority of evidence supports some actions on sensory nerves being largely responsible for early knockdown symptoms.

Knockdown by topical treatment was slower in the German cockroach, *Blattella germanica*, than knockdown by surface contact. Uptake and distribution of [^{14}C]permethrin showed that rapid knockdown by surface contact corresponded most closely to radiolabel within the legs (Scott et al., 1984). This supports the conclusion of Clements & May (1977) that knockdown is associated with the peripheral nervous system, and is compatible with the finding that houseflies "knocked down" by deltamethrin are still capable of flight but cannot maintain posture (J. R. Bloomquist, PhD dissertation work, 1983). Knockdown properties of pyrethroids are also said to be correlated with their ability to produce disruptions in cercal nerves of the American cockroach (Glickman & Casida, 1982).

Thus, although it is difficult to rule out completely actions on the CNS, actions on sensory nerves, particularly in the legs, suggest that knockdown might affect postural tonus (Zill & Moran, 1982) or some other key sensory feedback mechanism leading to postural failure.

Another conclusion one is forced to make is that the prostrate stage, as defined by overt behaviour, cannot accurately reflect the separate activities of the cercal sensory or ventral nerve cord interneuronal activity. Interneurons do not accurately reflect locomotory behaviour in insects, and the role of giant interneurons in ordinary behaviour is not understood (Camhi & Nolen, 1981). The dorsal giant axons are not involved in walking behaviour and the ventral giants are loosely correlated with walking in a perfunctory manner. Thus, it would be very difficult to conclude anything about the locomotory appendages by examining the ventral nerve cord.

The ventral nerve cord has provided a useful tool for studying the actions of insecticides as a model system. It was shown some time ago that the cercal–giant system of *Periplaneta americana* is still functional during the prostrate stage of poisoning by DDT (Heslop & Ray, 1959), and that pyrethrum poisoning progresses regardless of whether the abdominal ventral nerve cord is attached (Hutzel, 1942). Thus, all studies on the giant axons and the ventral nerve cords of the American cockroach must be considered as models of the poisoning process and as an indication that some nerve centres retain their function long after the advanced stages of poisoning.

Several studies have found a correlation between knockdown and polarity. Briggs et al. (1974) and Ford (1979) found that the optimum polarity for knockdown by a series of pyrethroids was greater that the optimum polarity for kill, and concluded that this was because greater polarity aided penetration to the site of action. Nishimura et al. (1982) found that knockdown rate was directly correlated with polarity for a series of substituted benzyl chrysanthemates. The data of Clements & May (1977) also show a correlation between knockdown and polarity, but these authors concluded that appropriate structural characteristics are relatively more important than polarity in determining knockdown activity. Clearly, structural features are of utmost importance in determining activity among widely different structures, but among a series of structurally similar pyrethroids, polarity can be seen to affect knockdown rate, most likely by affecting rate of penetration to the site of action. However, we do not know whether this effect is on penetration through the integument, penetration from the haemolymph to the nerve membranes or, once in contact with the nerve membranes, on penetration into the membrane to the site of action.

Central versus peripheral actions

There are six separate ideas concerning the actions of pyrethroids that need clarification: knockdown versus kill; peripheral versus central actions; type I vesus type II pyrethroids. These concepts have become somewhat blurred. Knockdown is an empirical concept, and means that a pyrethroid product causes rapid action, particularly in "knocking" flying insects out of the air in mid-flight. It has been known for many years that insects may or may not subsequently recover from this knockdown and therefore non-recovery (kill) is given a separate name. Some pyrethroids are said to have excellent knockdown and either little or no recovery or complete recovery, depending on the structures (Elliott & Janes, 1979; Ohsumi et al., 1981); thus these two properties were assumed to be somehow different.

What has never been clarified is whether recovery from knockdown is merely a matter of metabolism of the insecticide to non-lethal levels and whether those poisoning processes leading to knockdown are the same as those that eventually lead to kill. Since some potent pyrethroids (such as bioethanomethrin) have little or no knockdown (Miller & Adams, 1977), it may be that the two processes are indeed separate.

Peripheral actions are actions on the peripheral nervous system which in insects consists of sensory neurons and their axons, motor axons and their terminals, and all neurosecretory axons and neurohaemal organs that lie outside of the ventral nerve cord and paired ganglia. The central nervous system (CNS) is considered to be the ganglia, connectives, and commissures from the brain to the terminal abdominal ganglion. These definitions make the insect CNS analogous to the mammalian CNS, which consists of the brain and spinal cord. In mammals, the α-cyano-3-phenoxybenzyl pyrethroids were considered to be distinct as a group (type II pyrethroids) on the basis of the motor disturbances they produced (Verschoyle & Aldridge, 1980). Choreoathetosis was produced by deltamethrin (type II) (Ray & Cremer, 1979), while cismethrin (type I) produced fine tremors.

Until recently there has been little information on the underlying causes of the different motor symptoms. Mephenesin was found to prevent or reverse all motor symptoms in rats, and it was thought that deltamethrin acts at the brain and peripheral sites, but that the spine mediates the motor symptoms of poisoning (Bradbury et al., 1983). Similar information is not available for insects, and the impression remains that peripheral and central actions are somehow vaguely correlated with type I and type II actions, respectively. One must remember that the strict central or peripheral action has not yet been defined for insects because one needs to examine all elements involved and so far only a few nerve elements have been investigated. It is particularly rash to assume a compound is not acting on any part of the CNS simply because a sheath or barrier exists, without actually recording from central units.

Lowenstein (1942) in his very early studies of pyrethrin extracts on *Blatta orientalis* ventral nerve cord (VNC) showed greatly increased nerve activity followed by block of nerve conduction. In later studies this same procedure was repeated using the VNC from adult *Periplaneta americana* at various times after treatment with 0·5 µg (LD95, room temperature) of pyrethrin I (Burt & Goodchild, 1971) (Table 2.5).

Burt concluded from these studies that the VNC was not affected in poisoning, and that perhaps a ganglion might be. A major role in this understandable conclusion is that probably the metathoracic ganglion with the

Table 2.5. Condition of American cockroaches and of parts of the abdominal nervous system at various times after topical treatment with 0·5 µg pyrethrin I (LD95). Values are the means of 3 replicates. Reproduced from Burt & Goodchild (1971)

| Time after treatment | Condition of insects | Condition of giant fibres | | | Condition of cercal nerve-giant fibre synapses[a] | Spontaneous activity in the 6th abdominal ganglion | |
| | | Stimulus strength | Height of spike (mV) when stimulated at: | | | Counts per sec. | Abnormal symptoms |
			1 Hz	40 Hz			
0	CO_2	5·1	4·3	4·3	N	5·9	—
5 min	CO_2	3·4	1·0	0·8	N	7·6	—
15 min	N	3·4	2·8	2·3	AD	7·5	T
1 h	B	3·1	1·7	1·6	AD	6·9	T
2 h	P	4·4	3·2	3·0	N	6·9	T
4 h	P	6·2	1·7	1·6	S	6·5	—
10 h	P	6·0	1·1	1·0	N	3·6	—
16 h	M	6·5	1·5	1·1	Bl.	5·6	—
24 h	M	Bl.	Bl.	Bl.	Bl.	4·5	—

[a] Note that failure of cercal–giant synaptic transmission was the first sign of failure.
CO_2 = affected by CO_2 anaesthesia; N = normal; S = slightly affected; B = badly affected; P = prostrate; M = moribund; AD = after-discharge; Bl. = blocked; T = "trains".

motor output it controls is the key nerve centre to monitor. It is very difficult to determine why the insect has ataxia or uncoordinated walking based on the cercal–giant nervous pathway in the abdomen. Burt did show that, at least in terms of the VNC, block of nerve activity occurred long after prostration. It is clear from Table 2.5 that measurements of spontaneous activity of the sixth abdominal ganglion give no clue as to the poisoned condition of the CNS.

An observation that pyrethroids are more potent when administered close to the CNS of rats (Staatz et al., 1982) matches the same observation of Burt & Goodchild (1977) in cockroaches and represents the main evidence for an action of the CNS of animals. In fact, Burt & Goodchild (1977) carried this much further, and considered both the knockdown and lethal actions of pyrethroids to be the result of actions on the CNS. By way of contrast, Gammon (1978c) found that pretreatment of cockroaches with sublethal injections of tetrodotoxin (TTX), which because of barriers to penetration is thought to act on the peripheral nervous system, offered a measure of protection against subsequent allethrin poisoning. He concluded that allethrin was acting entirely peripherally, but because his assumption that TTX has only peripheral actions has never been verified this experiment is not conclusive (Burt, 1980).

Again from work with rats, non-cyano-containing pyrethroids such as permethrin are thought to produce symptoms mainly through actions on the peripheral nervous system (Verschoyle & Aldridge, 1980), whereas α-cyano-3-phenoxybenzyl pyrethroids, such as deltamethrin, produce symptoms thought to be mainly central in action, or at least to originate in higher nerve centres of the brain (Staatz et al., 1982). The same distinct set of symptoms are known to occur in insects (cf. Staatz et al., 1982).

Insects whose muscle activity is blocked do not display DDT symptoms of poisoning (Beard, 1958). Treatment of cockroaches with TTX, which caused general paralysis, did not protect cockroaches from toxic effects of DDT, but did offer significant protection against allethrin poisoning (Gammon, 1979). The surprising lack of protection by TTX against DDT action is difficult to understand if, as strongly concluded by Roeder & Weiant (1951), the sensory discharges are the key element in the initial and subsequent actions by DDT. To explain both the TTX experiment and the poisoning scheme of Roeder & Weiant (1951) one is forced to use accessibility arguments, suspecting that the highly lipophilic DDT has access to key compartments denied to the more water soluble TTX. This argument looks weak in the face of the prevention of poisoning from the highly lipophilic allethrin by TTX pretreatment.

In a series of extensive post-war experiments, Kenneth Roeder and co-workers examined the action of DDT on the cockroach. They showed that the action of DDT on the cercal nerves was much less marked than its action on the legs, and that campaniform sensillae were more sensitive than other sensillae, concluding that DDT tremors are due to an intense and patternless bombardment of motor neurons by trains of impulses originating in sensory endings (Roeder & Weiant, 1946, 1951). They estimated a concentration of DDT at the site of action in the leg of between 1·4 and 14 μM based on the injected dose of 5 μg. The solubility limit of DDT allowed to stand overnight in saline was estimated by bioassay to be 28–280 nM (Roeder & Weiant, 1946).

Eaton & Sternburg (1967a,b) reported a close correlation between the stage of poisoning and the degree of synaptic impairment in the thoracic ganglia after injecting American cockroaches with 12·5 or 25 μg of DDT. After waiting for the desired stage of poisoning, the cockroaches were mounted ventrum up and stimulated after recording electrodes were affixed to nerves leading from the metathoracic ganglion. As poisoning symptoms progressed from hyperactivity to prostration, conduction across the metathoracic ganglion changed from a prolonged after-discharge to block. The authors were able to reverse the initial block by raising the temperature from 20 to 35 °C, but the poisoning symptoms and ganglionic block in the late prostrate stage of poisoning could not be reversed.

There was much discussion about the cause of the synaptic block noted above. Eaton & Sternburg (1967a) concluded that block could have resulted from autointoxication (cf. Sternburg, 1963). Recently, however, Salgado et al. (1983a) showed that DDT or pyrethroids will eventually block neuromuscular junctions if allowed to poison sufficiently long, and initially this block could be reversed by raising the temperature (Salgado et al., 1983b). Thus, the synaptic block observed by Eaton & Sternburg was probably due to presynaptic depolarisation (Salgado et al., 1983a).

One assumes that Roeder interpreted the poisoning process to be entirely determined by the trains of sensory impulses arising from specific sensory neurons and that later stages of poisoning were merely a consequence of this increased discharge. He did not see synaptic blocking actions of DDT, but looked only at whether DDT produced discharges.

Roeder perfused the ventral nerve cord of the cockroach with millimolar emulsive concentrations of DDT and found no change in spontaneous activity. Suprisingly, if Roeder & Weiant (1946) injected an "effective" dose of DDT, they recorded an increase in nervous activity in the ventral nerve in an otherwise intact cockroach. By cutting posterior and anterior to the recording electrodes, they concluded that the increased activity was coming from the attached sensory structures innervating the posterior ventral nerve cord. Unfortunately they did not examine the function of reflex arcs. Roeder's results clearly implied that with action exclusively on the sensory structures, the reflex arc would be vital to the overt expression of symptoms.

Dresden (1948) injected cockroaches with a DDT emulsion and then recorded ascending crural nerve activity from the mesothoracic leg, or recorded descending activity from the crural nerve of the mesothoracic ganglion while isolated from the other thoracic ganglia and from all peripheral nerves. He found no difference in these preparations between treated and untreated cockroaches; however, when the leg and mesothoracic ganglion were left intact, crural nerve activity was elevated. Dresden concluded that DDT caused synaptic facilitation in the mesothoracic ganglion.

Using a spinal frog preparation with *Rana esculenta* to investigate central synaptic actions, Dresden concluded that later in DDT poisoning, reflex pathways became blocked and that this caused the death of the animal. Unfortunately, Dresden concluded that DDT had no action on sensory nerves of the cockroach leg. This result is difficult to understand unless one assumes that injection into the body cavity did not allow DDT to perfuse to the legs in sufficient concentration to cause effects.

So compelling were Roeder's papers on the potency of DDT in producing sensory discharges in the cockroach leg that Dresden's work was openly scoffed at (cf. O'Brien, 1967, p. 112). No weight was given to the possible synaptic (or neuromuscular) effects of DDT because none were seen during the time the Roeder experiments were conducted and those that were reported were considered to be due to overdoses (cf. Yeager & Munson, 1945). Even Clements & May (1977) did not find neuromuscular lesions in the locust extensor tibia preparation.

The kdr *gene*

Genetic evidence suggests a single mode of action of pyrethroids, and is compatible with the stated claim of a single site of action at the sodium channel to explain all pyrethroid actions (Vijverberg et al., 1982). The best genetic evidence in support of a single mode of action comes exclusively from insects, mainly from houseflies and mosquitoes. Farnham (1977) reported separation of a single gene for site-insensitivity resistance in housefly. This gene was shown to be homozygous recessive and located on the third chromosome. It was linked with green eye and brown body genetic markers in a genetically well defined strain, 538ge, which was cross-resistant to all pyrethroids tested and to DDT and all DDT-analogues tested, except possibly kelthane (Dicofol) (Sawicki, 1978), a specific acaricide.

Thus Farnham's results suggest that one gene or an isolated locus is responsible for conferring site insensitivity. We might conclude from this that there is indeed one mode of action which, as van den Bercken (Vijverberg et al., 1982) states, is the sodium channel. It is important to realise, however, that the genetic evidence does not prove that interference with the sodium channel is the single mode of action of pyrethroids and DDT because we don't know what the *kdr* gene controls.

Salgado et al. (1983a) found that presynaptic terminals of *kdr* insects were resistant to the depolarising actions of the pyrethroid deltamethrin, and another sodium channel modulator, aconitine. Similarly, *kdr* nerves have been shown to be resistant to the repetitive firing activity of pyrethroids (Gammon, 1980; Salgado et al., 1983b) which is also known to be due to action on the sodium channels. These findings suggest that *kdr* genes change the sensitivity of the sodium channel to such agents as DDT, pyrethroids and aconitine, but this has yet to be directly verified.

Chialiang & Devonshire (1982) measured the transition temperature of housefly head membranes from Arrhenius plots of the activity of membrane-bound acetylcholinesterase. They found that membranes from *kdr* and *super-kdr* flies had higher transition temperatures than those from normal flies (19 and 21 °C, respectively, versus 14 °C), and suggested that this difference in lipid environment could influence the function of the target membrane component, presumably the sodium channel. In conclusion, then, the best evidence to date indicates that *kdr* resistance is due to modified lipid composition of the nerve membranes.

Kdr resistance creates profound practical problems. Resistance to pyrethroids has been documented in ticks in Australia (Nolan et al., 1977). Those ticks, *Boophilus microplus*, being resistant to DDT by a *kdr*-type resistance mechanism would have had a built-in advantage in surviving pyrethroid treatment because of cross-resistance between DDT and pyrethroids. In fact, the occurrence of a *kdr*-resistance mechanism in species treated with DDT at various times in the post-World War II era spells potential problems for pest control strategies hoping to switch to pyrethroids. This is particularly true for the WHO malaria control programme which experienced serious resistance to pyrethroids in Anopheline mosquitoes in early tests (R. Pal, 1979, personal communication).

DDT was banned for most uses considerably before the photostable pyrethroids were introduced on a large scale. For this reason, most pest species should have had a lapse of *kdr* toward the susceptible condition. However, since DDT is suspected to act in a manner similar to pyrethroids, and since both series of compounds induce *kdr* resistance, one can make educated guesses about whether persistent and stable pyrethroid insecticides will be successful as pest control agents. Put another way, the development of resistance to DDT by a particular pest some 30 years ago might be expected to repeat itself in the development of resistance to pyrethroids. Because of this unique knowledge, it could have been possible for the first time to plan a strategy to deal with pyrethroid resistance before it started.

Pyrethroids should never have been used as the single available compound, and never on a massive scale. They should have been alternated if possible from one season to another with another completely different type of insecticide such as an organophosphate.

As this was written (summer, 1983), information concerning the outbreak of high resistance to pyrethroids in Queensland, Australia in January of 1983 is becoming available. *Heliothis armigera* in Australian cotton fields developed substantial *kdr* resistance along with an unknown amount of metabolic resistance in January 1983. The highest levels of *kdr* resistance in individual *H. armigera* range from 100- to 500-fold compared to susceptible *H. armigera* collected from the same area, including resistance to permethrin and cypermethrin and cross-resistance to DDT. These levels are high enough to suggest a *super-kdr* resistance factor such as that found and characterised in housefly (Nicholson & Sawicki, 1982).

Although it is not clear what the level of use of pyrethroids was in Queensland that might have led to the present predicament, the prediction of Elliott et al. (1978) that overuse will lead to a resistance problem, appears to have come true. Another point stressed by Elliott et al. (1978) worth keeping in mind is that, with judicious use, pyrethroids can be effective pest control agents for the forseeable future.

Strategies for preventing or dealing with the situation in Queensland must include ending reliance on one compound. Our experience with *Heliothis virescens* on cotton in the Imperial Valley of California, suggests that the first appearance of high levels of pyrethroid resistance is associated with poor viability and a phenomenon termed reversion (Brown, 1981). The resistant animals lay fewer eggs and survive laboratory rearing poorly. Thus, if the pyrethroids are immediately removed from use, or another type of insecticide to which there is no cross-resistance is introduced, there is every chance that the *super-kdr* will revert to susceptibility. Such a strategy is more likely to succeed if directed by chemical firms selling pyrethroids or government agencies licensing and registering pyrethroids for use.

A large number of insects and acaricines have now shown resistance or cross-resistance to pyrethroids (Brown, 1982) and the list grows with every season. The first reports described a resistance in diamondback moth, *Plutella xylostella* (L.) to DDT and the pyrethroids: permethrin (110-fold), cypermethrin (894-fold), deltamethrin (2235-fold) and fenvalerate (2880-fold) compared to a susceptible strain (Liu et al., 1982b). Synergism by DMC (1,1-di-(4-chlorophenyl) ethanol), piperonyl butoxide or DEF (*S,S,S*-tributyl-phosphorotrithioate) failed to alter the toxicity values appreciably (Liu et al., 1982a). Thus a *kdr* factor was suspected. If a *kdr* factor is present, the knockdown time should be greatly lengthened in resistant adults as compared to susceptible adults.

Recording miniature postsynaptic potentials (mEPSPs) from larval housefly or mosquitoes, Steve Irving was able to show that permethrin and deltamethrin were less effective at causing an increase in mEPSP rate leading to synaptic block in the *kdr*-resistant strain as compared to a susceptible strain (Miller et al., 1983) (Table 2.6). He was also able to provide direct proof of a suspected *kdr* phenotype in mosquito larvae (Omer et al., 1980).

Since the mEPSP technique requires only intracellular recording from muscles and since larval stages provide large numbers of uniform body wall muscles, the method is applicable to a large number of insects. To date a number of species have been examined for the presence of *kdr* factors by this method (S. N. Irving, 1983, personal communication, Table 2.7).

An important disadvantage of the mEPSP technique for measuring intrinsic

Table 2.6. Comparison of toxicity in three species with a nerve-insensitivity resistance factor. Taken from Miller et al. (1983)

Insect[a]	Compound	LC50 or LD50 (R : S ratio)		EC[b] (nM)	(R : S ratio)
Musca S	*Cis*-deltamethrin	0·6		1	
Musca R	*Cis*-deltamethrin	8·0	(13)	100	(100)
Culex S	*Trans*-permethrin	0·0021		0·5	
Culex R	*Trans*-permethrin	8·7	(4143)	1000	(2000)
Anopheles S	*Cis*-permethrin	0·0035		0·05	
Anopheles R	*Cis*-permethrin	0·082	(23)	1	(20)

[a] Species tested: *Musca* = *Musca domestica*, NAIDM S strain, and *kdr* R strain; *Culex quinquefasciatus*; *Anopheles stephensi*.
[b] EC = lowest concentration increasing mEPSP rate within 30 min.

activity of pyrethroids is its technical nature. Specialised training is necessary before one can use the equipment involved and such instrumentation is normally not available in the ordinary entomology laboratory, not even one specialising in insect toxicology. An experienced neurophysiologist could make these measurements, but such a person normally is not familiar with insecticides or their bioassay.

It should be possible to develop very rapid assessment methods based on the original observations of Busvine (1951) on the earliest *kdr* factors appearing in houseflies. We have attempted to assess the knockdown of susceptible and *kdr* houseflies (the same strains referred to in Table 2.6). Using 80 houseflies, a spray knocked down the susceptibles earlier than the *kdr* strain. Provided that there are no penetration factors involved, all metabolic factors should not determine knockdown behaviour – only recovery from poisoning.

Thus it should be possible to develop a sampling programme for pyrethroid resistance in adult moths. Population assessment by trap catches are common in agricultural regions. It should be possible to rescue a number of adult moths from such a trap and assess knockdown behaviour in the field in less than 20 minutes using a portable knockdown tower previously calibrated.

Pharmacokinetics and competition experiments

When applied on a leg of the housefly, 1% of the applied dose of [^{14}C]DDT was collected from the haemolymph at the cervical membrane (LeRoux & Morrison, 1954). If a dose of 1·5 μg was applied to the leg, 1% of this is 15 ng. The concentration of DDT in the haemolymph peaked a few minutes after treatment but then stayed constant for 24 hours, never increasing above

Table 2.7. Insects from which mEPSPs have been recorded as a measure of intrinsic activity of pyrethroids and where susceptible (S) and resistance (R) strains were available. The (R : S) ratio of toxicity is given

Insect	Compound	Resistance factor (R : S)	Source of insects
Heliothis armigera	Cypermethrin	50–500	Australia[a]
H. punctigera	Fenvalerate	1	Australia
H. viriscens	Permethrin, fenvalerate	10	California
H. zea	Permethrin, cypermethrin	1	California
Spodoptera littoralis	Permethrin	4	Egypt[a]
S. exigua	Permethrin	104	Holland[a]
Trichoplusia ni	Fenvalerate	1	California
Chilo partellus	Deltamethrin, permethrin, and cypermethrin	1	Jealott's Hill[a]
Plutella xylostella	Deltamethrin, permethrin, and cypermethrin	1	Jealott's Hill[a]
Mamestra brassicae	Deltamethrin, permethrin, and cypermethrin	1	Jealott's Hill[a]
Anopheles stephensi	Deltamethrin, permethrin and cypermethrin	20	California
Aedes aegypti	Deltamethrin, permethrin, and cypermethrin	1	Jealott's Hill[a]
Culex quinquefasciatus	Deltamethrin, permethrin, and cypermethrin	2000	California
Musca domestica	Deltamethrin, permethrin, and cypermethrin	100	New York[b]
M. domestica	Deltamethrin, permethrin, and cypermethrin	10	Rothamsted[c]
Diabrotica sp.	Deltamethrin, permethrin, and cypermethrin		Jealott's Hill[a]
Periplaneta americana	Deltamethrin, permethrin, and cypermethrin	1	Great Britain[a]

[a] Data from S. Irving, ICI, Jealott's Hill, England.
[b] Super *kdr*, J. Scott, 1983.
[c] 538ge *kdr* strain, A. Farnham, Rothamsted.

15 ng. A blood level of 15 ng of DDT represents about 4 μM, assuming perfect mixing and a blood volume of about 10 μl. This value of DDT, reached seconds after topical treatment, was considerably higher than the threshold dose of DDT needed to produce an increase in the rate of discharge of minature postsynaptic potentials from housefly larval muscles (Salgado et al., 1983b). Thus, the LeRoux & Morrison (1954) measurements represent reasonable concentrations in terms of those needed to begin producing a physiological lesion.

Table 2.1 shows that for toxic compounds, one expects increased nervous activity by perfusion of at least 100-nM concentrations, and usually 10 nM in the leg of the locust. This figure of 10 nM then should be compared to known concentrations of pyrethroids as they accumulate in the haemolymph and nervous tissues following topical treatment of toxic doses.

Soderlund's (1979) pharmacodynamic study of 2 × LD50 doses of NRDC 157 penetration into *Periplaneta americana* showed a consistent concentration of just over 150 nM achieved within 2 hours in the haemolymph and somewhat later in the nerve cord. Since the completely non-toxic 1S enantiomer of NRDC 157 achieved the same concentration, this figure of 150 nM should be used as a reference when considering toxic doses.

Burt et al. (1971) reported that they could not detect 20 nM pyrethrin I chromatographically in the haemolymph of cockroaches treated with LD95 doses (0·45 μg/g, *Periplaneta americana*). However, when haemolymph was collected 2 hours after topical treatment and bioassayed on the ventral nerve cord, responses were qualitatively and quantitatively comparable to perfusion of ≥ 200 nM of pyrethrin I on the same ventral nerve cord preparation. The value of 200 nM is close to the 150 nM concentration reported by Soderlund (1979) in the same insect for 2 × LD50 doses of NRDC 157. However, the inability of Burt to detect pyrethrin I is similar to the report of Blum & Kearns (1956). In the latter paper blood collected from pyrethrum-treated cockroaches produced poisoning responses that were attributed to an autoneurotoxin unrelated to pyrethrum. Thus blood assays with *Periplaneta* are complex.

Establishing a reasonable concentration of pyrethroid in the internal organs of a treated insect is vital to understanding the mode of action of pyrethroids. Pyrethroids have been shown to cause a variety of effects on nerves. Knowing concentrations of pyrethroids in the haemolymph of treated insects helps put primary actions into perspective.

Our measurements of [^{3}H]-labelled 1R-*trans*-permethrin accumulation in the haemolymph of topically treated female houseflies agree with the lower range of values noted above for the American cockroach (Figure 2.5). The data show that 2 hours after an LD85 dose (3·4 ng/female) of permethrin, the concentration of permethrin in the haemolymph was approaching 32 nM. Note from the behaviour of lower doses that a specific concentration is maintained over a considerable period and, as reported by Soderlund (1979), higher doses result in a decrease in the amount of collectable haemolymph, so that accurate concentrations are more difficult to measure later in poisoning.

Consideration of the competitive actions of pyrethroids is extremely important for practical reasons. Pyrethroids are often manufactured as mixed isomers. If the toxic 1R isomer were blocked by a non-toxic 1S isomer, either competitively or non-competitively, then various mixtures might be expected to be comparatively less toxic than the highest amount of the most toxic enantiomer. Casual inspection of toxicity data from pure of NRDC 157 (Soderlund, 1979) and various mixtures would argue that there is very little interaction or interference of this kind. Indeed, one must be extremely careful that where very large toxicity differences occur between pairs of enantiomers that the toxicity value of the inactive isomer is not due to extremely small ($< 0.001\%$) contamination of the active isomer.

Soderlund (1979) examined the effects of the highly insecticidal NRDC 157 (3-phenoxybenzyl (1*R*, *cis*)-3-(2,2-dibromovinyl)-2,2-dimethylcyclopropane-carboxylate), and its inactive 1*S*, *cis* enantiomer. Two hours following topical application of 0·17 µg/g of NRDC 157 (LD50 = 0·085 µg/g at 96 h, 20 °C) to adult American cockroaches, *Periplaneta americana*, Soderlund found that the enantiomers were distributed in similar patterns, with a steady state concentration of $1.2–1.7 \times 10^{-7}$ M in the haemolymph and nerve cord. He concluded that the difference in toxicity was due to a stereo-specificity of the site of action for the 1*R* configuration on the cyclopropane ring, all other portions of the molecules being identical.

For pretreatment "competition" experiments, Soderlund (1979) injected the inactive (1*S*, *cis*) enantiomer before treating with the insecticidal analogue (1*R*, *cis*) NRDC 157. If the 1*S* form competed with the 1*R* form, he would have measured a reduced effect of the 1*R*, but apparently the onset of symptoms of 0·17 µg/g of NRDC 157 was not altered at all by up to 5·1 µg of 1*S* or 30 times the dose of the 1*R* NRDC 157. This suggests a fairly specific recognition by the target site of the 1*R* configuration without interruption by the 1*S* enantiomer.

A similar conclusion was reached in a study of pyrethroids on crayfish claw opener muscle. Non-toxic stereoisomers of type II pyrethroids did not antagonise the action of toxic enantiomers, nor did pyrethroids causing type I symptoms antagonise the active type II compounds (Gammon & Casida, 1983).

Narahashi (1982) used 10–50 µM tetramethrin in his voltage-clamp studies, despite a 1–10 nM threshold needed for production of repetitive discharges. He explained the differences by calculating that less than 1% of the sodium channel population needs to be affected by pyrethroids to elevate the depolarising after-potential.

The result of an interesting "competition" experiment using stereoisomers of tetramethrin was also reported (Lund & Narahashi, 1982). Axons were voltage-clamped and the slow tail current induced by (1*R*)-*trans*-tetra-methrin was recorded. A 100% response was obtained in 100 µM (1*R*)-*trans*-tetramethrin. In the presence of 300 µM (1*S*)-*cis*-tetramethrin, the dose-response curve for (1*R*)-*trans*-tetramethrin was shifted towards higher concentrations and the peak response was markedly reduced, suggesting non-competitive interaction between (1*R*)-*trans*- and (1*S*)-*cis*-tetra-methrin. In a similar experiment the (1*S*)-*cis* form was found to antagonise the (1*R*)-*cis* form competitively.

These studies on the 1*R* versus 1*S* forms of pyrethroids on biological responses are in marked contrast with the result reported by Soderlund (1979) who found no inhibition of the active 1*R* form of NRDC 157 by its completely inactive 1*S* form. One must conclude that either the two biological tissues

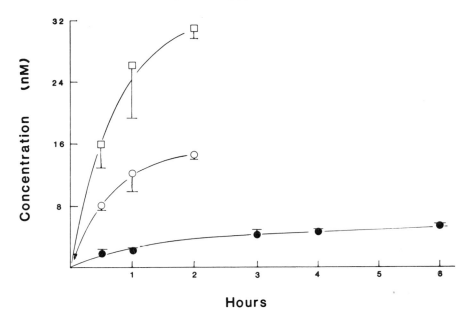

Figure 2.5. Concentration of [³H]1*R-trans*-permethrin in the haemolymph of female adult, *Musca domestica*, following topical treatment on the abdomen of 0·34 ng (LD2), filled circles; 1·2 ng (LD35), open circles; or 3·4 ng (LD85), open squares at 21 °C. All insects were pretreated with 4 µg of piperonyl butoxide before permethrin application (Collins, Miller & Kennedy, 1984, unpublished observation)

(squid giant axon voltage clamping versus cockroach poisoning symptoms) are so different as not to be comparable, or that the tail current measurements are not related to intoxication by pyrethroids in whole animals. The latter interpretation is probably correct. Lund & Narahashi (1982) found that the dose–response curve for tetramethrin and allethrin had two dissociation contrasts. The high affinity site has a K_d of 150 nM, but saturation of this site produced only 6% of the maximum response. This is the site most likely to be involved in poisoning. On the other hand, the competition Lund & Narahashi observed was for the lower affinity site and is probably not relevant to poisoning.

Pyrethroid action at the molecular level

Sodium channels

Repetitive firing

If the reader does not understand how the action potential is generated by the time-dependent sodium and potassium permeability of the membrane, he or

T. A. Miller and V. L. Salgado

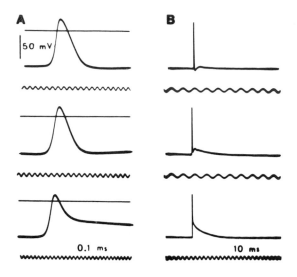

Figure 2.6. Changes in intracellularly recorded action potential of the cockroach giant axon after treatment with DDT 10^{-4} M. (A) from top to bottom, before, 38 min after, and 90 min after treatment with DDT. The horizontal lines indicate zero potential level. (B) as in (A), but with slower sweep. From Narahashi & Yamasaki, 1960a

she should read the original development of this remarkable theory by Hodgkin et al. (1952) and Hodgkin & Huxley (1952a,b,c,d), or a brief account such as those by Narahashi (1971, 1979).

Roeder & Weiant (1946) first observed that DDT caused repetitive firing of sensory cells of the cockroach trochanter. Since then, repetitive firing has been observed in nearly every type of nerve structure treated with DDT, DDT analogues, pyrethrins and type I pyrethroids. This includes studies on numerous sensory, motor and CNS neurons (see reviews by Narahashi, 1971; Wouters & van den Bercken, 1978), as well as neurosecretory axons (Orchard & Osborne, 1979).

The mechanism of induction of repetitive firing was first determined for DDT, but it is essentially the same for pyrethroids (see reviews by Narahashi, 1971, 1979). Repetitive firing is associated with a negative after-potential, as discovered by Shanes (1949) with extracellular recording and later confirmed by intracellular recording (Narahashi & Yamasaki, 1960a). Following an action potential in normal nerve, the membrane potential returns rapidly to the resting level and may even undershoot the resting level. When the undershoot occurs, as in the top of Figure 2.6, we call it a positive after-potential. It may seem odd to call this negative deflection a positive after-potential, but the terminology derives from extracellular recording, in which the action potential is recorded as a negative deflection and the positive after-potential is truly positive. The positive after-potential is due to the supernormal potassium conductance of the membrane following the action potential, as the potassium channels which open to repolarise the membrane after the action potential are still largely open at this time, up to 2 ms after the action potential (Narahashi & Yamasaki, 1960b). The first symptom which appears after treatment of the nerve with DDT or a type I pyrethroid is a decrease and reversal of the positive after-potential, or in other words, we say that the pyrethroid induces a negative after-potential (Figure 2.6). The negative after-potential gets progressively larger, and when it reaches a certain threshold amplitude, the nerve fibre fires repetitively. It is often incorrectly stated that the negative after-potential, when it reaches the threshold of the nerve, stimulates the nerve to fire repetitively. This is inaccurate for two reasons. First, the negative after-potential itself is not the stimulus; it represents a response of the membrane to underlying changes in membrane conductances, the latter of which are properly the causes of both the negative after-potential and the repetitive firing. Second, unpoisoned cockroach axons cannot fire repetitively in response to a steady depolarising current but can do so after DDT or pyrethroid poisoning. Thus DDT and pyrethroids change the state of the membrane so that it can respond repetitively to a long-lasting depolarising stimulus such as that which causes a negative after-potential (Narahashi & Yamasaki, 1960a). The nature of this depolarising stimulus is described below.

The mechanism responsible for generation of the negative after-potential was first studied by Narahashi & Haas (1968) in DDT-treated lobster giant axons. They found that the sodium inactivation and potassium activation were both slowed and suppressed. This would tend to decrease outward potassium current and increase inward sodium current following an action potential and could partially explain the negative after-potential. Later it was found that the synthetic pyrethroid allethrin caused similar changes in squid giant axons with the additional effect of suppressing the sodium current at higher concentrations (Narahashi & Anderson, 1967). The latter effect accounted for block of the action potential by allethrin. It was subsequently discovered that allethrin (Murayama et al., 1972) and DDT (Hille, 1968; Lund & Narahashi, 1981a) also slowed the turn-off of sodium current when the nerve was repolarised following a step depolarisation. This induces a so-called tail current which is necessary for producing the negative after-potential, as it allows the inward current to be maintained at fairly negative potentials. The DDT analogue EDO (Wu et al., 1980) and the pyrethroid tetramethrin (Lund & Narahashi, 1981b, c) had similar effects on sodium channels. The above studies led to the conclusion that the major site of action of DDT and pyrethroids in inducing the negative after-potential is the nerve sodium channel. The effect on the potassium channel is much less important in this respect.

The interactions of pyrethroids and DDT with sodium channels has been brought to a new level of sophistication by the recent studies of Lund & Narahashi (1981a,b,c). This important work will be considered in detail after we first describe the properties of sodium channels.

Sodium channel gating

The sodium channel is an integral membrane protein about which we have extensive functional knowledge but very little structural information. Thus we can describe the functional modifications by pyrethroids in exquisite detail, while knowing virtually nothing about the "pyrethroid receptor".

The sodium channel protein undergoes conformational changes between at least three functionally distinct states, depending upon transmembrane potential (Figure 2.7). In reality, there are more than three states, but three will suffice to illustrate the general features of sodium channel gating. The resting (R) state is the preferred conformation at potentials near resting, while the inactivated (I) state predominates at more depolarized potentials. The open (O) state is not stable at any potential, but exists transiently following depolarisation of the membrane, as channels move from R to O (activation) and then to I (inactivation). The open state of the channel is the only state that is open to the flow of sodium ions, and as the sodium concentration is normally about 10 times higher outside than inside the cell, it permits an inward flow of sodium ionic current. Thus, following a step depolarisation from the resting potential to around 0 mV, we record a transient inward current, with a rising phase due to the opening of channels and a decay phase due to the subsequent inactivation of these channels (Figure 2.7B).

If the membrane is repolarised after the decay is complete, the inactivated channels return to the resting state without reopening (Figure 2.7B). However, if the pulse duration is shortened so that repolarisation occurs while channels are still open, as in Figure 2.7D, there will be a discontinuity in the current trace while the potential is changing (seen as a downward blip in the current trace of Figure 2.7D) followed by decay of the current back to zero. This decay of current following repolarisation, known as the sodium tail current, is very important in pyrethroid studies. As shown in Figure 2.7D, the tail current decay is due to return of channels to the resting state (O–R) rather than to inactivation.

The complex voltage dependence of sodium channel gating described above implies that the forward transitions in the model (R → O, R → I and O → I) speed up with depolarisation, while the backward ones (O → R and I → R) speed up with hyperpolarisation. The backward transition I → O is negligibly slow at all potentials. The speeding up of the backward transition O → R with hyperpolarisation can be readily seen from the voltage dependence of tail current decay (Figure 2.9A).

To simplify the analysis of sodium currents, the assumption is usually made that inactivation is slow enough compared to activation that equilibrium between R and O is transiently attained during the peak of the inward current. The fraction of channels open during this transient equilibrium is denoted by m_∞ (m-infinity), and is very small at negative potentials but increases rapidly with depolarisation to saturation at around 0 mV (Figure

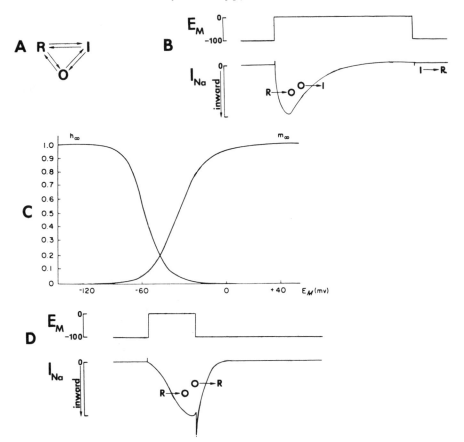

Figure 2.7. (A) Transitions between the R, resting; O, open; and I, inactivated states of the sodium channel. (B) Sodium current in response to a prolonged voltage step from -100 to 0 mV. (C) Computed values of the steady-state values of m and h as a function of membrane potential, E_M. Note that when the axon membrane is depolarised, h_∞ decreases to zero, while m_∞ increases towards one. Modified from Palti (1971). (D) Sodium tail current after a short duration voltage step from -100 to mV. See text for details

2.7C). Thus, a depolarisation to -60 mV would evoke a very small peak current, while one to 0 mV would give a nearly maximal one. m_∞ provides a very sensitive measure of the voltage dependence of sodium activation, and is often discussed in relation to pyrethroid action.

The equilibrium between R and I, on the other hand, is described by the variable h_∞ (h-infinity), which is defined as the fraction of channels in state R at equilibrium. Neurophysiologists call h_∞ the sodium inactivation, but as it is actually a measure of the fraction of channels that are in the resting state and thus available to open upon depolarisation, it would be more precise to call it availability, as cardiac physiologists do. The h_∞ curve for a squid axon is shown in Figure 2.7C, and from it we can see that all of the channels are available at a holding potential of -100 mV, while all are inactivated at 0 mV. h_∞ is 0.5 at -60 mV, indicating that the peak current evoked by a step from this potential to 0 mV would be only half as large as that evoked by a step to 0 mV from a holding potential of -100 mV.

The energy source for the gating transitions is one more point that needs to be made concerning the sodium channel before we consider its modification by pyrethroids. Hodgkin & Huxley (1952d), in their quantitative study of sodium and potassium conductances, first discussed the mechanism by which membrane potential could affect the ionic channels. They recognised the necessity of postulating that the gating structure possessed permanent charges or electrical dipoles which could move under an applied electric field. Simple statistical mechanical considerations allowed them to calculate that the movement of six electronic charges across the entire membrane field could account for the steepness of the voltage dependence of sodium activation (m_∞). They also predicted that the movement of these charges should produce a measurable "gating" current, which was eventually detected following improvements in techniques (Armstrong & Bezanilla, 1972). It should be emphasised that intramembrane charge movement alone can provide the energy for gating transitions, and no metabolic involvement need be postulated. In fact, under voltage-clamp conditions, sodium channels function normally for many hours after replacement of the cell contents by an artificial solution containing only inorganic ions and no ATP. In this case, energy for channel gating is supplied directly by the voltage-clamp circuit. In an intact functioning cell, phosphate bond energy is converted into potential energy of the resting potential by the Na^+-K^+ ATPase, and the gating of sodium channels is driven by this electrical potential energy.

The effect of pyrethroids upon sodium channels

Narahashi & Haas (1968) and Narahashi & Anderson (1967) described the effects of DDT and allethrin, respectively, on nerve sodium current. Hille (1968) made two important conceptual advances in interpreting these effects at a molecular level. First, he showed that DDT modified only a fraction of the channels, causing them to close slowly once opened; the remainder of the channels behaved normally. Second, he postulated that there is a "precursor–product" relationship between the peak current and the DDT-modified current, that is, that DDT modifies open channels. These observations were recently extended by Lund & Narahashi (1981a,b,c; 1982) in their studies with DDT and tetramethrin.

Figure 2.8 shows sodium currents before and after treatment with 2×10^{-5} M tetramethrin. The minimum concentration of tetramethrin needed to cause repetitive firing is 10^{-8} M in both squid and crayfish axons (Lund & Narahashi, 1982, 1981c), but at such a low concentration less than 0·25% of the sodium channels are affected. In order to perform voltage-clamp studies, we need to apply much higher concentrations to modify a larger fraction of the sodium channel population. After poisoning, the rising phase of the sodium current is normal, but the falling phase is not complete. However, as part of the

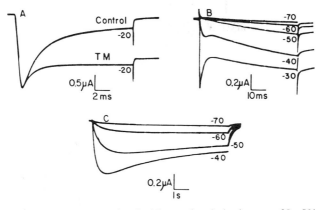

Figure 2.8. (A) Sodium currents associated with step depolarisations to −20 mV before and after internal treatment with 2×10^{-5} M tetramethrin (TM). (B,C) Sodium currents associated with step depolarisations to the levels indicated after treatment with TM. Note the different time scales in (A), (B), and (C). Temperature 5 °C. Taken from Lund & Narahashi, 1981c

peak current does inactivate with the same time course as normal, it seems that some channels still function normally and that the large steady state current is flowing through pyrethroid-modified channels which do not inactivate. The fraction of normal channels decreases with increasing concentration of tetramethrin, with no change in the normal channel kinetics. This suggests that tetramethrin interacts specifically with single channels, rather than having a non-specific effect on the membrane physical properties, which would be expected to change the kinetics of all of the channels in a graded manner.

If the membrane is repolarised during a pulse while sodium channels are still open, the channels close very rapidly, in less that 1 ms (Figures 2.7D, 2.9A). The pathway by which the channels close at these negative potentials is not inactivation, but the reverse of activation: they return to the resting state. The amplitude of this fast "tail current" is proportional to the sodium current at the time of repolarisation. The tail currents in pyrethroid- or DDT-poisoned axons are also proportional to the number of channels open at the moment of repolarisation, but now the decay of the tail current is much slower. If the membrane is repolarised during the first few milliseconds, while the normal sodium channels are open, one sees that a fraction of the tail current decays with the normal time constant (Figure 2.9B); this represents closing of the normal channels. However, in this figure it can be seen that after tetramethrin poisoning there is an additional slow component of tail current that does not decay significantly on the time scale shown in Figure 2.9. This slow tail current is due to modified channels, which return to the resting state in tens of milliseconds as opposed to less than 1 ms for normal channels.

The amplitude of the slow tail current is a very important parameter, as it is proportional to the number of modified channels. Thus by repolarising at various times during a pulse, the amplitude of the slow tail current shows the

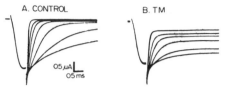

Figure 2.9. Families of tail currents (A) before and (B) after treatment with 2×10^{-5} M tetramethrin (TM) after conditioning the membrane at -20 mV for 0.75 ms. The repolarisation potentials were -30, -40, -50, -60, -70, -80, and -100 mV in A and -40, -50, -60, -70, -80 and -100 mV in B, with the fastest decay occurring at -100 mV in both cases. Temperature 5°C. Taken from Lund & Narahashi, 1981c

time course of formation of modified channels during a pulse. Hille (1968) used this method to measure the time course of formation of DDT-modified channels in frog axons. He found that modified channels were formed while normal channels were open, and that there was a distinct delay after the beginning of depolarisation until modified channels began to form. These observations suggested the already mentioned precursor–product hypothesis in which DDT or the pyrethroid reacts specifically with the open channel (O → O* in Figure 2.10).

Lund & Narahashi (1981a) made similar measurements with tetramethrin in squid axons. Unlike Hille, they saw no delay in the development of modified channels. Modified channels were formed with a dual exponential time course following the onset of depolarisation. The fast phase had a time constant of 0.8 to 1.8 ms, depending on pulse potential, and probably corresponded to modification of open channels, while the slow phase had a time constant of 14–25 ms. The slow phase was abolished by pronase treatment of axons, which destroys the ability of channels to inactivate, so it probably represents inactivated channels spontaneously opening and becoming modified. It should be pointed out that pronase treatment greatly potentiated the action of tetramethrin by removing the inactivation, as inactivation, competes with tetramethrin for the open channels. The pronase experiment also showed that tetramethrin modified some channels in the resting state which then activated more slowly than normal. This pathway (R → R* → O*) is shown in Figure

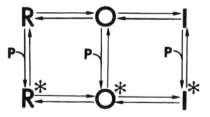

Figure 2.10. Hypothetical model for the kinetic interaction of pyrethroids with the sodium channel. R, resting state; O, open state; I, inactivated state; R*, pyrethroid modified state; O*, pyrethroid modified open state; I*, pyrethroid modified inactivated state; and P, pyrethroid. See text for details. Modified from Narahashi. (1982)

2.10. As we have mentioned above, modified open channels do close, but with a much slower time course than normal channels. It has not been established whether they close before ($O^* \to R^*$) or after ($O^* \to O \to R$) unbinding the pyrethroid.

Although it is now clear that DDT and pyrethroid-poisoned membranes retain some channels that appear to be normal, as of course is expected if the insecticides interact directly with individual channels, it is important to establish just how normal these channels are. Narahashi & Haas (1967) and Hille (1968) found no shift of the m_∞ or h_∞ curves by DDT, whereas a DDT analogue, EDO, shifted both curves towards more negative potentials (Wu et al., 1980). Wu et al. (1980) postulated that the potential shift by EDO was due to its large dipole moment as compared with DDT, which might allow the molecules to orient in the membrane field, thereby reducing the field. On the other hand, Lund & Narahashi (1981c), using short (8 ms) pulses to selectively observe normal channels, found no shift in m_∞ by DDT in squid axons, but using long (200 ms) pulses they did find that DDT shifted the h_∞ curve 6-8 mV in the negative direction. It should be recognised that in the above experiments h_∞ of both normal and modified channels were lumped together, so the shift of h_∞ does not require the voltage dependence of normal inactivation be shifted, but may mean that the inactivation of only the modified channels is shifted in the negative direction. These processes are difficult to separate with DDT, as the inactivation of the modified channels is slowed only about 15-fold. With tetramethrin, which slows the inactivation of modified channels 1000-fold, Lund & Narahashi (1981a) found that inactivation occurring during 30 ms pulses was shifted only 5 mV, while that occurring during 160 ms pulses was shifted 10 mV in the negative direction, supporting the idea that tetramethrin-modified channels inactivate at more negative potentials than normal channels, and suggesting that the voltage dependence of inactivation of normal channels may not be shifted at all.

Type I versus type II

Although the above model was developed for tetramethrin, it can obviously account for DDT and type I pyrethroids with minor changes in rate constants. Lund & Narahashi (1983) have examined the time course of channel modification and recovery for DDT analogues and type I pyrethroids, and found that there is a wide spectrum of kinetics. DDT analogues, in general, had the fastest kinetics with slow tail current time constants, ranging from 9·5 to 86 ms at -100 mV from DDT to EDO, respectively, while type I pyrethroid-induced tail current time constants ranged from tetramethrin (620 ms) to GH401 (3·4 s).

There appears to be a qualitative difference between type II pyrethroids, which depolarise nerves without repetitive firing, and the type I compounds. However, at the molecular level, the difference is merely quantitative. Lund &

Narahashi (1983) studied several type II pyrethroids, and found that they depolarised crayfish axons and that the depolarisation was enhanced by stimulation. Under voltage clamp, they found that depolarisation resulted in modification of a small number of channels, resulting in a barely detectable tail current, but these channels recovered so slowly after depolarisation (tail current time constants of several minutes) that the tail currents summated upon repetitive depolarisation. This is the mechanism by which stimulation enhances the depolarisation.

Salgado & Narahashi (1982) have studied the action of the type II pyrethroid fenvalerate on voltage-clamped crayfish axons. They confirmed these findings, also finding that fenvalerate induces a steady-state sodium current at potentials less negative than -100 mV. At these potentials, normal channels that spontanteously open can become modified by fenvalerate, and the modified open state is stable with no inactivation. Without voltage clamp, the small number of channels that become modified at the resting potential would depolarise the axon, permitting even more channels to open and become modified. In this way, fenvalerate can depolarise the membrane, and the mechanism appears to differ only quantitatively from type I compounds in that the modified open state is much more stable.

It seems certain that the crayfish results and the model derived from them are directly applicable to insects, because even though voltage-clamp studies have not been done on insect nerves, other types of neurophysiological studies on insects are consistent with these findings. Thus, Salgado et al. (1983a) found that α-cyano pyrethroids were potent in depolarising insect motor nerve terminals and that the depolarisation was enhanced by nerve stimulation, as found by Lund & Narahashi (1983) for crayfish axons. They further obtained evidence that the nerve terminal depolarisation was due to prolonged sodium current. Salgado et al. (1983b) found that type I pyrethroids and DDT all increased mEPSP rate (shown to be due to nerve terminal depolarisation) at higher concentation than needed to cause repetitive firing. This corresponds to the observation by Lund & Narahashi (1983) that all of the type I pyrethroids depolarised the crayfish axon at 100–1000 times the concentration needed to cause repetitive firing.

Single channel and gating current studies

As the model for pyrethroid effects on sodium channels is derived somewhat indirectly from sodium current measurements, it is of interest to obtain more direct experimental confirmation of some of its features. This comes from single channel and gating current studies.

The recently developed single channel recording method (Hamill et al., 1981) allows us to seal a blunt pipette tip a few microns in diameter to a patch of membrane and record the current through any channels which happen to be in the patch. Yamamoto et al. (1983) have used this technique to study sodium channels modified by tetramethrin. In control patches, sodium channels open following a step depolarisation, but then inactivate. The channel, once open at -50 mV, stays open for an average of 1·7 ms. In tetamethrin-poisoned patches, channels which open at -50 mV fall into two distinct classes based on the length of time they remain open. There is a group resembling normal channels, with the same average open time found in control patches, and there is a group showing the expected property of modified channels with a much longer average lifetime of 11 ms. This result confirms the validity of separating the macroscopic currents into normal and modified fractions. Another important result from this study is that the single channel conductance of the modified channels was identical to that of the normal channel.

As has already been mentioned, the conformational changes associated with opening and closing of the sodium channels involve intramembrane charge movement known as gating current. Because in the presence of pyrethroids, the channels are thought to open normally and then become modified while open, the pyrethroid should not affect the ON gating current associated with channel opening, but should decrease the OFF gating current which is associated with channel closing. Dubois & Bergman (1977) demonstrated that DDT has this effect on gating current in frog myelinated nerves. On the other hand, Vijverberg (1982) found that allethrin did not have a specific effect on the OFF gating current in squid giant axons, but decreased the ON and OFF components equally. Furthermore, in preliminary studies on frog myelinated nerve, Vijverberg (1982) found that cismethrin also depressed the ON and OFF components equally, and slowed both components. Further investigation is needed in this area.

Pyrethroid effects on vertebrate sodium channels

The only vertebrate nerves in which the actions of DDT and pyrethroids have been studied in detail are the frog myelinated fibres. As in invertebrate nerves, sodium inactivation is slowed down in a fraction of the channels and tail current persists after repolarisation (Hille, 1968; Vijverberg et al., 1983). However, there are two important differences. First, with pyrethroids, the inactivation is not greatly affected, so the conductance declines more or less normally during the pulse, but after repolarisation the conductance increases, as though modified channels are opening. This gives rise to a so-called hooked tail current (Vijverberg et al., 1983). Small hooks were also seen with DDT (Hille, 1968). The simplest interpretation of the hooked tail current is that the modified channel can inactivate almost as rapidly as normal ($O^* \rightarrow I^*$ in Figure 2.10), so that the sodium current is not prolonged, but following repolarisation the modified inactivated channel returns to the modified open state ($I^* \rightarrow O^*$) from which it returns slowly to the resting state.

The second major difference between frog and invertebrate nerve is that the effect of α-cyano pyrethroids is much less dramatic on the frog. Whereas in crayfish axon, tail currents induced by α-cyano compounds decay with time constants of several minutes (Lund & Narahashi, 1983), in frog the time constants range from 0·5 to 1·5 seconds (Vijverberg et al., 1983). Thus, in frog, the α-cyano pyrethroids are much less effective in stabilising the open state than in invertebrates. This is probably also true for mammals, as deltamethrin did not depolarise mouse neuroblastoma cells (Jacques et al., 1980). Although the tail current induced by α-cyano pyrethroids in frog nerve does not decay as slowly as that in crayfish, it nevertheless decays two orders of magnitude more slowly than the tail current caused by type I pyrethroids in frog. This difference may be the basis for the unique neurotoxic effects of α-cyano pyrethroids in rats.

Potassium channels

Although the primary site of action of pyrethroids seems to be the nerve sodium channels, there may be other membrane components directly affected by pyrethroids. Internal perfusion of 10–100 μM allethrin in squid axons suppressed the potassium current without changing its kinetics (Narahashi & Anderson, 1967; Wang et al., 1972). DDT also suppressed the potassium current in lobster axons (Narahashi & Haas, 1968). On the other hand, tetramethrin (3×10^{-4} M) or fenvalerate (10^{-5} M) did not affect the potassium current in crayfish axons (Lund & Narahashi, 1979; V. L. Salgado & T. Narahashi, unpublished observations). The toxicological significance of the effect on potassium channels has not been studied, as it is generally too small to be of importance in itself. However, when combined with the tendency of pyrethroids to keep sodium channels open, suppression of the potassium current can enhance the ability of some pyrethroids to induce negative after-potentials, repetitive firing and nerve depolarisation.

ATPases

It has recently been reported that pyrethroids inhibit calcium-dependent ATPase activity in squid nerve. Allethrin and pyrethrins primarily inhibited Ca^{++}-ATPase activity, whereas "highly modified type" pyrethroids (deltamethrin and cypermethrin) mainly inhibited Ca^{++}, Mg^{++}-ATPase (Clark & Matsumura, 1982). The difference was not correlated with the presence of an α-cyano group, as permethrin, which lacks a cyano group, strongly inhibited Ca^{++}, Mg^{++}-ATPase, while fenvalerate, which contains the cyano group, was not very potent in this respect. The toxicological significance of ATPase inhibition by pyrethroids is wholly unknown. The Ca^{++}, Mg^{++}-ATPases are thought to be involved in sequestering intracellular calcium, so their inhibition should increase intracellular calcium. It was suggested that this could account for the increase of spontaneous transmitter release by pyrethroids (Clark & Matsumura, 1982), but Salgado et al. (1983) showed that the increase of spontaneous transmitter release by pyrethroids was due to depolarisation of the nerve terminals by pyrethroid-induced TTX-sensitive sodium influx.

DDT also inhibits ATPases, specifically the mitochondrial oligomycin-sensitive Mg^{++}-ATPase which catalyses the terminal step of oxidative phosphorylation (Patil & Koch, 1979). As with pyrethroids, this effect has not been found to have any toxicological significance (Jackson & Gardner, 1978).

GABA receptor–ionophore system†

A 3–10 mg/kg i.p. pretreatment dose of diazepam protected mice against mortality due to i.c.v. doses of either deltamethrin or permethrin; diazepam delayed the onset of symptoms due to deltamethrin or fenvalerate but not permethrin or allethrin (Staatz et al., 1982; Gammon et al., 1982). This diazepam effect was claimed to support the contention that some pyrethroids are acting at the GABA–receptor complex. This evidence cannot be taken seriously, however, because the diazepam dose used is excessive. The therapeutic i.v. dose for status epilepticus in human beings is 5 mg/person (70 kg) according to standard references, suggesting that penetration to the GABA site in the brain is not a problem. Using the volume of distribution of diazepam in man of 0·7 l/kg, the dose of 3–10 mg/kg would be expected to give a blood diazepam concentration of 15–50 μM. However, 50 μM diazepam has been found to block 50% of the sodium current in myelinated nerve (Courtney & Etter, 1983), underlining the difficulties one encounters using excessive doses and assuming a specific action.

Symptoms of permethrin or deltamethrin poisoning may also be partially relieved by i.p. treatment with cycloheximide or aminooxyacetic acid (Staatz et al., 1982); however, this is not compelling evidence to implicate the GABA system in pyrethroid poisoning (Staatz et al., 1982). Quite obviously, much more information is needed on the action of pyrethroids on the central nervous system.

Leeb-Lundberg & Olsen (1980) reported that deltamethrin inhibited binding of α-dihydropicrotoxinin (DHP) to rat brain extracts with an IC50 of 0·1 μM, although in Olsen et al. (1980) the figure was reported at 0·05 μM. The binding of labelled DHP was 50% inhibited by cold DHP at 1·1 μM, by picrotoxinin at 0·5 μM, by the convulsant barbiturate DMBB (dimethylbutylbarbiturate) 0·05 μM, and by *t*-butylbicyclophosphate at 2·0 μM. Thus, deltamethrin was similar in potency to DMBB, one of the most potent

---

† GABA, γ-aminobutyric acid.

compounds inhibiting DHP binding to rat brain. The αR isomer of delta-methrin, which is without comparable insecticidal activity, only inhibited binding at greater that 10 μM (Leeb-Lundberg & Olsen, 1980).

Using the ^{35}S-labelled ligand t-butylbicyclophosphorothioate (TBPS), Lawrence & Casida (1983) reported stereospecific inhibition of TBPS binding by a variety of α-cyanophenoxybenzyl pyrethroids. Ability to inhibit binding in rat brain extracts was correlated with toxicity values obtained by intracerebral injection (i.c.v.) into mice. The inhibition of the binding, however, is relatively weak, with 37% inhibition being produced by 5 μM concentrations of (1R, αS)-cis-cypermethrin, and no other values approaching this. When compared to the reported inhibition of 50% DHP binding by 0·05 μM deltamethrin in rat brain, one sees a 100-fold difference if, indeed, it is legitimate to compare inhibition of TBPS and DHP binding.

Gammon & Casida (1983) recently reported that insecticidal isomers of the type II pyrethroids deltamethrin, cypermethrin, fenpyrithrin and fenvalerate blocked the GABA-activated conductance in crayfish claw opener muscles, while two inactive αS isomers of cypermethrin and three insecticidal type I pyrethroids (permethrin, resmethrin and S-bioallethrin) had no effect. However, the IC50 values for inhibition of the GABA-activated conductance were 79, 310 and 29 μM for deltamethrin, fenpyrithrin and fenvalerate, respectively. As we know (see p. 73) that pyrethroids reach levels of only 30–200 nM during poisoning, effects on the GABA system are probably not relevant to poisoning, at least for invertebrates.

Pyrethroid binding studies

Measurement of binding of labelled pyrethroids and DDT has recently been achieved by two different groups. Soderlund et al. (1983) found saturable binding of NRDC 157 (3-phenoxybenzyl [1R-cis-3-(2,2-dibromovinyl]-2,2-dimethylcyclopropanecarboxylate) that was stereospecific for the 1R isomer. This stereospecific binding was half-saturated at 40 nM and 100% saturated above 100 nM which is near the concentration expected for the site of action.

DDT, permethrin and cypermethrin were found to bind to a receptor in housefly brain membranes, though DDT did not bind at precisely the same site as the pyrethroids (Chang & Plapp, 1983a,b). Receptor sites were reported to be saturated at 45–90 nM [^{14}C]DDT and had a dissociation constant of 12·2 nM. These figures from housefly brain membranes were of the same order of magnitude as similar binding studies on mouse brain (Soderlund et al., 1983). The data of Chang & Plapp (1983b) showed that DDT was bound to membranes from kdr houseflies two to three times less tightly than to similar extracts from susceptible houseflies. Since the kdr factor is thought to be responsible for site insensitivity in the nervous system, one would expect such a difference at the site of action.

These binding studies are still very new, and the binding components have not yet been identified. However, we should expect to see much important information from binding studies in the near future.

Summary

There is no general agreement on what happens after target organisms are treated with pyrethroids. Opinions may be conveniently divided between those who feel all poisoning is a direct result of interactions with the function of the sodium channel, and those others who feel that other sites of action are equally or more important. At this time of writing there is no universally held theory of pyrethroid mode of action, despite the commitments anyone in particular has to one view.

The symptoms of pyrethroid poisoning and their underlying causes are at the root of most mode-of-action studies. Everyone agrees that the symptoms are indicative of an attack on the insect nervous system; beyond this there is little or no consensus of opinion.

For a complete understanding of the mode of action of a compound, we must identify the molecular site or sites of action and show how these primary effects lead to death of the animal. In mammals the point of death is sharply defined by respiratory or cardiac failure. Pyrethroid-poisoned mice and rats die during seizures within one or two hours after treatment (Ray & Cremer, 1979; Lawrence & Casida, 1982). Insects, however, are not dependent on continuous nervous control of respiration and circulation, and it is therefore difficult to define the point of death. In fact, death of insects seems to be due to irreversible damage to the nervous system occurring when poisoning lasts for more than a few hours. Because of these differences in physiology, different aspects of pyrethroid action may be important for insect and mammalian toxicity. For instance, mammals may be sensitive to any disruptive effect on the nervous system, while insects will be relatively more sensitive to those effects that cause irreversible nerve damage.

Several molecular effects of pyrethroids have been identified. There is no question that pyrethroids interact with sodium channels, potassium channels, GABA-activated chloride channels and membrane-bound ATPases. The effects on potassium channels and membrane-bound ATPases do not appear to be important for poisoning. There is some evidence that the effects on the GABA system may be involved in the poisoning of mammals by type II pyrethroids, but this does not appear to be the case in insects. The most general effect of pyrethroids, which seems to account for all their diverse actions in insects, is their interaction with sodium channels. The function of sodium channels is complex and, therefore, the interactions of pyrethroids with them can be expressed in different ways. Thus, type I pyrethroids tend to cause repetitive firing, while type II pyrethroids depolarise the nerves. Also, some compounds, allethrin for example, can block nerve conduction without depolarisation.

Pyrethroids have been shown to be active on some sensory nerve structures, motor nerve structures, presynaptic terminals, and neurosecretory axons. The actions on the sensory and motor axons are not known to lead to irreversible lesions; however, actions at the presynaptic terminal lead to irreversible damage, when examined over sufficient time. The block of neuromuscular junctions explains why exaggerated movements gradually decline in later stages of poisoning, possibly coinciding with the prostrate and moribund stages.

A theory of a common mode of action for all pyrethroids and DDT analogues was offered recently by Salgado et al. (1983b), who claimed that the depolarising action of pyrethroids was the key feature of their action. This claim suggests that, no matter how severe, burst discharge is not fatal; that while hyperactivity and convulsions are the most obvious events in pyrethroid poisoning, other more important and fatal events occur that determine survival or death. Although the depolarisation theory was developed from studies on the neuromuscular junction, it badly needs to be tested on central nervous tissues to see if irreversible lesions occur there too.

Pyrethroids and DDT are two categories of insecticide that share with several other compound (veratridine, aconitine, batrachotoxin, grayanotoxin and scorpion toxins) an ability to alter indirectly the kinetics of the sodium channel. These compounds are called sodium channel modulators. All of these compounds share the property that their actions are reversed or eliminated by perfusion with tetrodotoxin.

If the sodium channel is indeed the site of action, some of the other sodium channel modulators might be expected to be less active on the *kdr* strains of insects. In fact, we need to study the sodium channel in *kdr* strains of insects above all to see whether this site is indeed transformed compared to susceptible insects.

References

Adams, M. E. & Miller, T. A., 1979, Site of action of pyrethroid: repetitive "backfiring" in flight motor units of housefly. *Pestic. Biochem. Physiol.*, **11**, 218–31.

Adams, M. E. & Miller, T. A., 1980, Neural and behavioral correlates of pyrethroid and DDT-type poisoning in the house fly, *Musca domestica* L. *Pestic. Biochem. Physiol.*, **13**, 137–47.

Armstrong, C. M. & Bezanilla, F., 1974, Charge movement associated with the opening and closing of the activation gates of the sodium channels. *J. Gen. Physiol.*, **63**, 533–52.

Beard, R. L., 1958, Secondary physiological effects of DDT in Galleria larvae. *Entomol. Exp. Appl.*, **1**, 200–10.

Beeman, R. W., 1982, Recent advances in mode of action of insecticides. *Ann. Rev. Entomol.*, **27**, 253–81.

Blum, M. S. & Kearns, C. W., 1956, Temperature and the action of pyrethrum in the American cockroach. *J. Econ. Entomol.*, **49**, 862–5.

Bradbury, J. E., Sorshaw, P. J., Gray, A. J. & Ray, D. E., 1983, The action of mephenesin and other agents on the effects produced by two neurotoxic pyrethroids in the intact and spinal rat. *Neuropharmacol.*, **22**, 907–14.

Briggs, G. G., Elliott, M., Farnham, A. W. & Janes, N. F., 1974, Structural aspects of the knockdown of pyrethroids. *Pestic. Sci.*, **5**, 643–50.

Brown, T. M., 1981, Countermeasures for insecticide resistance. *Bull. Entomol. Soc. Amer.*, **27**, 198–202.

Brown, T. M., 1982, Prevention of pest resistance to synthetic pyrethroid insecticides. *Chemical Times and Trends*, **5**, 33–6.

Burt, P. E., 1980, The effects of neurotoxicants on the nervous systems of intact insects. In *Insect Neurobiology and Pesticide Action (Neurotox 79)* (London: Society of Chemical Industry), pp. 407–14.

Burt, P. E. & Goodchild, R. E., 1971, The site of action of pyrethrin I in the nervous system of the cockroach, *Periplaneta americana. Entomol. Exp. Appl.*, **14**, 179–89.

Burt, P. E. & Goodchild, R. E., 1974, Knockdown by pyrethroids: its role in the intoxication process. *Pestic. Sci.*, **5**, 625–33.

Burt, P. E. & Goodchild, R. E., 1977, The action of pyrethroids in the insect central nervous system. I. Features of molecular structure associated with toxicity to cockroaches and to their giant fibre axons. *Pestic. Sci.*, **8**, 681–91.

Busvine, J. R., 1951, Mechanism of resistance to insecticide in houseflies. *Nature, Lond.*, **168**, 193–5.

Camhi, J. M. & Nolen, T. G., 1981, Properties of the escape system of cockroaches during walking. *J. Comp. Physiol.*, **142**, 339–46.

Casida, J. E., Gammon, D. W., Glickman, A. H. & Lawrence, L. J., 1983, Mechanisms of selective action of pyrethroid insecticides. *Ann. Rev. Pharmacol. Toxicol.*, **23**, 413–38.

Chang, C. P. & Plapp, F. W., 1983a, DDT and pyrethroids: receptor binding and mode of action in the house fly. *Pestic. Biochem. Physiol.*, **20**, 76–85.

Chang, C. P. & Plapp, F. W., 1983b, DDT and pyrethroids: receptor binding in relation to knockdown resistance (*kdr*) in the house fly. *Pestic. Biochem. Physiol.*, **20**, 86–91.

Chialiang, C. & Devonshire, A. L., 1982, Changes in membrane phospholipids, identified by Arrhenius plots of acetylcholinesterase and associated with pyrethroid resistance (*kdr*) in houseflies (*Musca domestica*). *Pestic. Sci.*, **13**, 156–60.

Christensen, T. A. & Carlson, A. D., 1981, Symmetrically organized dorsal unpaired median (DUM) neurones and flash control in the male firefly, *Photuris versicolor. J. Exp. Biol.*, **93**, 133–47.

Clark, J. M. & Matsumura, F., 1982, Two different types of inhibitory effects of pyrethroids on nerve Ca and Ca+Mg−ATPase activity in the squid, *Loligo pealei. Pestic. Biochem. Physiol.*, **18**, 180–90.

Clements, A. N. & May, T. E., 1977, The actions of pyrethroids upon the peripheral nervous system and associated organs in the locust. *Pestic. Sci.*, **8**, 661–80.

Courtney, K. R. & Etter, E. F., 1983, Modulated anticonvulsant block of sodium channels in nerve and muscle. *Eur. J. Pharmacol,*, **88**, 1–9.

Cutkomp, L. K. Koch, R. B & Desaiah, D., 1982, Inhibition of ATPases by chlorinated hydrocarbons. In *Insecticide Mode of Action* (New York: Academic Press), pp. 45–69.

Dresden, D., 1948, Site of action of DDT and cause of death after acute DDT poisoning. *Nature, Lond.*, **162**, 1000–1.

Dubois, J. M. & Bergman, C., 1977, Asymmetrical currents and sodium currents in Ranvier nodes exposed to DDT. *Nature, Lond.*, **266**, 741–2.

Eaton, J. L. & Sternburg, J., 1964, Temperature and the action of DDT on the nervous system of *Periplaneta americana* (L.). *J. Insect Physiol.*, **10**, 471–85.

Eaton, J. L. & Sternburg, J. G., 1967a, Temperature effects on nerve activity in DDT-treated American cockroaches. *J. Econ. Entomol.*, **60**, 1358–64.

Eaton, J. L. & Sternburg, J. G., 1967b, Uptake of DDT by the American cockroach central nervous system. *J. Econ. Entomol.*, **60**, 1699–703.

Elliott, M. & Janes, N. F., 1979, Recent structure–activity correlations in synthetic pyrethroids. In *Advances in Pesticide Science* (Oxford: Pergamon Press), pp. 166–73.

Elliott, F. M., Janes, N. F. & Potter, C., 1978, The future of pyrethroids in insect control. *Ann. Rev. Entomol.*, **23**, 443–69.

Farley, J. M., Narahashi, T. & Holan, G., 1979, The mechanism of action of a DDT analog on the crayfish neuromuscular junction. *Neurotoxicol.*, **1**, 191–207.

Farnham, A. W., 1977, Genetics of resistance of houseflies (*Musca domestica* L.) to pyrethroids. I. Knock-down resistance. *Pestic. Sci.*, **8**, 631–6.

Ford, M., 1979, Quantitative structure–activity relationships of pyrethroid insecticides. *Pestic. Sci.*, **10**, 39–49.

Gammon, D. W., 1978a, Neural effects of allethrin on the free walking cockroach, *Periplaneta americana*: an investigation using defined doses at 15 and 32 °C. *Pestic. Sci.*, **9**, 79–91.

Gammon, D. W., 1978b, Effects of DDT on the cockroach nervous system at three temperatures. *Pestic. Sci.*, **9**, 95–104.

Gammon, D. W., 1978c, The action of tetrodotoxin on the cockroach, *Periplaneta americana*: a toxicological and neurophysiological study. *Physiological Entomol.*, **3**, 37–42.

Gammon, D. W., 1979, An analysis of the temperature-dependence of the toxicity of allethrin to the cockroach. In *Neurotoxicology of Insecticides and Pheromones* (New York: Plenum Press), pp. 97–117.

Gammon, D. W., 1980, Pyrethroid resistance in a strain of *Spodoptera littoralis* is correlated with decreased sensitivity of the CNA *in vitro*. *Pestic. Biochem. Physiol.*, **13**, 53–62.

Gammon, D. & Casida, J. E., 1983, Pyrethroids of the most potent class antagonize GABA action at the crayfish neuromuscular junction. *Neurosci. Lett.*, **40**, 163–8.

Gammon, D. W., Brown, M. A. & Casida, J. E., 1981, Two classes of pyrethroid action in the cockroach. *Pestic. Biochem. Physiol.*, **15**, 181–91.

Gammon, D. W., Lawrence, L. J. & Casida, J. E., 1982, Pyrethroid toxicology: protective effects of diazepam and phenobarbital in the mouse and the cockroach. *Toxicol. Appl. Pharmacol.*, **66**, 290–6.

Gerolt, P., 1983, Insecticides: their route of entry, mechanism of transport and mode of action. *Biol. Rev.*, **58**, 233–74.

Glickman, A. H. & Casida, J. E., 1982, Species and structural variations affecting pyrethroid neurotoxicity. *Neurobehav. Toxicol. Teratol.*, **4**, 793–9.

Hamill, O. P., Marty, A., Neher, E., Sakmann, B. & Sigworth, F. J., 1981, Improved patch-clamp techniques for high-resolution current recording from cells and cell-free membrane patches. *Pflugers Arch.*, **391**, 85–100.

Heslop, J. P. & Ray, J. W., 1959, The reaction of the cockroach, *Periplaneta americana* L., of bodily stress and DDT. *J. Insect Physiol.*, **3**, 395–401.

Hille, B., 1968, Pharmacological modifications of the sodium channels of frog nerve. *J. Gen. Physiol.*, **51**, 199–219.

Hodgkin, A. L. & Huxley, A. F., 1952a, Currents carried by sodium and potassium ions through the membrane of the giant axon of *Loligo*. *J. Physiol.*, **116**, 449–72.

Hodgkin, A. L. & Huxley, A. F., 1952b, The components of membrane conductance in the giant axon of *Loligo*. *J. Physiol.*, **116**, 473–96.

Hodgkin, A. L. & Huxley, A. F., 1952c, The dual effect of membrane potential on sodium conductance in the giant axon of *Loligo*. *J. Physiol.*, **116**, 497–506.

Hodgkin, A. L. & Huxley, A. F., 1952d, A quantitative description of membrane current and its application to conduction and excitation in nerve. *J. Physiol.*, **117**, 500–44.

Hodgkin, A. L., Huxley, A. F. & Katz, B., 1952, Measurement of current–voltage relations in the membrane of the giant axon of *Loligo*. *J. Physiol.*, **116**, 242–48.

Holan, G., O'Keefe, D. F., Virgona, C. & Walser, R., 1978, Structural and biological link between pyrethroids and DDT in new insecticides. *Nature, Lond.*, **272**, 734–6.

Hrdina, P. D., Singhal, R. L. & Ling, G. M., 1975, DDT and related chlorinated hydrocarbon insecticides: pharmacological basis of their toxicity in mammals. *Adv. Pharmacol. Chemother.*, **12**, 31–88.

Hutzel, J. M., 1942, Action of pyrethrum upon the German cockroach. *J. Econ. Entomol.*, **35**, 933–7.

Irving, S. N. & Fraser, T. E. M., 1984, Insecticidal activity of tralomethrin: electrophysiological assay reveals that it acts as a propesticide. *J. Agric. Food Chem.*, **32(1)**, 111–3.

Irving, S. N. & Miller, T. A., 1980, Aspartate and glutamate as possible transmitters at the "slow" and "fast" neuromuscular juctions of the body wall muscles of *Musca* larvae. *J. Comp. Physiol.*, **135**, 299–314.

Jackson, D. A. & Gardner, D. R., 1978, *In vitro* effects of DDT analogs on trout brain Mg^{++}-ATPases. 1. Specificity and physiological significance. *Pestic. Biochem. Physiol.*, **8**, 113–22.

Jacques, Y., Romey, G., Cavey, M. T., Kartalovski, B. & Lazdunski, M., 1980, Interaction of pyrethroids with the Na^+ channel in mammalian cells in culture. *Biochim. Biophys. Acta.*, **600**, 882–97.

Läuger, P., 1946, Mechanism of intoxication of DDT insecticides in insects and warm-blooded animals. Joint meeting of the Army Committee for Insect and Rodent Control, pp. 1–24.

Lawrence, L. J. & Casida, J. E., 1982, Pyrethroid toxicology: mouse intracerebral structure–toxicity relationships. *Pestic. Biochem. Physiol.*, **18**, 9–14.

Lawrence, L. J. & Casida, J. E., 1983, Stereospecific action of pyrethroid insecticides in the γ-aminobutyric acid receptor-ionophore complex. *Science*, **221**, 1399–401.

Leeb-Lundberg, F. & Olsen, R. W., 1980, Picrotoxinin binding as a probe of the GABA postsynaptic membrane receptor–ionophore complex. In *Psychopharmacology and Biochemistry of Neurotransmitter Receptors*, edited by H. I. Yamamura, R. W. Olsen & E. Usdin (New York: Elsevier/North-Holland), pp. 593–606.

LeRoux, E. J. & Morrison, F. O., 1954, The adsorption, distribution, and site of action of DDT in DDT-resistant and DDT-susceptible house flies using carbon-labelled DDT. *J. Econ. Entomol.*, **47**, 1058–66.

Lindquist, A. W., Wilson, H. G., Schroeder, A. O. & Madden, A. H., 1945, Effect of temperature on knockdown and kill of houseflies exposed to DDT. *J. Econ. Entomol.*, **38**, 261–4.

Liu, M. Y., Sun, C. N. & Huang , S. W., 1982a, Absence of synergism of DDT by piperonyl butoxide and DMC in larvae of the diamondback moth (Lepidoptera: Yponomeutidae). *J. Econ. Entomol.*, **75**, 964–5.

Liu, M. Y., Tzeng, Y. J. & Sun, C. N., 1982b, Insecticide resistance in the diamondback moth. *J. Econ. Entomol.*, **75**, 153–5.

Lockskin, R. A., 1971, Programmed cell death: nature of the nervous signal controlling breakdown of intersegmental muscles. *J. Insect Physiol.*, **17**, 149–58.

Lowenstein, O., 1942, A method of physiological assay of pyrethrum extracts. *Nature, Lond.*, **150**, 760–2.

Lund, A. E. & Narahashi, T., 1979, The effect of the insecticide tetramethrin on the sodium channel of crayfish giant axons, *Soc. Neurosci. Abstr.*, **5**, 293.

Lund, A. E. & Narahashi, T., 1981a, Interaction of DDT with sodium channels in squid giant axon membranes. *Neurosci.*, **6**, 2253–58.

Lund, A. E. & Narahashi, T., 1981b, Kinetics of sodium channel modification by the insecticide tetramethrin in squid axon membranes. *J. Pharmacol. Exp. Therap.*, **219**, 464–73.

Lund. A. E. & Narahashi, T., 1981c, Modification of sodium channel kinetics by the insecticide tetramethrin in crayfish giant axons. *Neurotoxicol.*, **2**, 213–29.

Lund, A. E. & Narahashi, T., 1982, Dose-dependent interaction of the pyrethroid isomers with sodium channels of squid axon membrane. *Neurotoxicol.*, **3**, 11–24.

Lund, A. E. & Narahashi, T., 1983, Kinetics of sodium channel modification as the basis for the variation in the nerve membrane effects of pyrethroids and DDT analogs. *Pestic. Biochem. Physiol.*, **20**, 203–16.

Merrill, R. S., Savit, J. & Tobias, J. M., 1946, Certain biochemical changes in the DDT poisoned cockroach and their prevention by prolonged anesthesia. *J. Cell. Comp. Physiol.*, **28**, 465–76.

Miller, T. A. & Adams, M. E., 1977, Central vs. peripheral action of pyrethroids on the house fly nervous system. *ACS Symp. Ser.*, **42**, 98–115.

Miller, T. A., Salgado, V. L. & Irving, S. N., 1983, The *kdr* factor in pyrethroid resistance. In *Pest Resistance to Pesticides* (New York: Plenum Press), pp. 353–66.

Mordue, W., 1982, Neurosecretory peptides and biogenic amines. In *Neuropharmacology of Insects*, edited by D. Evered, M. O'Connor & J. Whelan, Ciba Foundation Symposium 88 (London: Pitman), pp. 88–101.

Moriarty, F., 1969, The sublethal effects of synthetic insecticides on insects. *Biol. Rev.*, **44**, 321–57.

Munson, S. C. & Yeager, J. F., 1945, DDT-like effects from injection of other compounds into cockroaches. *J. Econ. Entomol.*, **38**, 618.

Muryama, K., Abbott, N. J., Narahashi, T. & Shapiro, B. I., 1972, Effects of allethrin and *Condylactis* toxin on the kinetics of sodium conductance of crayfish axon membranes. *Comp. Gen. Pharmac.*, **3**, 391–400.

Narahashi, T., 1963, The properties of insect axons. In *Advances in Insect Physiology* (New York: Academic Press), pp. 175–256.

Narahashi, T., 1971, Effects of insecticides on excitable tissues. *Adv. Insect Physiol.*, **8**, 1–93.

Narahashi, T., 1979, Nerve membrane ionic channels as the target site of insecticides. In *Neurotoxicology of Insecticides and Pheromones*, edited by T. Narahashi (New York: Plenum), pp. 211–43.

Narahashi, T., 1982, Cellular and molecular mechanisms of action of insecticides: neurophysiological approach. *Neurobehav. Toxicol. Teratol.*, **4**, 753–8.

Narahashi, T. & Anderson, N. C., 1967, Mechanism of excitation block by the insecticide allethrin applied externally and internally to squid giant axons. *Toxicol. Appl. Pharmacol.*, **10**, 529–47.

Narahashi, T. & Haas, H. G., 1968, Interaction of DDT with the components of lobster nerve membrane conductance. *J. Gen. Physiol.*, **51**, 177–98.

Narahashi, T. & Yamasaki, T., 1960a, Mechanism of increase in negative after-potential by dicophanum (DDT) in the giant axons of the cockroach. *J. Physiol.*, **152**, 122–40.

Narahashi, T. & Yamasaki, T., 1960b, Mechanism of the after-potential production in the giant axons of the cockroach. *J. Physiol.*, **151**, 75–88.

Nicholson, R. A. & Sawicki, R. M., 1982, Genetic and biochemical studies of resistance to permethrin in a pyretheroid-resistant strain of the housefly (*Musca domestica* L.). *Pestic. Sci.*, **13**, 357–66.

Nishimura, K., Okajima, N., Fujita, T. & Nakjima, M., 1982, quantitative structure–activity studies of substituted benzyl chrysanthemates. 4. Physicochemical properties and the rate of progress of the knockdown symptom induced in house flies. *Pestic. Biochem. Physiol.*, **18**, 341–50.

Nishimura, K., Hayashi, K., Keno, A. & Fujita, T., 1983, Effect of pyrethroids and DDT analogs on the frequency of spontaneous discharges in crayfish central nerve cord. *J. Pestic. Sci.*, **8**, 283–91.

Nolan, J., Roulston, W. J. & Wharton, R. H., 1977, Resistance to synthetic pyrethroids in a DDT-resistant strain of *Boophilus microplus*. *Pestic. Sci.*, **8**, 484–6.

O'Brien, R. D., 1967, *Insecticides Action and Metabolism* (New York: Academic Press), p. 112.

Ohshumi, T., Hirano, M., Itaya, N. & Fujita, Y., 1981, A new pyrethroid of high insecticidal activity. *Pestic. Sci.*, **12**, 53–8.

Olsen, R. W., Leeb-Lundberg, F. & Napias, C., 1980, Picrotoxin and convulsant binding sites in mammalian brain. *Brain Res. Bull.*, **5**, Suppl. **2**, 217–21.

Omer, S. M., Georghiou, G. P. & Irving, S. N., 1980, DDT/pyrethroid resistance inter-relationships in *Anopheles stephensi*. *Mosquito News*, **40**, 200–9.

Orchard, I. & Osborne, M. P., The action of insecticides of neurosecretory neurons in the stick insects, *Carausius morosus*. *Pestic. Biochem. Physiol.*, **10**, 197–202.

O'Shea, M. & Evans, P. D., 1979, Potentiation of neuromuscular transmission by an octopaminergic neuron. *J. Exp. Biol.*, **79**, 169–90.

Page, A. B. P., Stringer, A. & Blackith, R. E. 1949, Bioassay system for the pyrethrins. I. Water base sprays against *Aedes aegypti* L. and other flying insects. *Ann. Appl. Biol.*, **36**, 225–49.

Palti, Y., 1971, Description of axon membrane ionic conductances and currents. In *Biophysics and Physiology of Excitable Membranes*, edited by W. J. Adelman (New York: Van Nostrand Reinhold), pp. 168–82.

Patil, T. N. & Koch, R. B., 1979, Differential inhibition responses caused by DDT on oligomycin-sensitive Mg^{++}-ATPase activity: nature of the requirements for DDT sensitivity. *Pestic. Biochem. Physiol.*, **12**, 205–15.

Pearson, K. G., 1981, Interneurones and locomotion. *Trends Neurosci.*, **4**, 128–31.

Pringle, J. W. S., 1974, Locomotion: flight. In *Physiology of Insecta* (New York: Academic Press), pp. 433–76.

Raabe, M., Baudry, N., Grillot, J. P. & Provansal, A., 1974, The perisympathetic organs of insects. In *Neurosecretion — the Final Neuroendocrine Pathway*, edited by F. Knowles & L. Vollrath (Berlin: Springer-Verlag), pp. 59–71.

Ray, D. E., 1982, The contrasting actions of two pyrethroids (deltamethrin and cismethrin) in the rat. *Neurobehav. Toxicol. Teratol.*, **4**, 801–04.

Ray, D. E. & Cremer, J. E., 1979, The actions of decamethrin (a synthetic pyrethroid) on the rat. *Pestic. Biochem. Physiol.*, **10**, 333–40.

Richards, A. G. & Cutkomp, L. K., 1945, Neuropathology in insects. *J. N. Y. Entomol. Soc.*, **53**, 313–55.

Roeder, K. D. & Weiant, E. A., 1946, The site of action of DDT in the cockroach. *Science*, **130**, 304–6.

Roeder, K. D. & Weiant, E. A., 1951, The effect of concentration, temperature, and washing on the time of appearance of DDT-induced trains in sensory fibers of the cockroach. *Ann. Entomol. Soc. Amer.*, **44**, 372–80.

Romey, G., Chicheportiche, R. & Lazdunski, M., 1980, Transition temperatures of the electrical activity of ion channels in the nerve membrane. *Biochem. Biophys. Acta*, **602**, 610–20.

Salgado, V. L., Irving, S. N. & Miller, T. A., 1983a, Depolarization of motor nerve terminals by pyrethroids in suscepible and *kdr*-resistant house flies. *Pestic. Biochem. Physiol.*, **20**, 100–14.

Salgado, V. L., Irving, S. N. & Miller, T. A., 1983b, The importance of nerve terminal depolarization in pyrethroid poisoning of insects. *Pestic. Biochem. Physiol.*, **20**, 169–82.

Salgado, V. L. & Narahashi, T., 1982, Interactions of the pyrethroid fenvalerate with the nerve membrane sodium channel. *Soc. Neurosci. Abstr.*, **8**, 251.

Sawicki, R. M., 1978, Unusual response of DDT-resistant houseflies to carbinol analogues of DDT. *Nature, Lond.*, **275**, 443–44.

Scott, J. G. & Matsumura, F., 1983, Evidence for two types of toxic actions of pyrethroids on susceptible and DDT-resistant German cockroaches. *Pestic. Biochem. Physiol.*, **19**, 141–50.

Scott, J. G., Ramaswamy, S. B. & Matsumura, F., 1984, Effect of temperature and method of application on resistance to pyrethroids in *kdr* resistant *Blattella germanica*, unpublished.

Shanes, A. M., 1949, Electrical phenomena in nerve. II. Crab nerve. *J. Gen. Physiol.*, **33**, 75–102.

Singh, G. J. P. & Orchard, I., 1982, Is insecticide-induced release of insect neurohormones a secondary effect of hyperactivity of the central nervous system? *Pestic. Biochem. Physiol.*, **17**, 232–42.

Soderlund, D. M., 1979, Pharmacokinetic behavior of enantiomeric pyrethroid esters in the cockroach, *Periplaneta americana* L. *Pestic. Biochem. Physiol.*, **12**, 38–48.

Soderlund, D. M., Ghiasuddin, S. M., & Helmuth, D. W., 1983, Receptor-like stereospecific binding of a pyrethroid insecticide to mouse brain membranes. *Life Sciences*, **33**, 261–67.

Sparks, T. C., Shour, M. H. & Wellemeyer, E. G., 1982, Temperature–toxicity relationships of pyrethroids on three lepidopterans. *J. Econ. Entomol.*, **75**, 643–46.

Sparks, T. C., Panloff, A. M., Rose, R. L. & Clower, D. F., 1983, Temperature–toxicity relationships of pyrethroids on *Heliothis virescens* (F.) (Lepidoptera: Modiuae) and *Anthonomus grandis grandis* Boheman (Coleoptera: Curculionidae). *J. Econ. Entomol.* **76**, 243–46.

Staatz, C. G., Bloom, A. S. & Lech, J. J., 1982, A pharmacological study of pyrethroid neurotoxicity in mice. *Pestic. Biochem. Physiol.*, **17**, 287–92.

Starkus, J. G. & Narahashi, T., 1978, Temperature dependence of allethrin induced repetitive discharges in nerves. *Pestic. Biochem. Physiol.*, **9**, 225–30.

Sternburg, J. G., 1963, Autointoxication and some stress phenomena. *Ann. Rev. Entomol.*, **8**, 19–38.

Sternburg, J. & Hewitt, P., 1962, *In vivo* protection of cholinesterase against inhibition by TEPP and its methyl homologue by prior treatment with DDT. *J. Insect Physiol.*, **8**, 643–63.

van den Bercken, L., Akkermans, L. M. A. & van der Zalm, J. M., 1973, DDT-like action of allethrin in the sensory nervous system of *Xenopus laevis*. *Eur. J. Pharmacol.*, **21**, 95–106.

Verschoyle, R. D. & Aldridge, W. N., 1980, Structure–activity relationships of some pyrethroids in rats. *Arch. Toxicol.*, **45**, 325–29.

Vijverberg, H. P. M., 1982, Interaction of pyrethroids with the sodium channels in myelinated nerve fibres. PhD dissertation, University of Utrecht.

Vijverberg, H. P. M., Ringt, G. S. F., & van den Bercken, J., 1982, Structure-related effects of pyrethroid insecticides on the lateral-line sense organ and on peripheral nerves of the clawed frog, *Xenopus laevis*. *Pestic. Biochem. Physiol.*, **18**, 315–24.

Vijverberg, H. P. M., van der Zalm, J. M., van Kleef, R. G. D. M. & van den Bercken, J. 1983, Temperature- and structure-dependent interaction of pyrethroids with the sodium channels in frog node of ranvier. *Biochim. Biophys. Acta*, **728**, 73–82.

Vinson, E. B. & Kearns, C. W., 1952, Temperature and the action of DDT on the American cockroach. *J. Econ. Entomol.*, **45**, 484–90.

Wang, C. M., Narahashi, T. & Scuka, M., 1972, Mechanism of negative temperature coefficient of nerve blocking action of allethrin. *J. Pharmacol. Exp. Therap.*, **182**, 442–53.

Wouters, W. & van den Bercken, J., 1978, Action of pyrethroids. *Gen. Pharmacol.*, **9**, 387–98.

Wu, C. H., van den Bercken, J. & Narahashi, T., 1975, The structure–activity relationship of DDT analogs in crayfish giant axons. *Pestic. Biochem. Physiol.*, **5**, 142–9.

Wu, C. H., Oxford, G. S., Narahashi, T. & Holan, G., 1980, Interaction of a DDT analog with the sodium channel of lobster axon. *J. Pharmacol. Exp. Therap.*, **212**, 287–93.

Yamamoto, D., Quandt, F. N. & Narahashi, T., 1983, Modification of single sodium channels by the insecticide tetramethrin. *Brain Res.*, **274**, 344–9.

Yaeger, J. F. & Munson, S. C., 1945, Physiological evidence of a site of action of DDT in an insect. *Science*, **102**, 305–7.

Zill, S. N. & Moran, D. T., 1982, Suppression of reflux postural tonus: a role of peripheral inhibition in insects. *Science*, **216**, 751–3.

Zlotkin, G., Lester, D., Lazorovice, P. & Pelhate, M., 1982, The chemistry and axonal action of two insect toxins derived from the venom of the scorpion *Buthotus judaicus*. *Toxicon.*, **20**, 332–31.

Zushi, S., Miyagawa, J. I., Yamamoto, M., Kataoko, K. & Seyama, I., 1983, Effect of grayanotoxin on the frog neuromuscular function. *J. Pharmacol. Exp. Ther.*, **226**, 269–75.

3. Toxicity to mammals

M. H. Litchfield

Introduction

The mammalian toxicology of the pyrethroids has been a developing subject over a number of years. The predecessors of the modern synthetic pyrethroids are the active ingredients of pyrethrum which has been used since the last century for insect control. Because these were naturally derived products it was always assumed that they were inherently safe to use. Their very early introduction precluded the comprehensive toxicological assessment which has been accorded new pesticides or new uses of pesticides in the last 20 years. Their toxicological evaluation has been built up slowly, and is still not equivalent to the present-day requirement for a new insecticide. However, the experience of the very long usage of pyrethrum products has apparently vindicated the term "safe" bestowed upon them. The amount of toxicological information on the early synthetic pyrethroids such as allethrin was also relatively small although some aspects of their mode of action and effects upon mammals had been established. Since their greater commercialisation from the mid-1970s, the modern synthetic pyrethroids have been the subject of an intensive investigation of their toxicological properties which has resulted in more data being generated within a relatively short time period, about 5 years, than for any other class of pesticide previously.

The review of the toxicology of a class of compounds can be undertaken in various ways, for example on a compound-by-compound basis, by assessment of different toxicological features (acute toxicity, chronic toxicity and so forth) or by some combination of these two models. The model followed in this review is to compare and contrast the various toxicological features of the pyrethroids and, where possible, to evaluate the overall toxicological potential of the pyrethroid molecule. The ultimate aim of toxicological studies on a compound is to be able to predict the effect in man, and this review is built up towards this end, also taking into account any experience on man himself. Although the review encompasses a wide range of the scientific literature on the subject there are certain publications which have provided a core of background information. The chapters on toxicology in the book *Pyrethrum – The Natural Insecticide* edited by J. E. Casida were particularly interesting and instructive in delineating the earlier studies on that material. For the synthetic pyrethroids the reviews from the Joint Meetings on Pesticide Residues, held jointly by the Food and Agricultural Organisation (FAO) and the World Health Organisation (WHO) have been a fruitful and up-to-date source of information, and the reader is referred to these for a compound-by-compound toxicological assessment.

Nomenclature and structures

A detailed explanation of pyrethroid structure and nomenclature is given in Chapter 1, but some information is repeated here in order to facilitate reference to the major compounds covered in this review.

The insecticidally active compounds present in natural pyrethrum are the esters of two acids (chrysanthemic acid and pyrethric acid) and three alcohols (pyrethrolone, cinerolone and jasmolone). Six esters arise from these combinations of which pyrethrins I and II represent the largest component (67%). The natural pyrethrum powder is usually extracted to give a dark viscous material containing about 30–35% pyrethrins, known as oleoresin concentrate. This may be diluted to give a standardised concentration of pyrethrins, called oleoresin extract. The oleoresin can be processed further to yield a more refined extract containing just over 50% pyrethrins. In the reports on toxicological studies on the pyrethrins it is not always clear which of these materials is being tested. Therefore in this chapter the extracts from the natural product will be referred to collectively as "pyrethrins" and their purity will be qualified if the information is available.

The synthetic pyrethroids were developed by modifying the basic pyrethrin I structure in order to improve insecticidal effectiveness and photostability. The resulting compounds, as discussed in Chapter 1, are usually a mixture of a number of isomeric forms. Since the isomeric composition of a pyrethroid significantly affects its biological activity, especially the *cis* : *trans* ratio for cyclopropane derived pyrethroids, it is important to know this isomeric composition when describing the results of toxicological studies.

Fortunately most of the commercially important synthetic pyrethroids are produced to a consistent isomer content, for example 40 : 60 *cis–trans* ratio for permethrin, 50 : 50 *cis–trans* ratio for cypermethrin, so that it will only be necessary to indicate significant deviations from the commonly used material when addressing the various compounds in toxicological studies. In this chapter the synthetic pyrethroids will be referred to universally as "pyrethroids".

Metabolism, distribution and excretion

The mechanism and rate of metabolism of a compound, the distribution of parent compound and metabolites in animal tissue and the overall rate of excretion usually have a major bearing on the toxicology of that compound. A comprehensive review of the metabolism of the pyrethroids in mammals is given in Chapter 5. It is relevant however to highlight the major features of their metabolism in the present chapter and in particular to detail their excretion and retention characteristics.

Table 3.1. The major events in the metabolism of permethrin after oral administration to rats

Metabolic process	Major products
1. Ester cleavage (esterase and oxidase enzymes)	
2. Oxidation	
3. Hydroxylation	
4. Conjugation	Glucuronide conjugates of carboxylic acids Sulphate conjugates of phenols

The pyrethroids are metabolised very rapidly in mammals, and the major steps in their metabolic pathway are illustrated in Table 3.1 using permethrin as the example (Gaughan et al., 1977). When the compound is administered orally to rats (1·6–4·8 mg/kg) it is rapidly metabolised via ester cleavage to give the acid and alcohol fragments which are then oxidised to the corresponding cyclopropane carboxylic acid and the 3-phenoxybenzoic acid. These acids are further oxidised to the 2-hydroxy derivative in the former case and the 2' and 4' hydroxy derivatives for the latter. The final metabolic process is conjugation, giving glucuronides of the carboxylic acids or sulphates of the phenols which are excreted in the urine. Similar routes of metabolism in rats are shown by cypermethrin (Crawford et al., 1981a,b), deltamethrin (Ruzo et al., 1978), fenvalerate (FAO, 1980) and resmethrin (Ueda et al., 1975). In the metabolic pathway for the pyrethrins (as shown in Chapter 5) oxidative attack at various points on the molecule is the major reaction, and in contrast to the pyrethroids ester cleavage is not a significant feature (Casida et al., 1971; Yamamoto et al., 1971; Elliott et al., 1972). The metabolites arising from this oxidative attack are found in the urine and faeces together with unmetabolised pyrethrins in the faeces only. It should be noted that the pyrethroid allethrin with its close structural relationship to the pyrethrins is also subject mainly to oxidative metabolism (Elliott et al., 1972).

The metabolism of the pyrethroids has been studied in several species in addition to the rat. There are no notable differences from the main pathway and products of metabolism as outlined for the rat except for the species-specific conjugative reactions with the 3-phenoxybenzoic acid metabolite (see Chapter 5 for details). A point arising from the metabolism studies of several pyrethroids is the difference between the degree and rate of reaction of the *cis* and *trans* isomers. Generally the *trans* isomers are metabolised at a faster rate than the *cis* as observed in animal and *in vitro* studies (Gaughan et al., 1977; Soderlund & Casida, 1977; Shono et al., 1979).

The site and mechanism of action for the metabolism of the pyrethroids and pyrethrins have been studied *in vitro*. The major system for cleaving the pyrethroids is located in the hepatic microsomes and is probably a carboxylesterase. The hydroxylating enzymes are also located in liver microsomes and are probably part of the cytochrome P-450 mediated mixed-function oxidase system (Soderlund & Casida, 1977; Hutson, 1979). Other *in vitro* studies on the location and kinetics of the reactions of the pyrethrins also indicate that the cytochrome P-450 oxygenase system may be implicated in their oxidative metabolism (Hutson, 1979).

The rapid rate of metabolism referred to above is reflected by the rate of appearance of excretion products in the urine and faeces. Oral doses of 1·6–4·8 mg/kg permethrin, labelled with ^{14}C on the acid and alcohol moieties, resulted in about 90% of the dose appearing in the urine and faeces within three days of administration to rats. A large proportion of this appeared in the first 24 hours. Urinary products were essentially permethrin metabolites but the faeces contained some unchanged parent compound as well (Gaughan et al., 1977). A very similar picture was shown by the study of ^{14}C-labelled cypermethrin in rats. After oral doses of 1·2–2·2 mg/kg cypermethrin, 85–90% of the label appeared in urine and faeces within three days (Crawford et al., 1981a). Deltamethrin at similar doses also shows the same rate of excretion in the rat (Ruzo et al., 1978), and both cypermethrin and deltamethrin have similar excretory rates in the mouse (Hutson et al., 1981;

Ruzo et al., 1979). After the oral administration of resmethrin to rats, the rate of elimination of excretory products is rather slower than for the other pyrethroids, about 50–70% of the dose appearing after six days (Ueda et al., 1975, Miyamoto, 1976). In most of the studies on these compounds the products of the breakdown of the *trans* isomer appeared in urine in relatively greater proportions than those from the *cis* isomer, while in faeces the opposite was the case. The *cis* isomer metabolites generally were excreted more slowly than the *trans*. Thiocyanate, arising from the metabolism of the α-cyano-pyrethroids, is eliminated slowly in the urine (Crawford et al., 1981b; Cole et al., 1982). The metabolic products from the pyrethrins are excreted fairly rapidly, with 50–60% of the labelled material appearing within 48 hours.

The corollary of the fast and generally complete elimination of the pyrethroids following oral administration to animals is low retention in the body, and this is borne out by actual measurement of tissue levels. After a single oral dose of pyrethroid the tissue concentration of radiolabelled material (apart from fat) usually reaches a maximum within one or two days, and is progressively and rapidly eliminated thereafter. Typical tissue concentrations one day after the dose of a pyrethroid are illustrated in the work on cypermethrin by Crawford et al (1981a). The levels, as microgram per gram equivalents of cypermethrin, ranged between 0·9–1·5 μg/g in fat and 0·005–0·024 μg/g in brain. The typical residue levels a few days after administration of a single dose of pyrethroid are shown by the values in Table 3.2. These values represent the microgram per gram equivalents of parent compound after administration of the ^{14}C-alcohol labelled material to rats or mice. The tissue concentrations remaining after dosage with ^{14}C-acid-labelled pyrethroid are usually somewhat lower than from the ^{14}C-alcohol label. Labelling with ^{14}C on the CN group of the α-cyano-pyrethroids shows a greater retention of material in the body tissues particularly in the skin and hair. This is in accordance with the known retention characteristics of the metabolite thiocyanate ion and accounts for the relatively slow excretion of the [14]CN label for these pyrethroids (Crawford et al., 1981b; Cole et al., 1982)

The tissue retention characteristics have been examined in more detail with some pyrethroids. In particular the retention characteristics in fatty tissue have been followed. Permethrin was dosed orally to rats at 1 mg/kg/day for 11 weeks and then the animals were maintained for a further 7 weeks undosed. The residues in the adipose tissue attained a plateau concentration within 3 weeks and did not exceed 2 μg/g. The adipose tissue levels declined steadily after the last dose and nothing could be detected at the end of 7 weeks. The half-life of the adipose tissue residues was calculated to be about two weeks. A similar study was undertaken with fenvalerate with similar results (FAO, 1980).

The rapid metabolism and elimination of the pyrethroids after administration to animals must have a significant impact upon their subsequent toxic action. The rapid decrease in acute toxic effect after the administration of the pyrethroids to be described in the following sections, shows that the metabolic

Table 3.2. Tissue retention[a] in rodents following the single dose oral administration of ^{14}C-alcohol labelled pyrethroids

Pyrethroid	Species	Dose (mg/kg)	Day	Blood	Brain	Fat	Kidney	Liver	Reference
Cypermethrin (*cis*)	Rat	1·7–2·5	8	0·01	0·001	1·15	0·02	0·05	Crawford et al. (1981a)
Cypermethrin (*cis*)	Mouse	8	8	0·02	<0·002	1·80	0·07	0·07	Hutson et al. (1981)
Deltamethrin	Mouse	1·7	8	0·005	n.d.	0·11	0·03	0·02	Ruzo et al. (1979)
Resmethrin (*trans*)	Rat	0·79–1·32	6	0·46	<0·01	1·56	0·16	0·11	Ueda et al. (1975)

[a]Expressed in terms of μg/g equivalents of parent compound.

reaction is essentially a detoxifying process. The influence of metabolism and tissue levels upon the toxicological properties of these compounds will be noted at the appropriate occasions in the following sections.

Mode of action

The signs of toxic action of the pyrethroids in insects and mammals are very similar. Hyperexcitation, tremoring and convulsions are the earlier signs followed by paralysis and then death at lethal levels of the insecticide. It is apparent therefore that the nervous system is the target in both insects and mammals. The investigations into the mode of action in insects have been described in Chapter 2. Although the site of action of the pyrethroids is similar in insects and mammals, there is a large differential in the dose level at which the effects occur. Obviously this is the basis for the favourable comparative toxicological profile of the pyrethroids whereby the target species, the insects, are killed by incredibly low concentrations of a compound (housefly topical LD50, deltamethrin, 0·01 mg/kg) whereas the mammalian non-target species are unaffected by such a dose (rat oral LD50, deltamethrin, 130 mg/kg).

Isolated organ studies

It is fairly easy to undertake mode of action studies using *in vitro* tissue preparations for compounds such as the organophosphates and carbamates where the reactions of a key enzyme can be followed. The pyrethroids do not appear to act upon any convenient enzyme system and therefore mode of action studies have to be undertaken on the target sites, that is the nervous tissue, directly. Detailed studies using isolated nervous tissue have been performed, notably by Narahashi and his co-workers utilising crayfish abdominal nerve cord and by van den Bercken and co-workers using isolated frog nerves. The work from the latter group is summarised here to describe the main features of the pyrethroid mode of action of relevance to mammalian systems. The experiments were carried out on the isolated peripheral system of the clawed frog *Xenopus laevis*. A brief electrical stimulus of the excised frog nerves evokes a single action potential in the control situation but when exposed to low concentrations of a pyrethroid (e.g., 1×10^{-5} M allethrin) they quickly show repetitive activity, that is the production of a train of action potentials. Sense organs such as the cutaneous touch receptor and the lateral line organ appear to be particularly sensitive to the pyrethroid action. Motor nerve fibres showed little or no repetitive activity even when higher concentrations of pyrethroids were used. This mode of action of the pyrethroids is very similar to that of DDT (Vijverberg, van der Zalm & van den

Bercken, 1982). DDT induces intense repetitive activity in frog sense organs and it has been shown that this activity increases with decrease in temperature. This inverse repetitive activity–temperature relationship has been shown to operate for the action of the pyrethroids on isolated nerve preparations and parallels a similar negative temperature coefficient for the toxicity of DDT and pyrethroids in insects, fish and mammals (Vijverberg & van den Bercken, 1982).

Differences in the manifestation of the action on isolated nerve preparations have been shown among the pyrethroids. Thus the pyrethroids that do not have an α-cyano group on the alcohol moiety, for example allethrin, permethrin and bioresmethrin, induce short trains of impulses in the lateral line sense organ. By contrast the pyrethroids containing an α-cyano group, such as cypermethrin, deltamethrin and fenvalerate, induce very long trains of nerve impulses which last for seconds and contain hundreds or even thousands of impulses (Vijverberg, Ruigt & van den Bercken, 1982). It would seem, therefore, that the α-cyano group is responsible for this difference in action between these two types of pyrethroids. The mechanism of the action of the pyrethroids on isolated frog nerves has been elucidated by the application of the voltage clamp technique (Vijverberg et al., 1983). This has shown that the pyrethroids cause prolongation of the sodium current associated with the depolarisation of the membrane. This has been explained as a delay in the closing of the sodium channels upon repolarisation of the membrane. The duration of these sodium tail currents correlates well with the duration of the nerve impulses induced in the lateral line sense organ, and both are subject to the negative temperature coefficient described previously. There is also a large difference in the sodium tail current duration between the α-cyano and non-α-cyano pyrethroids, which reflects their differences on lateral line repetitive activity. Despite these apparent differences it is postulated that the basic mode of action of all pyrethroids is on the sodium channels of the nerve membrane (Vijverberg & van den Bercken, 1982).

Mammalian studies

It would appear that the pyrethroids can be divided into two classes on consideration of their signs of toxicity following their administration to mammals by a variety of routes (Verschoyle & Aldridge, 1980). There are those pyrethroids which quickly produce an aggressive sparring behaviour in rats and an increased sensitivity to external stimuli. This is followed by the onset of fine tremor progressing to whole body tremor and, with lethal doses, occasionally rigor just before death. This was designated as the T-syndrome by Verschoyle & Aldridge (1980) and is produced by pyrethroids such as allethrin, cismethrin (*cis* isomer of resmethrin) and permethrin, which do not possess an α-cyano group on their alcohol moiety. The toxic action of these pyrethroids is similar to that for DDT (Joy, 1982). The toxic action of the pyrethroids that possess an α-cyano group in their alcohol moiety, such as cypermethrin and deltamethrin, is somewhat different. The initial signs of toxicity are character-ised by pawing and burrowing behaviour, followed within minutes by profuse salivation, coarse tremor, enhanced startled response and abnormal hind limb gait. These signs progress to sinuous whole body writhing, choreoathetosis, and with lethal doses clonic seizures are occasionally observed prior to death. This was designated as the CS-syndrome by Verschoyle & Aldridge (1980).

A similar division of the pyrethroids into two classes on the basis of toxicology has been noted by other investigators. Lawrence & Casida (1982) prefer to designate the classes as Type I (= T) and Type II (= CS) syndromes since differences also can be produced in systems where the effects are not described in terms of mammalian signs of toxicity. Thus differences in action between the two types of pyrethroids have been described for the cockroach (Gammon et al., 1981) and for isolated nervous tissue from the frog (Vijverberg, Ruigt & van den Bercken, 1982). Lawrence & Casida (1982) point out that there can be exceptions to this division into two classes, with the α-cyano pyrethroid fenproparthrin appearing to have some elements of the Type I syndrome. Verschoyle & Aldridge (1980) also noted exceptions in their studies where two compounds showed salivation associated with tremors. Lawrence & Casida (1982) noted that the Type I pyrethroids have a high *cis/trans* differential in toxic effect (*cis* > *trans*) whereas the Type II pyrethroids generally have a smaller differential between the two isomers.

The action of the pyrethroids directly on mammalian nerves has been studied using methodology previously utilised on humans. The compound nerve action potential was recorded proximally from the rat tail following the distal stimulation of the tail (Parkin & LeQuesne, 1982). The supernormal nerve excitability lasting from 4 to 30 ms in control animals was increased and extended to as much as 400 ms in animals dosed intravenously with 1·5 mg/kg deltamethrin. The supernormal nerve excitability effect showed a clear relationship with dose and was closely correlated with clinical signs of toxicity (Takahashi & LeQuesne, 1982). An intravenous dose of 0·3 mg/kg deltamethrin showed no response for both parameters. The non-α-cyano pyrethroid cismethrin at an intravenous dose of 3 mg/kg caused an increased excitability of only short duration, in contrast to deltamethrin.

There have been several investigations to try to locate the site of action of the pyrethroids in mammalian nervous tissue. Verschoyle & Aldridge (1980) hypothesised on the basis of their observations in rats that the T-syndrome exhibited by some pyrethroids originated from action upon the peripheral system. They could not decide upon the primary site of action of either type of pyrethroid however. Lawrence & Casida (1982) highlighted the probable importance of the role of the brain for the Type II (= CS) syndrome as a result of their investigations. In order to try to elucidate the site of action, Staatz et al. (1982) undertook studies in mice using intracerebroventricular (i.c.v.)

injections of either permethrin (Type I) or deltamethrin (Type II), comparing the response with those evoked by i.v. injection of the pyrethroids. Both compounds were much more toxic when administered i.c.v. than i.v. It was concluded that the toxic action of both pyrethroids was predominantly due to a central mechanism because the amounts injected into the brain were similar to those found by analysis at the time of toxic effects in other studies (Ruzo et al., 1979; Glickman & Lech, 1982). Other investigators using direct injection techniques into the central nervous system did not come to such a firm conclusion. Gray & Rickard (1982a), using cismethrin or deltamethrin, found that the toxic signs produced by injection into the lateral ventricles of the brain or the subarachonoid spaces around the spinal cord were very similar to those found by p.o., i.v. or i.p. injection. The use of radiolabelled pyrethroid showed that there was little movement from the spinal region after lumbar injection. The effects arising from this injection could be equated with those from studies in spinal rats or rabbits, further indicating that the involvement of the brain was not necessary. Since no deaths occurred after the lumbar or intraventricular injection of cismethrin, even with high doses, then it could be implied that a peripheral rather than a central action could be adduced to this pyrethroid. The argument for a spinal site of action to account for the motor signs produced by pyrethroids was further supported by studies involving the influence of pharmacological agents, such as mephenesin, on pyrethroid action (Bradbury et al., 1983). Similar modification of the pyrethroid motor signs by mephenesin was obtained using either the intact conscious rat or the spinal rat. Whatever the differences that might occur between the two types of pyrethroid or at the possible sites of action it is likely that the underlying mechanism of action is mediated via changes in the sodium permeability of nerve membranes (Bradbury et al., 1983; Gray & Rickard 1982a).

Another approach to investigating the mode or site of action of the pyrethroids has been taken by other investigators. Brodie & Aldridge (1982) measured cyclic GMP levels in different parts of the brain after the i.p. injection of deltamethrin to rats. They found that the time course of cerebellar cyclic GMP changes reflected the changes in motor activity as the signs of toxicity developed and not the dose of compound administered. There were no significant changes in cortical or striatal cyclic GMP which indicated that deltamethrin was acting as a locomotor stimulant rather than as a convulsant, in which case cyclic GMP levels would have been increased in all three regions of the brain. Cerebellar cyclic GMP was also elevated after the i.v. administration of cypermethrin and the change did not appear to correlate with the convulsive state of the rats (Lock & Berry, 1981). No increase in cyclic GMP was seen when cypermethrin was incubated with slices of the cerebellum *in vitro*. It was surmised that the primary site of action of the pyrethroids occurs outside the cerebellum. Another possible mechanism of action has been indicated by the reactions to the administration of drugs prior to pyrethroid injection (Gammon et al., 1982). Diazepam delayed the onset of the toxic signs of deltamethrin and fenvalerate (Type II pyrethroids) in the mouse but not those of allethrin and permethrin (Type I pyrethroids). By contrast phenobarbitone pretreatment only slightly delayed the onset of the toxic signs of deltamethrin. The higher potency of diazepam compared with phenobarbitone as an anticonvulsant for the Type II pyrethroid parallels the findings with GABAergic drugs. Therefore a possible mechanism for this type of pyrethroid could involve an action at the GABA receptor complex.

A feature of the mammalian action of pyrethroids is the capacity for recovery from toxic effects. Ray & Cremer (1979) noted that animals which survived potentially lethal doses of deltamethrin recovered within two to eight hours of the onset of athetosis, and were apparently normal for an observation period of two weeks afterwards. Monitoring of the cardiovascular and respiratory systems during early choreoathetosis showed that their status was relatively well maintained and that this was consistent with the good recovery made by the animals which survived.

Summary of present views

The signs of acute toxicity of the pyrethroids in mammals have been described in detail by several investigators. There is universal recognition that the pyrethroids can be put into two categories on the basis of the signs of toxicity in mammals as well as associated differences such as the protective influence of drugs and the action on isolated nerves of frogs. There has been less unanimity on deciding the site of action of the pyrethroids in the nervous system and it is possible that they act whenever presented in sufficient concentration to an available nerve tissue. It seems fairly certain however that the involvement of the brain is not fundamental for their action. Whatever site of action is involved it would appear that the mechanism at the nerve membrane level is by an effect on sodium permeability. This can account also for the differences between the toxic actions of those pyrethroids possessing or not possessing an α-cyano group on the alcohol moiety. These are probably a reflection of the differences shown in the duration of the transient increase in sodium permeability in the nerve membrane (Gray & Rickard, 1982a; Vijverberg & van den Bercken, 1982).

Acute and subacute toxicity

Acute toxicity studies provide the means for quantifying the outcome of the mode of action of a compound and enables comparisons of potency to be made between compounds. The subacute toxicology (repeated dosing of 14 days or less) of the pyrethroids is described also in this section since it affords more detailed study of their toxic action at the acute level. The quantitative and comparative acute toxicology of a class of compounds is difficult to review because acute toxicity results are very much dependent upon the dosing vehicle used, the environmental conditions of testing, the strain and sex of the animal and its dietary status at the time of dosing. With the pyrethroids there is also the additional factor of the isomer content of the test material to take into account. Thus the description of studies and results in this section are qualified as far as possible by the relevant factors detailed above.

Acute toxicity

Oral and parenteral administration

The difficulties of making strict comparisons of the acute toxicity of the pyrethroids are exemplified by a brief survey of the influence of the dosing

Table 3.3. Influence of the vehicle on the acute oral LD50 of pyrethroids

Pyrethroid (isomer ratio)		Species	Sex	Vehicle	LD50 (mg/kg)
Permethrin	(40 : 60)	Rat	M	Water	2949
Permethrin	(40 : 60)	Rat	M	DMSO	1500
Permethrin	(40 : 60)	Rat	M	Corn oil	500
Permethrin	(40 : 60)	Mouse	F	Water	4000
Permethrin	(40 : 60)	Mouse	F	Corn oil	540
Cypermethrin	(40 : 60)	Mouse	F	Aq. Susp	779
Cypermethrin	(50 : 50)	Mouse	F	DMSO	138
Cypermethrin	(50 : 50)	Mouse	F	Corn oil	82
Fenvalerate		Rat	—	DMSO	451
Fenvalerate		Rat	—	PEG:H$_2$O	3200
Deltamethrin[a]		Rat	M	Sesame oil	129
Deltamethrin[a]		Rat	M	PEG 200	67

Source: All FAO (1980) except [a]Glomot (1982).

vehicle and the compound isomer ratio upon the oral LD50. Table 3.3 shows that the vehicle has a marked influence on the LD50 values of a pyrethroid when the isomer ratio is constant. Aqueous suspensions of a pyrethroid usually provide the highest LD50 values (least toxic) whereas oil based dosing solutions generally give rise to the lowest values (most toxic). A ten-fold difference in toxicity using different vehicles can be seen in some cases. Similarly the isomer content of a pyrethroid has a profound influence upon the oral LD50 value (Table 3.4). The *cis* isomers are more toxic than the *trans*, and

Table 3.4. The influence of isomer content on the acute oral LD50 of pyrethroids

Pyrethroid	Cis (%)	Trans (%)	Species	Sex	Vehicle	LD50 (mg/kg)
Permethrin[a]	80	20	Rat	F	Corn oil	224
	60	40	Rat	F	Corn oil	445
	50	50	Rat	F	Corn oil	1000
	40	60	Rat	F	Corn oil	1260
	30	70	Rat	F	Corn oil	1684
	20	80	Rat	F	Corn oil	6000
Cypermethrin[b]	90	10	Rat	F	Corn oil	367
	40	60	Rat	F	Corn oil	891
Resmethrin[c]	—	100	Mouse	F	Corn oil	800
	100	—	Mouse	F	Corn oil	160

Source: [a]James (1980); [b]FAO (1980); [c]Miyamoto (1976).

Table 3.5. The comparative oral LD50 of several pyrethroids (Miyamoto et al., 1976)

Pyrethroid (racemic mixtures)	LD50[a] in male mice (mg/kg)
Allethrin	500
Phenothrin	>5000
Permethrin	490
Resmethrin	690
Tetramethrin	1920
Pyrethrins[b]	370

[a]Corn oil vehicle.
[b]Calculated as active ingredient.

with some pyrethroids are very much more toxic. Thus it is important to define the isomer ratio of a pyrethroid when alluding to its toxicological properties and this applies to aspects in addition to acute toxicity. This survey emphasises that it is difficult to make strict comparisons between the potency of the pyrethroids. The most useful comparisons are made where all the work has been undertaken in one laboratory as exampled in Table 3.5. The LD50 values shown in Tables 3.3–3.5 indicate that the current synthetic pyrethroids have a wide range of acute toxicities.

It is interesting to note that the pyrethrins derived from natural sources do not have a more favourable acute toxicity profile in the above comparisons with the synthetic pyrethroids. Admittedly it is even more difficult to obtain a precise figure, given the additional complexity of correctly identifying the test material used in some studies. Hayes (1982) illustrates this point when

Table 3.6. Rat and mouse acute oral LD50 of pyrethroids

Pyrethroid	LD50 (mg/kg)		Source
	Rat (sex)	Mouse (sex)	
Cypermethrin	250 (M)	82 (M)	FAO (1980)
Deltamethrin	129 (M)	33 (M)	Glomot (1982)
	139 (F)	34 (F)	Glomot (1982)
Fenvalerate	451 (M)	200–300 (M)	FAO (1980)
Pyrethrins	584 (M)	273 (M)	Barthel (1973)

quoting values of 584–900 mg/kg for the rat oral LD50 from a study of several grades of extracts with pyrethrin contents ranging from 20 to 78% (Malone & Brown, 1968). Barthel (1973) also demonstrates the wide range of values reported by several investigators in his survey of pyrethrin toxicology.

The rat and mouse are the most studied species, with the mouse generally showing the lower acute oral LD50 values for the pyrethroids as exemplified in Table 3.6. The quality and preciseness of the data from the study of other species does not allow for such good comparisons. It is noted however that fenvalerate is rather more toxic to the Chinese hamster (acute oral LD50 98 mg/kg) than to the mouse (acute oral LD50 200–300 mg/kg) whereas this order is the reverse for cypermethrin (FAO, 1980). A survey of the acute toxicity of other species such as rabbit, guinea pig, dog and Syrian hamster revealed that most pyrethroids have oral LD50 values in these species within the range set by the rat and mouse (James, 1980; FAO, 1980, 1982). With regard to routes of parenteral administration it is pertinent to note that i.v. dosing gives rise to a much greater order of toxicity than oral dosing (Table 3.7). This demonstrates the protective defences available after oral dosing where slower absorption combined with the rapid rate of detoxifying metabolism, as described previously in this chapter, dissipate the inherent toxicity of the pyrethroid molecule.

Dermal application

The oral route of administration provides relevant information for determining the mode of toxic action of a compound and gives some insight into the anticipated symptoms and toxic dose in man. However, apart from accidental or intentional ingestion man is not usually at acute risk to the oral intake of pesticides. The more relevant assessment for their normal usage is for possible dermal absorption. The experimental model to assess the risk is by acute dermal tests in the rat or rabbit. The pyrethroid molecule universally has a low order of toxicity in this respect, and in nearly every case the dermal LD50 is

Table 3.7. Rat oral and i.v. acute LD50 of pyrethroids

Pyrethroid	Oral LD50 (sex)	I.v. LD50 (sex)	Source
Deltamethrin (vehicles)	31 (F) (peanut oil)	4 (F) (acetone)	Kavlock et al. (1979)
Deltamethrin (vehicles)	67 (M) (PEG 200)	3 (M) (PEG 200)	Glomot (1982)
Resmethrin (vehicles)	1347 (F) (DMSO)	160[a] (F) (glycerol formal)	Verschoyle & Barnes (1972)
Pyrethrin I (vehicles)	260–420 (M) (DMSO)	5[a] (F) (glycerol formal)	Verschoyle & Barnes (1972)
Pyrethrin II (vehicles)	>600 (M) (DMSO)	1[a] (F) (glycerol formal)	Verschoyle & Barnes (1972)

[a]Approximate lethal doses.

Table 3.8. Acute dermal LD50 of pyrethroids

Pyrethroid	Species (sex)	LD50 (mg/kg)	Source
Permethrin	Rat (F)	>4000	FAO (1980)
	Rat (M)	>2500	FAO (1980)
	Rabbit (F)	>200	FAO (1980)
Cypermethrin	Rat(F)	>4800	FAO (1980)
	Rabbit (F)	>2400	FAO (1980)
Bioresmethrin	Rat (F)	>10000	FAO (1977)
Fenvalerate	Rat	5000	FAO (1980)
	Rabbit	>2500	FAO (1982)
Deltamethrin	Rat	>800	Kavlock et al. (1979)
	Rat (M)	>2940	FAO (1982)
	Rabbit (M)	>2000	FAO (1982)

greater than the maximum amount of compound that can be applied to the test animals under the conditions of the test (Table 3.8).

Inhalation exposure

The assessment of inhalation hazard has been a relevant factor since the earlier uses of the pyrethrins. Barthel (1973) refers to their usage for mosquito-borne disease control during World War II. Aerosol bombs containing the pyrethrins were the means of application and it is possible that fine respirable particles were generated during the process. The studies of Carpenter et al. (1950) were the first recorded work on the inhalation toxicity of the pyrethrins and pyrethroids. In a part of their programme fogs were generated to give concentrations of allethrin of about 19 g/m³ for a two-hour exposure or 13·8 g/m³ for a four-hour exposure with a mean particle size of 2–5 μm. In the first study one rat out of ten died and in the second four out of ten. It was pointed out that these concentrations were several thousand times greater than any exposure expected for man (Carpenter et al., 1950).

Several other pyrethroids have been subjected to acute inhalation testing. Miyamoto (1976) using standard conditions for each compound tried to determine the LC50 for allethrin, phenothrin, permethrin, resmethrin and tetramethrin. Exposure periods were for two, three or four hours and the particle size of the mists generated by an atomiser was 1–2 μm. No deaths occurred with any compound and the LC50s were therefore greater than the concentrations tested, ranging from 686 mg/m³ for permethrin to 2500 mg/m³ for tetramethrin in rats and mice. Fenvalerate was tested as an aerosolised formulation generated from an aqueous suspension containing 3 g/litre (FAO, 1980). The mean particle size of the atmosphere was 77 μm and the exposure to rats was for four hours. Acute signs of toxicity were noted for a short time but there were no deaths after exposure ceased. The same source also quotes a three-hour LC50 >101 mg/m³ for fenvalerate in the rat and mouse. The LC50 for deltamethrin, in aerosols generated from 10% DMSO solutions was determined to be 940 mg/m³ and 785 mg/m³ respectively in male and female Sprague-Dawley rats for exposures of 2–2·5 h (Kavlock et al., 1979). A similar value, 600 mg/m³ for a six-hour LC50 was determined in another rat study with deltamethrin aerosol (Glomot, 1982). The particle size distribution in the deltamethrin studies was not recorded.

Signs of toxicity

The signs of toxicity noted in the acute studies have generally been very similar to those described by Verschoyle & Aldridge (1980). The descriptions have not always fallen neatly into the two categories outlined by Verschoyle & Aldridge but the most characteristic signs, tremoring for non-α-cyano pyrethroids and splayed limb gait for α-cyano pyrethroids, have featured prominently in each case. For the pyrethrins the acute toxic signs in rats have been described as laboured or rapid breathing, ataxia, sprawling limbs and tremors (Barthel, 1973 quoting Weir, 1966), aggressiveness, sensitivity to external stimuli, tremor, convulsive twitching and prostration (Verschoyle & Barnes, 1972). The constant feature of the acute toxic signs with the pyrethrins and pyrethroids is their rapidity of onset and recovery. Typically, after acute oral dosing, the signs occur within 1–3 h, may last up to 24 h and most surviving animals have recovered by 48–72 h. Administration by the i.v. route produced the signs within minutes, with a correspondingly faster recovery within a few hours.

Attempts have been made to correlate the signs of toxic action with the levels of pyrethroid attained in the brain. Cismethrin was injected i.v. to rats either at doses producing tremors (2·5 mg/kg) or not producing toxic signs (0·5 mg/kg). The compound was cleared from blood very rapidly (99% in 3 min) and reached a maximal concentration in the brain within 2 min. The tremoring was directly related to the brain level of cismethrin, and it appeared that a threshold of 3·5 nmol/g had to be reached before toxic signs appeared. The pyrethroid was present mainly as the parent compound (Gray et al., 1980). By contrast it seems that deltamethrin injection does not achieve the same correlation between peak brain concentrations and toxic signs (Gray & Rickard, 1982b). When injected i.v. at 1·75 mg/kg to rats, a toxic but not lethal dose, the peak brain concentration of 0·5 nmol/g was attained within 1 min. This and subsequent concentrations did not correlate with the severity of toxic action. The clearance from the brain was slower than that for cismethrin, and generally it appeared that the concentration of deltamethrin was about 10% of that required for cismethrin. A similar difference has been shown by the direct adminstration of these compounds in to the brain (Gray & Rickard, 1982a).

Eye and skin tests

The acute toxicological assessment of the pyrethroids has included several studies for eye and skin irritation and skin sensitisation potential. Mostly these have been undertaken using standard methods, with the rabbit eye, rat or rabbit intact or abraded skin (irritation) and guinea pig skin (sensitisation) as the models.

Carpenter et al. (1950) used several methods to test for skin irritation on the rabbit with technical allethrin or pyrethrin concentrates. These involved single or repeated application and resulted in moderate erythema. Neither undiluted technical allethrin or an oleoresin concentrate of pyrethrins caused any significant damage to rabbit eye. The application of 0·1 ml of barthrin or dimethrin to the rabbit eye did not result in any adverse reaction (Ambrose et al., 1963, 1964). The acute and repeated dermal application of barthrin to rabbits showed no visible skin reactions and this observation was confirmed microscopically. Similar results were obtained from dimethrin after a single (5 ml/kg) or repeated (0·5 ml/kg) application to rabbit skin. Permethrin elicited only minimal effect in the rabbit eye (Metker, 1978; FAO, 1980) but did not show any indication of skin irritation in the same species. Cypermethrin (FAO, 1980) and deltamethrin (Glomot, 1982) have mildly irritant effects on rabbit eye but bioresmethrin (FAO, 1977) had no effect. The same sources of information showed that cypermethrin was a moderate skin irritant to the rabbit whereas deltamethrin had no effect.

Skin sensitisation studies in the guinea pig have demonstrated that the pyrethroids have little potential in this respect. No reaction has been ascribed to allethrin (Carpenter et al., 1950), permethrin (Metker, 1978; FAO, 1980) or deltamethrin (Glomot, 1982), whereas bioresmethrin and cypermethrin have been described only as weak sensitisers (FAO, 1977, 1980).

Subacute toxicity

The acute toxicity studies have provided little information on the toxic effects of the pyrethroids other than clinical signs and lethality. Gross pathology and microscopic findings, when reported, have generally shown non-specific or secondary effects such as congestion of the lungs. It is only comparatively recently that attention has been directed to examining the target organs, that is nervous tissue, in the acute and sub-acute studies on the pyrethroids. Rats given very high oral doses of pyrethroids at levels resulting in severe toxic signs and mortality within 24 hours have showed histopathological lesions in the sciatic nerve (FAO, 1980). There was no such evidence for an effect on the sciatic nerve at lower levels at which no toxic signs were observed (200 mg/kg fenvalerate or permethrin; 100 mg/kg cypermethrin).

Subacute studies have been undertaken on the pyrethroids to examine these changes in more detail, and to elicit no-effect levels and their capacity for reversibility. In a study with fenvalerate, rats were fed diet containing 3000 p.p.m. fenvalerate for 10 days during which time 60% of the animals died. The survivors of the 10-day administration were maitained on control diet for up to 12 weeks to study recovery aspects. The animals surviving during the treatment period showed swelling and disintegration of the axons of the sciatic nerve which persisted for up to three weeks into the recovery period. Sciatic nerves examined six weeks after the end of fenvalerate treatment were normal (FAO, 1980). Rats given 2000 p.p.m. fenvalerate in diet for 8–10 days did not show any adverse changes in the sciatic nerve, even though the animals exhibited the typical signs of acute toxicity.

Cypermethrin and permethrin were examined in 14-day subacute studies in the rat (FAO, 1980; Litchfield, 1983). The animals were observed throughout the studies and the sciatic nerves were taken at post mortem for microscopic and electron-microscopic examination after special slide preparation and staining techniques. Cypermethrin was studied at dose levels from 1250 to 5000 p.p.m. in diet, and deaths occurred at the two higher levels of 2500 and 5000 p.p.m. Clinical signs of acute toxic effect were seen at all levels in a dose-related manner and diminished noticeably during the course of the study at the two lower levels of 1250 and 2500 p.p.m. cypermethrin. Examination of the sciatic nerves from animals on the 2500 p.p.m. dose level showed that only one rat had evidence for microscopic and ultrastructural changes due to cypermethrin. The dose levels for permethrin were 2500 to 7500 p.p.m. in diet. Deaths occurred at the higher doses of 5000 and 7500 p.p.m. permethrin and signs of acute toxic action were seen at all doses. However, even for animals surviving 5000 p.p.m. for 14 days these toxic signs had disappeared before the end of the study. Microscopic evidence for an increase in the number of degenerating nerve fibres was seen only in two rats surviving the full 14-day treatment with 5000 p.p.m. permethrin. Nerves from the rats given 2500 p.p.m. permethrin showed no microscopic or ultrastructural changes compared with their paired controls.

Some attempts have been made to extend the assessment of the toxic action of the pyrethroids beyond clinical observations and examination of nervous tissue. Estimates of neurological deficit by measurement of the mean slip angle have shown variable results and have not proved more sensitive to the action of the pyrethroids than the examinations described above (FAO, 1980). The assay of enzyme activity, such as β-glucuronidase and β-galactosidase in peripheral nerve tissue, have shown dose-related increases in studies with cypermethrin and fenvalerate (FAO, 1980) but have not shown correlation with neuromuscular dysfunction produced by pyrethroids (Rose & Dewar, 1983). The value of such assays in providing meaningful data for the evaluation of pyrethroid toxicity is uncertain and would require further extensive evaluation.

Electrophysiological studies in the rat have shown that cypermethrin, at doses that induce severe signs of acute toxicity, did not have any effect on maximal motor conduction velocity or conduction velocity of the slower motor fibres in peripheral nerve (FAO, 1980).

Some pyrethroids have been tested in the hen for delayed neurotoxicity. The tests were undertaken in the same way as for the organophosphorus compounds whereby the test material was given orally at high dose level either after single or multiple doses, the birds observed for 21 days and then, in most cases, redosed. Cypermethrin and fenvalerate (1000 mg/kg/day for 5 days), permethrin (9000 mg/kg single dose or 1000 mg/kg/day for 5 days) and deltamethrin (5000 mg/kg single dose) showed no toxic signs and produced no histopathologicial changes in nervous tissue in the hen when subjected to such a regime of testing (FAO, 1980, 1981). It is not suprising that a test devised for the organophosphates does not show any effects for the pyrethroids which have a different mechanism of action, and the test cannot be considered appropriate for this class of compounds.

There is a small amount of data available on subacute toxicology in the dog. Bioresmethrin was administered orally at 500 mg/kg/day for 7 days followed by 1000 mg/kg/day for 14 days. There were no effects on mortality, clinical condition, blood or urine biochemistry, haematology and electrocardiology. There were also no effects when the dose was raised to 2000 mg/kg/day for a further 7 days (FAO, 1977). There were no cardiovascular effects in dogs given fenvalerate at dose levels sufficient to induce toxic signs (FAO, 1980).

Summary

The numerous studies described above have adequately characterised and quantified the acute toxic action of the pyrethroids. Most of the pyrethroids reviewed have a moderate or low acute toxicity in several species. The signs of toxicity and their time course in rodents are well recognised, and surviving animals appear to recover quickly and completely. Examination of nervous tissue from acute and subacute studies show that microscopic or ultrastructural changes occur only at lethal levels of pyrethroid administration and there is no evidence for changes at lower non-lethal doses. The assessment of more prolonged administration on these and other parameters is addressed in the next section.

Subchronic and chronic toxicity

In the previous section the studies described as subacute were defined as those using daily administration up to 14 days duration. The present section is concerned with studies of longer duration and is divided between subchronic (4–26 week) and chronic (1 year and longer) administration. These studies are used for assessing the toxicology of a compound over prolonged periods to see if the constant administration changes the degree or mechanism of action, alters the level of effect or leads to delayed effects. They also serve to set toxicological no-effect levels for regulatory requirements. The pyrethroids have been the subject of numerous studies of this kind and there are several ways of approaching their review. The approach taken here is to split the subject matter into three parts:

(*a*) Description of studies by dietary administration, their design and results.
(*b*) Description of studies by other routes of administration.
(*c*) Commentary on the findings from the above studies.

The review in this section does not cover carcinogenicity, this subject being addressed separately later in this chapter.

Dietary administration

Rodent studies

The pyrethrins and most of the pyrethroids have been evaluated in subchronic and chronic rodent studies, and descriptions of numerous studies have appeared in the literature. The duration of such studies has ranged from four weeks to two years. The list of subchronic studies reported so far is shown in Table 3.9. The typical experimental details for such studies are exampled by a 13-week rat study on cypermethrin (FAO, 1980). Groups of 20 male and 20 female rats were fed cypermethrin in diet at concentrations of 0 (control), 75, 150 and 1500 p.p.m.. The material used had a 44 : 56 *cis–trans* ratio. The animals were observed daily, bodyweight and food consumption measured weekly, and haematology, blood and urine biochemistry assays undertaken during the course of the study. At termination a full post mortem was undertaken on all animals, the weight of several organs were measured and a histopathological examination was carried out on a wide range of tissues. There were no signs of toxic action and no mortality in the study. Bodyweight gain and food consumption was reduced at the top dose of 1500 p.p.m. cypermethrin. There was no effect on any of the haematological or blood and urine biochemical parameters, nor upon organ weights or any tissues examined by light microscopy. Electron microscopic examination of the liver showed that there was an increase in smooth endoplasmic reticulum proliferation in males and females at 1500 p.p.m. and in males only at 150 p.p.m. cypermethrin. This increase had substantially reverted to normal in rats retained for a four-week recovery period after the termination of the main study.

The other pyrethroids listed in Table 3.9 showed very similar toxicological profiles in their subchronic studies. Some pyrethroids elicited signs of acute toxicity at the highest dose levels used, for instance, permethrin at 3000 p.p.m. in diet in the 26-week study (Miyamato, 1976; FAO 1980), deltamethrin at 10 mg/kg/day by gavage in the 13-week study (FAO, 1981) and resmethrin at 5000 p.p.m. in diet in the 24-week study (Miyamoto, 1976). It was usual for these effects to regress or disappear as the study proceeded. Permethrin, resmethrin, allethrin, barthrin and dimethrin showed liver enlargement with minor histopathological changes at the top dose in their subchronic studies.

Long-term studies of one year or more in rat and mouse have been performed with most of the commercially important pyrethroids; the list of rat studies appearing in the scientific literature is shown in Table 3.10. These are usually of a comprehensive design involving large numbers of animals as exampled by the two-year rat study on permethrin (FAO, 1980). Groups of 60 male and 60 female rats were fed diets containing 0, 500, 1000 or 2500 p.p.m. permethrin. At one year 12 males and 12 females per group were killed for an interim evaluation. The animals were observed daily, bodyweights being recorded weekly for the first 12 weeks and then at' two-weekly intervals thereafter. Food consumption was measured weekly for the first 12 weeks and then at regular less-frequent intervals. Ophthalmological examination was undertaken during the study and haematological, blood and urine biochemical assays at intervals throughout the experimental period. Animals killed or dying during the study and all those at the scheduled kills at one or two years were

Table 3.9. Rat subchronic studies on pyrethroids

Pyrethroid	Dietary levels (p.p.m.)	No. rats/group		Duration (weeks)	Source
		M	F		
Allethrin	1000, 5000, 15 000	12	12	12	Miyamoto (1976)
Barthrin	1000, 3000, 10 000, 20 000	5	6	16	Masri et al. (1964)
Cypermethrin	75, 150, 1500	20	20	13	FAO (1980)
	25, 100, 400, 1600	12	12	13	FAO (1980)
Deltamethrin	0·1, 1, 2·5, 10[a]	20	20	13	FAO (1981, 1982)
Dimethrin	2000, 6000, 15 000, 30 000	5	6	13	Masri et al. (1964)
Fenvalerate	125, 500, 1000, 2000	12	12	13	FAO (1980)
Permethrin	9, 27, 85, 270, 850[b]	10	10	13	Metker et al. (1978)
	20, 100, 500	10	10	13	FAO (1980)
	375, 750, 1500, 3000	16	16	26	FAO (1980)
Pyrethrins	360[a]	10	10	13	Bond et al. (1973)
Resmethrin	400, 1200, 4000/8000[c],	18	18	13	FAO (1977)
	500, 1500, 5000	20	20	24	Miyamoto (1976)

[a]Mg/kg/day by gavage.
[b]Mg/kg/day in diet.
[c]The concentration was changed from 4000 to 8000 p.p.m. during the study.

subject to full post mortem and a histopathological examination which involved more than 30 tissues. Weights were measured on a variety of organs. Tremors and hypersensitivity were noted in rats at the top dose of 2500 p.p.m. during the first two weeks of the study but were not seen thereafter. There were no compound-related effects on mortality, growth, food consumption, haematology, blood and urine biochemistry. There was also no effect on organ weights except for the liver which increased to some extent at each dose level. Histopathological changes in the liver were observed as hypertrophy at 1000 and 2500 p.p.m. permethrin and as hepatocyte vacuolation at 2500 p.p.m. No histopathological changes due to treatment were seen in any of the other organs examined. The adaptive changes in the liver are discussed in more detail later.

The results from the long-term rat studies on the other pyrethroids (Table 3.10) generally followed a similar pattern except that the effects on liver weight or histopathology were not seen with cypermethrin, deltamethrin or fenvaler-

Table 3.10. Rat chronic studies on pyrethroids

Pyrethroid	Dietary levels (p.p.m.)	No. rats/group M	No. rats/group F	Duration (weeks)	Source
Allethrin	500, 1000, 2000	30	30	80	Miyamoto (1976)
Barthrin	500, 1000, 5000, 10 000, 20 000	10	10	52	Ambrose et al. (1963)
Cypermethrin	1, 10, 100, 1000	48	48	104	FAO (1980)
Deltamethrin	2, 20, 50,	90	90	104	FAO (1981)
Dimethrin	500, 1000, 5000, 10 000, 20 000	10	10	52	Ambrose et al. (1964)
Fenvalerate	1, 5, 25, 250	93	93	104	FAO (1980)
	50, 150, 500, 1500	80	80	104(M) 119(F)	FAO (1982)
Permethrin	20, 100, 500	60	60	104	FAO (1980)
	500, 1000, 2500	60	60	104	FAO (1980)
Phenothrin	200, 600, 2000, 6000	50	50	104	FAO (1981)
Pyrethrins	200, 1000, 5000	12	12	104	Williams (1973)
Resmethrin	500, 2500, 5000	60	60	104	Federal Register (1983)

ate, although fenvalerate increased the incidence of giant cell infiltration in the spleen and liver at 1500 p.p.m. (FAO, 1982).

Long-term mouse studies have been undertaken on deltamethrin, fenvalerate, permethrin, phenothrin and resmethrin, mainly for carcinogenic evaluation, and these are described in the more relevant section on carcinogenicity later in this chapter. However it should be noted that they were undertaken at similar dose levels to the corresponding rat studies described above and that in terms of non-neoplastic findings the chronic toxicity profile was generally similar to that of the rat.

Non-rodent studies

The dog is used universally as the non-rodent species for the evaluation of subchronic and chronic toxicity of compounds. The studies performed on the pyrethroids are listed in Table 3.11 and represent a range of evaluations from 13 to 104 weeks in duration. The design and outcome of these studies is represented by the following descriptions of the 13-week and two-year studies with cypermethrin (FAO, 1980, 1982). For the 13-week study, four male and four female dogs were fed cypermethrin in diet at dose levels of 0 (control), 5, 50, 500 and 1500 p.p.m. The animals were observed daily and bodyweight measured weekly. Haematological and clinical biochemical investigations were undertaken during the study. At termination all dogs were submitted to a full post mortem, a variety of organs weighed and a wide range of tissues submitted for histopathological examination. There were clinical signs of toxicity at the highest dose of 1500 p.p.m. cypermethrin but not at 500 p.p.m. Apart from a lowering of the kaolin–cephalin time (a clotting factor) at 500 p.p.m. cypermethrin in females there were no other haematological or biochemical changes of note. There were no treatment-related changes in any of the tissues examined, apart from a non-specific bronchopneumonia in the lungs of the animals at 1500 p.p.m. cypermethrin. There were no effects on organ weights. This overall pattern of effect was seen also in the two-year dog study in which four males and four females were fed diets containing 0 (control), 3, 30, 300 and 600 p.p.m. cypermethrin. The highest dose initially was 1000 p.p.m. but was reduced to 600 p.p.m. because of marked persistent signs of toxicity. There was a reduction in bodyweight in these top dose animals, probably due to inappetence in the earlier stages of the study. No treatment-related differences were seen in the other parameters examined.

The effect of other pyrethroids on the dog was basically similar with signs of toxicity seen in the top dose of most of the studies shown in Table 3.11. Permethrin caused small increases in liver weight in dogs on the top dose levels of the two studies reported (FAO, 1980). Deltamethrin given by gavage over a 13-week period produced several effects, including liquid faeces, growth depression and dilatation of the pupils. However these changes were not seen in the two-year study in which deltamethrin was administered in the diet at dose levels up to 40 p.p.m. (FAO, 1982).

Other routes of administration

Dermal or inhalation routes of administration have also been used for the evaluation of pyrethroids in subchronic studies. The rabbit has been used as the species of choice for the dermal studies, while several species have been utilised for the inhalation studies.

The evaluation of permethrin in a three-week dermal study on the rabbit has been described comprehensively by Metker et al. (1978). The compound was applied daily to the clipped skin of New Zealand White rabbits for 21 days. Groups of four male and four female rabbits received 0 (control), 100, 320 or 1000 mg permethrin/kg/day. The application site was abraded on the first day of the study for half the animals in each group. Several biochemical assays of the blood were undertaken during the study. On the tenth day after the last application the animals were necropsied and several tissues, including skin, liver, kidney, brain, heart and testes were taken for histopathological examination. Moderate primary irritation of the skin was observed by the eighteenth day of the study and a mild irritation was present 10 days after the

Table 3.11. Dog subchronic and chronic studies on pyrethroids

Pyrethroid	Dose levels	No. dogs/group		Duration (weeks)	Source
		M	F		
Cypermethrin	5, 50, 500, 1500 (p.p.m.)[a]	4	4	13	FAO (1980)
	3, 30, 300, 600 (p.p.m.)[a]	4	4	104	FAO (1982)
Deltamethrin	0·1, 1, 2·5, 10 (mg/kg/day)[b]	3–5	3–5	13	FAO (1981)
	1, 10, 40 (p.p.m.)[a]	8	8	104	FAO (1982)
Fenvalerate	0·05, 0·25, 1·25, 12·5 (mg/kg/day)[a]	4	4	13	FAO (1980)
Permethrin	5, 50, 500 (mg/kg/day)[c]	4	4	13	FAO (1980)
	10, 100, 2000 (mg/kg/day)[c]	4	4	13	FAO (1980)
Pyrethrins	5000 (p.p.m.)[a]	3	3	13	Barthel (1973)
Resmethrin	25, 80, 250 (mg/kg/day)[c]	3	3	13	FAO (1977)

[a]Dietary administration.
[b]By gavage.
[c]Capsule administration.

last application. No systemic effects were noted on the evidence of the results for growth rate, organ weights, clinical chemistry or histopathology.

Fenvalerate was evaluated similarly in two dermal studies in rabbits, one with 14 applications of pyrethroid over a 22-day period, and the other with 15 applications over 3 weeks (FAO, 1980). The dosages ranged from 30 to 400 mg fenvalerate/kg/day over the two studies. There were clinical signs of toxicity at the highest doses of 300 and 400 mg/kg/day and there was some mortality. A dose 100 mg/kg/day showed no signs of systemic toxicity although there was a mild irritant effect on the skin. A duration of three weeks was used also for the evaluation of a 1% pyrethrins formulation at a dose of 10 mg pyrethrins/kg/day on intact or abraded skin. There were no observable effects in this study (WHO, 1973). Barthrin and dimethrin were subjected to longer periods of evaluation in their dermal studies with rabbits. The animals received the pyrethroids for 5 days/week for 13 weeks at dose levels of 0·25 or 0·5 ml/kg/day for each compound (Ambrose et al., 1963, 1964). No effects were observed in either study.

The earliest subchronic inhalation studies on the pyrethrins and pyrethroids were undertaken by Carpenter et al. (1950). They used a regime of 30-min exposures once or twice a day for up to 67 days in rats or dogs at air concentrations of approximately 20 mg/m^3 pyrethrin or allethrin. No effects were observed on the basis of toxic signs, bodyweight gain, liver and kidney weight, haematology or histopathology. The most comprehensively described subchronic inhalation study on a pyrethroid is again from the work of Metker and his associates (Metker, 1980). Permethrin was tested on three species: the rat, dog and guinea pig. The animals were exposed to the aerosolised material for 6 h/day for 5 days/week for 13 weeks at exposure levels of 0 (control), 125, 250 or 500 mg/m^3. The group sizes were 10 males and 10 females (rat), 10 males (guinea pig), two males and two females (dog). The animals were observed during the study, bodyweights and organ weights were measured and several tissues were examined microscopically at termination. Additional evaluations included haematology, blood and urine biochemistry and pulmonary function studies on the dogs, an intradermal challenge on the guinea pigs for assessment of sensitisation, and oxygen consumption measurements on the rats. The only effect recorded from these evaluations was the appearance of severe tremor in the rats exposed to 500 mg/m^3 permethrin. These toxic signs disappeared in the second week of the study.

Fenvalerate (FAO, 1980), allethrin and resmethrin (Miyamoto, 1976) have been evaluated in four-week inhalation studies in the rat and mouse, and phenothrin in a four-week study in the rat (FAO, 1981). The study designs were very similar in each case, exposure for 3 or 4 h/day for 5 days/week. The exposure levels were 2, 7 and 20 mg/m^3 for fenvalerate (1–2 μm particle size), 23, 47 and 210 mg/m^3 for resmethrin, 43 and 220 mg/m^3 for phenothrin and up to 123 mg/m^3 for allethrin. Apart from clinical signs of toxicity at the highest exposure levels for fenvalerate and allethrin there were no other effects arising from the series of experimental evaluations, including haematology, clinical biochemistry and histopathology.

Commentary on findings

General

The subchronic and chronic toxicology of the pyrethroids in several species by different routes of administration can be summarised by one word – undramatic. At the highest doses in some studies signs of acute toxicity have been observed. These have usually been transient in appearance and have considerably regressed or disappeared as a study progressed. This finding is not unexpected since it follows the pattern seen in the subacute studies described in the previous section. The other most notable finding at the highest dose levels was growth depression either as a result of direct toxic action or due to inappetence early in a study. No effect levels were established in studies in which these changes were noted at the higher levels. There was no consistent effect on any of the many haematological or clinical biochemical parameters measured in the studies described. Pathological examination in these studies has shown only one tissue affected with any consistency, the liver, and this did not occur with all pyrethroids.

There does not seem to be any species difference in the general pattern of response nor does the route of administration alter the basic toxicological profile. We are left with two findings worthy of further discussion, the clinical signs of toxicity, together with any other manifestation thereof, and the hepatic changes seen with some of the pyrethroids.

Nervous system

Because the findings from acute studies have shown toxic effects associated with the nervous system, particular attention has been paid to the examination of nervous tissue from most of the subchronic and chronic studies described above. The tissues of the central and peripheral nervous systems have been examined routinely by light microscopy and in some cases by electron microscopy. Apart from the conventional methods of slide preparation, special techniques also have been used in many cases to ensure that any perceptible changes in nervous tissue structure did not go undetected. For example in some studies the left sciatic nerve of rats was fixed in 3% v/v glutaraldehyde in 0·1 M phosphate buffer and embedded in "Araldite" resin and semi-thin (1 μm) sections were stained with Toluidine blue. The right sciatic nerve was fixed in formol saline and sections stained with Luxol fast blue-cresyl violet and in some cases by Palmgren's silver technique. Studies where special emphasis has been placed upon the examination of nervous tissue included a 13-week rat study with cypermethrin and a two-year rat study on permethrin. In each case there was no evidence for an effect on nervous tissue at dose levels up to 1500 p.p.m. with cypermethrin and 2500 p.p.m. with permethrin (Litchfield, 1983). Similarly a two-year rat study with fenvalerate at dose levels up to 500 p.p.m. revealed no effects on the sciatic nerves which had been subjected to special examination (FAO, 1980). Virtually all the subchronic and chronic rodent studies on the pyrethroids have shown no evidence for histopathological changes in the nervous tissue, nor have there been any other manifestations of effects on the nervous system other than transitory clinical signs at high dose levels in some studies. The results from the dog studies have shown a similar picture. Even where pronounced signs of toxicity were seen for a few weeks, as in the 13-week dog study on deltamethrin at the highest dose level of 10 mg/kg/day, there were no changes in the nervous system tissues attributable to treatment (FAO, 1981; Glomot, 1982).

Another evaluation of the effect of pyrethroids on the nervous system has been provided by the studies of Takahashi & LeQuesne (1982). They measured the nerve excitability changes from the tails of rats fed 0 (control), 25, 50 100 or 200 p.p.m. deltamethrin in diet over a period of eight weeks. The electrophysiological measurements were undertaken at intervals during

the study and also on animals that were returned to control diet after cessation
of deltamethrin administration. Increased excitability was recorded from rats
at 50 p.p.m. deltamethrin and above during dosing but the effect became less
marked as the study progressed. No effect remained in those rats returned to
control diet for 24–48 h after cessation of deltamethrin administration.

The evidence arising from the subchronic or chronic studies of the
pyrethroids in a variety of species using different routes of administration
shows that the basic mode of action observed acutely is not modified by the
more prolonged administration. There has been no indication of a delayed
effect and no chronic or cumulative effect has been displayed as the result of
prolonged administration at high dose level.

Hepatic changes

Liver enlargement in rodents had been shown in studies with the earliest
pyrethroids: allethrin, barthrin, and dimethrin. For example, dimethrin fed to
rats at a dose level of 30 000 p.p.m. in diet for 16 weeks increased the
liver : bodyweight ratio from 3·26 in the controls to 5·39 in the test group.
This was accompanied by histopathological findings in the form of eosinophilic
changes in the liver (Masri et al., 1964). Similar findings were shown for the
pyrethrins at 1000 p.p.m. in diet by Kimbrough et al. (1968). Springfield et al.
(1973) studied these hepatic changes from a biochemical standpoint with a
23-day rat study using a dose of 200 mg pyrethrins/kg/day. The
liver : bodyweight ratio of the treated animals was increased 25% above
control level and several hepatic microsomal drug-metabolising enzyme
activities were increased. Hepatic hexabarbitone oxidase activity was doubled
by pyrethrin administration and there was a corresponding decrease in
hexabarbitone sleeping time. In a separate experiment it was shown that the
liver weight and increases in hepatic microsomal enzyme activity were fully
reversible within seven days of cessation of dosing.

More recently Carlson & Schoenig (1980) demonstrated that permethrin
(80 : 20 cis-trans ratio) was capable of inducing hepatic cytochrome P-450
levels in rats dosed 50 mg/kg/day p.o. for up to 12 days. The effect was
comparatively weak when compared to phenobarbitone administration at the
same dose level i.p. (Table 3.12). In the same study cypermethrin (40 : 60
cis-trans ratio) did not induce cytochrome P-450 at a dose of 50 mg/kg/day.
The reversibility of the liver changes due to permethrin was shown in a study
designed for this purpose. Rats were fed 2500 p.p.m. permethrin in diet for
four weeks, at which time half the animals were terminated for hepatic studies
and the remainder were returned to control diet. Liver weight, hepatic
aminopyrine-N-demethylase activity, cytochrome P-450 content and smooth
endoplasmic reticulum proliferation were significantly increased in the
permethrin-treated animals after four weeks. After the recovery period of four
weeks these parameters had returned to control values (Litchfield, 1983).

Table 3.12. The effect of permethrin (80 : 20 *cis–trans* ratio) on rat liver microsomal cytochrome P-450

Administration (12 days)	Cytochrome P-450 (nmol/mg protein). Mean ± SE
Control	1·00 ± 0·05
Permethrin – 50 mg/kg/day p.o.	1·33 ± 0·05[a]
Phenobarbitone – 50 mg/kg/day i.p.	2·15 ± 0·18[a]

[a]Significantly different from control mean($P<0.05$).
Source: Extracted from Table 2 of Carlson & Schoenig (1980).

Because the hepatic changes described above have not been associated with toxic liver damage and have been shown to be rapidly reversible they may be regarded as an adaptive response to the administration of a compound (Feuer et al., 1965; Gilbert & Goldberg, 1965; Crampton et al., 1977). It is also relevant to note that this condition does not alter as the result of prolonged administration. The effect of permethrin on a number of these hepatic parameters did not change over a two-year period in the rat (Table 3.13). There were only minor histopathological changes associated with these hepatic effects after two years in the form of hypertrophy and hepatocyte vacuolation (Litchfield, 1983).

Not all pyrethroids appear to have the capacity to cause liver enlargement. Cypermethrin at dose levels equitoxic to permethrin did not cause an increase in liver weight but did increase hepatic microsomal enzyme activity and smooth endoplasmic reticulum proliferation in a 13-week rat study (Litchfield, 1983). Fenvalerate increased liver : bodyweight ratio at 500 p.p.m. and above in a 13-week rat study, but did not cause increases in hepatic microsomal enzyme activity at 1000 p.p.m. in diets (FAO, 1980). No effect on liver weight or histopathology has been reported from several deltamethrin rodent studies (FAO, 1981, 1982; Glomot, 1982).

In summary the pyrethrins and some pyrethroids have been shown to cause weak adaptive responses in rodent liver. Such effects are fully reversible and are not altered by prolonged administration at high dose level.

Mutagenicity

The pyrethroids, in common with many other pesticides, have been tested extensively for mutagenic potential in recent years. When assessing the results from such programmes of testing it is important to identify the experimental procedures and method for calculating the results. Such is the proliferation of procedures, particularly among the *in vitro* tests, that unless comparability is

Table 3.13. Hepatic assays on permethrin rat studies: measurements over two years

Assay	Liver weight (g)			APDM[a] activity (μmol AP/h/g)			SER[b] Index		
Period (weeks)	4	52	104	4	52	104	4	52	104
Dietary 0	8·6	10·4	12·8	0·03	0·04	0·05	115	127	101
permethrin 2500	9·3	11·8	14·0	0·16	0·17	0·16	159	172	136
(p.p.m.)									

[a] APDM: aminopyrine-*N*-demethylase activity.
[b] SER: smooth endoplasmic reticulum proliferation.

maintained then it is difficult to relate the results from different investigators. Because of this, most emphasis has been given in this review to reports of studies that have clearly defined the conditions of testing and the basis of assessing a negative or positive result.

Submammalian systems

The most common mutagenic tests applied to the pyrethroids have been bacterial assays utilising *Salmonella typhimurium* or *Escherichia coli*. Typically the methods used have been similar to those described by Waters et al. (1982). For the *S. typhimurium* assays, up to five tester strains, TA 1535, TA 1537, TA 1538, TA 98, TA 100, have been utilised and the test procedure undertaken according to Ames et al. (1975). The assays were carried out in the presence or absence of metabolic activation provided by homogenate preparations obtained from the livers of Aroclor-1254-induced rats or mice. Several concentrations of pyrethroid have been tested and a positive result has been ascribed to those assays which produce a dose-related increase for at least three concentration levels in at least one tester strain. Positive controls were run alongside the test materials to demonstrate the sensitivity of the system. The *E. coli* assays commonly utilised strain WP2 and followed procedures similar to that of Bridges (1972). Tests were undertaken with or without metabolic activation.

The results for these two assays for several pyrethroids from studies undertaken by different investigators are shown in Table 3.14. The repeatability of the results should be noted, where permethrin (nine separate studies), deltamethrin (four separate studies) and cypermethrin (three separate studies) consistently showed negative, that is non-mutagenic, results. Of the 12 pyrethroids tested only allethrin has shown positive results (Waters et al., 1982; Moriya et al., 1983). In the studies by Waters et al. (1982) the positive result for allethrin was obtained in the presence of metabolic activation only,

Table 3.14. Results of *S. typhimurium* and *E. coli* assays on pyrethroids

Pyrethroid	*S. typhimurium* references	*E. coli* references
A) *Studies which showed negative (non-mutageniuc) responses*		
Allethrin	Miyamoto (1976)	Miyamoto (1976), Waters et al. (1982) Moriya et al. (1983)
Cypermethrin	FAO (1980) Waters et al. (1982)	FAO (1980), Waters et al. (1982)
Chrysanthemic acid	Waters et al. (1982)	Waters et al. (1982)
Deltamethrin	Kavlock et al. (1979) Bartsch et al. (1980) FAO (1981)	Kavlock et al. (1979), FAO (1981)
Ethyl chrysanthemate	Waters et al. (1982)	Waters et al. (1982)
Fenvalerate	FAO (1980)	
Permethrin	Miyamoto (1976) Metker et al. (1978), FAO (1980), Bartsch et al. (1980), Waters et al. (1982), Moriya et al. (1983)	Miyamoto (1976), Metker, et al. (1978), FAO (1980), Waters et al. (1982), Moriya et al. (1983)
Phenothrin/Sumithrin	Miyamoto (1976), FAO (1981), Waters et al. (1982)	Miyamoto (1976), FAO (1981), Waters et al. (1982)
Pyrethrins	Miyamoto (1976)	Miyamoto (1976), Ashwood-Smith et al. (1972)
Resmethrin	Miyamoto (1976), Waters et al. (1982)	Miyamoto (1976), Waters et al. (1982)
Tetramethrin	Miyamoto (1976)	Miyamoto (1976)
B) *Studies which showed positive (mutagenic) responses*		
Allethrin	Waters et al. (1982), Moriya et al. (1983).	

while Moriya et al. (1983) found their positive result with the TA 100 tester strain only. Apart from the studies reported in Table 3.14 there have also been brief references in the literature to negative results from fenproparthrin (Kadry & Abu-Hadeed, 1982) and resmethrin (Swentzel et al., 1978) for tests utilising *S. typhimurium*. Several pyrethroids (allethrin, cypermethrin, deltamethrin, permethrin and resmethrin) have also been shown to be non-mutagenic in yeasts utilising *Saccharomyces cerevisiae* (FAO, 1980; Kavlock et al., 1979; Waters et al., 1982).

Bacterial tests for pyrethroid mutagenic potential have been undertaken utilising the host-mediated assay. In this procedure the pyrethroid was administered orally to mice followed by an i.p. injection of an appropriate strain of microorganism such as *S. typhimurium*. After a few hours the bacterial cells were harvested from the abdominal cavity of the animals and their reversion frequency determined. Allethrin, phenothrin, permethrin, resmethrin, tetramethrin, pyrethrins (Miyamoto, 1976), cypermethrin and fenvalerate (FAO, 1980) were shown to be non-mutagenic in this assay. The use of a battery of tests to act as a screening procedure for the assessment of mutagenic potential has been undertaken by some workers in this field. In an impressive programme undertaken on behalf of the Genetic Toxicology Division of the US Environmental Protection Agency, the genotoxic potential of 65 pesticides was examined, involving several hundred assays (Waters et al., 1982). Six assays were used routinely on most of the test compounds. These assays tested for different types of mutagenic action viz. point/gene mutations in prokaryotes, primary DNA damage in prokaryotes and primary DNA damage in eukaryotes. The tests used and the results for the seven pyrethroids investigated are shown in Table 3.15. In their summary and conclusions, Waters et al. (1982) designate chemicals that elicit positive responses in several kinds of genetic bioassays as giving the greatest concern. On the evidence of the results in Table 3.15 it can be seen that these pyrethroids have little or no potential for mutagenic action.

Mammalian systems

The studies in microorganisms described above give an indication of the mutagenic potential of chemicals in those systems and have some predictive value for that potential in mammalian systems. The use of mammalian systems themselves obviously gives a better indication of this potential, with *in vivo* tests providing better evidence than those *in vitro*. Among the *in vitro* mammalian tests used for the pyrethroids the results of the human lung fibroblast unscheduled DNA synthesis assay have already been quoted for the pyrethroids in Table 3.15 and show them to be non-mutagenic (Waters et al., 1982). Deltamethrin has been shown to be negative in a mammalian assay with Chinese hamster ovarian cells in the presence or absence of metabolic

Table 3.15. The results of a battery of mutagenicity assays for seven pyrethroids

Pyrethroid	Point/gene mutation		Prim. DNA damage prokaryotes		Prim. DNA damage eukaryotes	
	S. typhimurium	*E. coli*	*E. coli pol A*	*B. subtilis rec*	*S. cerevisiae*	Human lung fibroblast
Allethrin	+"	−		−	−	−
Chrysanthemic acid	−	−	−	+	−	−
Cypermethrin	−	−			−	−
Ethyl chrysanthemate	−	−	−	−	+	
Permethrin	−	−		−	−	
Resmethrin	−	−		−	−	
Phenothrin (sumithrin)	−	−			−	−

+ = positive response; − = negative response; " = + only with metabolic activation.
Source: Information) extracted from Table 3 of Waters et al. (1982).

activation (FAO, 1981) while allethrin was shown to be positive in Chinese hamster lung cell in the presence of metabolic activation (Matsuoka et al., 1979).

The most relevant tests involve the administration of pyrethroids to animals and examination for evidence of mutagenic changes *in vivo*. Two such tests which have been applied routinely for some pyrethroids are the *in vivo* cytogenic assay and the dominant lethal test. In the former assay, groups of hamsters, mice or rats were administered the pyrethroid by single or multiple (usually five daily) doses and the bone marrow cells sampled from animals killed at a specific time (6–48 h) after the last dose. The bone marrow preparations were examined for chromosome aberrations by standard techniques. Cypermethrin, fenvalerate (in hamsters), deltamethrin (in mice) and permethrin (in rats) all failed to increase the number of chromosomal abnormalities above control levels in this type of study (FAO, 1980, 1981; Polakova & Vargova, 1983). Two briefly reported studies indicated that fenpropathrin caused chromosomal damage in rat bone marrow (Kadry & Abu-Hadeed, 1982), and that cypermethrin and deltamethrin did not cause bone marrow chromosomal changes in mice although permethrin induced changes of borderline significance (Paldy, 1982). In the dominant lethal test, male mice were dosed orally with the pyrethroid at one or more dose levels daily for five days after which each male was mated with two new females each week over an eight week period. Fifteen days after pairing, the females were killed and their uteri examined for numbers of live implantations and early or late deaths. This standard test, or slight variations of it, has shown no consistent effect on pregnancy, numbers of implantations, early or late deaths for deltamethrin, cypermethrin, fenvalerate or permethrin, providing further evidence for the lack of mutagenic potential in these compounds (FAO, 1980, 1981).

The mutagenic assessment of compounds or classes of compounds is recognised as a difficult process. Even if the results from different assays on a chemical are uniformly negative or positive then allocation of non-mutagenic or mutagenic properties respectively to other species is not necessarily a foregone conclusion. The assessment of mutagenic potential is even more difficult when a series of tests show a composite of negative and positive results for a compound. In any event is is always necessary to take into account all aspects of the toxicology of a compound to place its mutagenic potential into perspective. Most of the pyrethroids reviewed above have shown uniformly negative, that is non-mutagenic, results in a variety of assays. The exception is allethrin which has shown positive responses in the *S. typhimurium* bacterial assay (Waters et al., 1982; Moriya et al., 1983) and the Chinese hamster lung assay (Matsuoka et al., 1979). Contrastingly it showed negative results in five other types of mutagenicity assay (Miyamoto, 1976; Waters et al., 1982; Moriya et al., 1983). Interestingly Kimmel et al. (1982) have isolated mutagenic photoproducts from allethrin after the pyrethroid was exposed as a thin film to sunlight or ultra-violet light (*S. typhimurium* TA 100 assay). The significance of this finding in relation to the reaction of allethrin in mammalian systems is uncertain. Even including the position with allethrin the overall evidence of mutagenic assessments indicates that the pyrethroid molecule has a very low potential for mutagenic action.

The results from the mutagenicity assays have also been used for predicting the carcinogenic potential of compounds or classes of compounds. This has been based upon the observed correlation between positive or negative mutagenicity findings with the demonstrated carcinogenic or non-carcinogenic properties of several compounds (Purchase et al., 1978; Bartsch et al., 1980). The role of mutagenic assessments in the evaluation of the carcinogenic potential of the pyrethroids is developed in the next section.

Carcinogenicity

It is possible to give some prediction of the carcinogenic potential of a compound or class of compounds from consideration of chemical structure or reaction in mutagenicity assays. If the chemical structure is very similar to that of known carcinogens then a carcinogenic potential may be suspected. Mutagenicity evaluations have been shown to have good correlation with carcinogenic potential in the case of certain compounds, for example the azo-benzenes, polycyclic hydrocarbons and aromatic amines. With other types of compounds, although the association is not always as good, the presence or lack of mutagenic potential is often taken as an indicator for carcinogenic potential. The mutagenic potency of a chemical can also provide an insight into its mode of carcinogenic action. It is likely that carcinogens which are able to react directly with DNA will have what is known as a genotoxic, that is direct carcinogenic, action in mammalian species (Weisburger & Williams, 1980). By contrast, a compound which does not react directly with DNA and yet is capable of carcinogenic action is said to be non-genotoxic. The distinguishing factor is that in the latter case the chemical must act via a secondary mechanism, for example, continuous toxic action upon an organ, to induce carcinogenic action. In this case there is a definite level at which the chemical is unable to induce an increased incidence in tumours. With a genotoxic compound this level is not so clearly distinguished and may be difficult to define in practice.

When applying the above considerations to the pyrethroids the first point to note is that on structural grounds they would not appear to be associated with known carcinogens. Secondly, the review of mutagenicity tests in the previous section revealed an almost total lack of mutagenic potential in submammalian and mammalian systems for the pyrethroids. Thus the prediction for this class of compounds is that they should have little or no genotoxic potential. However, since these are predictive assessments for animals and man the next step is the practical assessment in animal models. Animal studies, even with rodents, are necessarily long term (two years or more) and involve large numbers of animals to ensure statistical acceptance of the findings. Nevertheless, at present they remain the best available means of assessing carcinogenic potential in the animals themselves and for providing the most realistic prediction of that potential in man. What follows is a review of the long-term rodent studies undertaken on several pyrethroids with an assessment of their carcinogenic potential as individual compounds and as a class.

Rodent bioassays

The earliest studies were undertaken on the pyrethrins in the US Food and Drug Administration laboratories in the 1950s. These have been referenced several times over the years, the most detailed account being given by Clara Williams (1973). Groups of 12 male and 12 female rats were fed 0 (control), 200, 1000 and 5000 p.p.m. pyrethrins for two years. There was no obvious effect on growth or mortality with the control group having the highest number of deaths. Forty per cent of the animals in the study survived to termination. There was no effect on tumour incidence due to the pyrethrins. In another study at the US Food and Drug Administration Laboratories pyrethrins were tested in combination with a number of flavouring substances and pesticides in a long-term rat study (Fitzhugh, 1966). The objective of this experiment was to determine if such a combination altered the intrinsic toxicological properties of the individual components. There were five test groups which contained varying combinations of eight flavouring agents and six pesticides, one group with the six pesticides only and a control group. These were fed to groups of 35 male and 35 female rats for two years. The highest concentration of pyrethrins tested was 10 p.p.m. There was no evidence for increased tumour incidence in the treated animals overall nor in any individual organ.

   The above studies must be judged inadequate for testing the carcinogenic potential of the pyrethrins because of the small numbers of animals involved in the first case and the low concentration of compound tested in the second. A rather more searching evaluation was undertaken with the synthetic pyrethroid allethrin. Groups of 30 male and 30 female rats were fed concentrations of 0, 500, 1000 or 2000 p.p.m. allethrin for 80 weeks (Miyamoto, 1976). A detailed histopathological examination was undertaken on 26 tissues and there was no indication of an increase in tumour formation in the allethrin-treated animals. The carcinogenic evaluations of five other pyrethroids have been undertaken in long-term rodent studies to modern protocols.

   In the most recently described study fenvalerate was administered in diet to groups of 50 male and 50 female B6C3F1 mice for two years (Parker et al., 1983). The mice, seven to nine weeks old at commencement, were fed 10, 50, 250 or 1250 p.p.m. fenvalerate and were compared with two control groups.

There was an increased mortality and decreased growth compared to the controls in the top dose group and some decrease in growth in animals given 250 p.p.m. fenvalerate. A very detailed histopathological examination of several tissues including lung, liver, reproductive organs, skin, pituitary, adrenals and thyroid did not reveal any increase in the number or type of neoplasms in the treated mice compared with the control groups. Support for these findings was shown in a different strain (ddY) where fenvalerate was administered at levels of 100, 300, 1000 or 3000 p.p.m. for 78 weeks. These groups and a control group each had 35–47 male or female mice (FAO, 1980). Increased mortality occurred at 1000 or 3000 p.p.m. fenvalerate and there was depression of growth at 3000 p.p.m. There was no indication from histopathological examination of increased tumour formation due to the administration of the compound. Fenvalerate was also evaluated for carcinogenic potential in the two long-term rat studies shown in Table 3.10 (FAO, 1980, 1982). In the one study, groups of Sprague-Dawley rats were fed diets containing up to 250 p.p.m. fenvalerate and in the other study the Wistar strain of rat was used with dose levels up to 1500 p.p.m. An effect on bodyweight gain was noted at the highest dose level of 1500 p.p.m. There was no evidence from either study for a compound-related effect on tumour incidence.

Cypermethrin was evaluated in a two-year rat study at dose levels of 1, 10, 100 and 1000 p.p.m. Each experimental group contained 48 males and 48 females while a control group had twice the number of animals (FAO, 1980). Mortality was unaffected by cypermethrin treatment but growth rate was depressed in the 1000 p.p.m. group. There was no significant increase in the incidence of tumour formation over the course of the study. The carcinogenic potential of deltamethrin was evaluated in two long-term rodent studies (FAO, 1981). In the mouse study, groups of animals of the Charles River CD-1 strain were fed deltamethrin at concentrations of 0 (control), 1, 5, 25, or 100 p.p.m. for 24 months. Sixty males and 60 females per group, including two control groups, were designated for treatment for the two-year period. Tumour incidence was unaffected by the administration of deltamethrin. The rat study was undertaken to the design shown in Table 3.10 using the Charles River CD strain. There was a slight depression of growth rate at the top dose of 50 p.p.m. but overall there was no evidence for a carcinogenic response.

Permethrin has been evaluated in seven long-term rodent studies at three different laboratories. Five of the studies were carried out with the 40 : 60 *cis–trans* isomer ratio and two with the 25 : 75 isomer combination. Four studies with 40 : 60 *cis–trans* ratio have been described in some detail as a result of the Joint Meeting on Pesticide Residues review in 1979 (FAO, 1980). Two rat studies, each of two years duration, were undertaken both using initial group sizes of 60 males and 60 females. In the study with the Alderley Park Wistar derived strain the dietary dose levels were 0, 500, 1000 and 2500 p.p.m. permethrin. The highest dose level represented one-half of the dietary LD50 and transient clinical signs of toxicity were noted in the early stages of the study. A detailed histopathological examination of a large range of organs on all animals in the study did not reveal any increased incidence of tumours due to permethrin. In the other study, using the Long Evans strain, the dietary levels were 0, 20, 100 and 500 p.p.m. permethrin. A full histopathological examination on a wide range of tissues did not reveal any carcinogenic potential due to permethrin.

Of the two mouse studies one was undertaken with the Alderley Park (Swiss derived) strain and the other with the Charles River CD-1 strain. The Alderley Park study contained groups of mice (70 males, 70 females per group) fed 0, 250, 1000 and 2500 p.p.m. permethrin for up to 98 weeks. Mortality was slightly increased and bodyweight slightly decreased at the top dose. An extensive histopathological examination was undertaken on all animals in a wide range of tissues. There was a high incidence of lung adenomas in all groups including controls, typical of the Alderley Park mouse, with a slightly higher incidence at 2500 p.p.m. This was not considered to be related to treatment with permethrin (FAO, 1980, 1983). The other study with CD-1 mice was carried out with group sizes of 75 males and 75 females and dietary concentrations of 0, 20, 500 or 4000 p.p.m. permethrin. The study was maintained for two years overall. Mortality and growth rate were affected at 4000 p.p.m. but a detailed histopathological examination did not reveal any effect on the number or type of tumours due to permethrin administration.

Three other long-term rodent studies on permethrin have been referred to, but not described in any detail, in the public domain (Federal Register, 1982). A two-year mouse study was undertaken with the 40 : 60 *cis–trans* material and a 92-week mouse and a two-year rat study on the 25 : 75 isomer combination. A statistically significant increase in lung and liver tumours was observed at the mid- and high-dose levels (2500 and 5000 p.p.m.) of the female mice in the study with the 40 : 60 *cis–trans* permethrin. The debate concerning the interpretation of these findings in terms of the carcinogenic potential of permethrin has been a lengthy one in the USA. It concerns the familiar story of assessing the relevance of mouse lung and liver tumours in a carcinogenic evaluation. In most strains these tumours are present at a high and variable incidence in ageing mice and are susceptible to change due to factors (e.g. environment, diet) other than the administration of a test material (Sher, 1974, 1982). It is well recognised that less significance is placed upon changes in the incidence of these tumours than for neoplasms occurring at other sites when evaluating the carcinogenic potential of chemicals (IARC, 1980). In the case of permethrin it is necessary to place the findings of the one mouse study into the context of all the other available evidence on the compound. The outcome of this particular debate was the conclusion by the US Environmental Protection Agency (Federal Register, 1982) and its Scientific Advisory Panel (Gray, 1981) that although a low oncogenic potential was shown by mice none was shown by rat studies, and that the oncogenic potential for humans was non-existent or extremely low. The Joint Meeting on Pesticide Residues in

1982 also concluded that the long-term rodent studies on permethrin did not indicate any oncogenic risk (FAO, 1983).

The fifth pyrethroid for which there are published descriptions of long-term rodent studies is phenothrin (FAO, 1981). An 18-month study was carried out in Swiss mice at dose levels of 300, 1000 or 3000 p.p.m. There were 50 male and 50 female mice per experimental group and in the control group. There was a slight effect on growth at 3000 p.p.m. There was no evidence for carcinogenic potential upon a detailed histopathological examination. Rats were studied over two years at dose levels of 0, 200, 600, 2000 or 6000 p.p.m. phenothrin in diet. Fifty male and 50 female rats were present in each group. Growth rate was retarded at the top dose of 6000 p.p.m. A description of the pathological findings is not given but the Joint Meeting on Pesticide Residues evaluation stated that there was no carcinogenic potential for phenothrin shown by this study (FAO, 1981). One other pyrethroid which has received mention of a carcinogenic evaluation is resmethrin (Federal Register, 1983). An 85-week mouse study at dose levels up to 1000 p.p.m. and a two year rat study with a highest dose level of 5000 p.p.m. resmethrin did not show the compound to be oncogenic.

Summary

In summary, several pyrethroids have been evaluated for carcinogenic potential by rodent studies caried out to modern protocols. A feature of these studies was the very high dose levels used, for example up to 3000 p.p.m. for fenvalerate, 5000 p.p.m. for permethrin and 6000 p.p.m. for phenothrin, thus fully testing the carcinogenic potential of the compounds. Some toxic effect was shown at the top dose in these studies for all the compounds tested. The studies involved several strains of mice and rats overall. For one compound, permethrin, there was an increased incidence of lung and liver tumours in one sex of one mouse study, and for reasons already given this indicated little or no carcinogenic potential when the totality of the evidence from rat, mouse and mutagenicity studies on permethrin were considered. From the overall lack of evidence shown by the long-term studies on several pyrethroids it would appear that the pyrethroid molecule does not have carcinogenic properties. This is in line with the prediction from the assessment of chemical structure and mutagenicity assays.

Reproductive toxicity

The reproductive toxicity of a compound is a complex process involving the separate and interlinking assessments of male and female fertility, overall reproductive performance and the effect on the developing foetus. These assessments are covered in the toxicological evaluation of a compound by a range of studies which usually include a dominant lethal test, teratology studies in two species and a multigeneration study in the rat. All the commercially important pyrethroids have been subjected to this battery of studies and this represents the main subject of this section.

Dominant lethal test

The primary purpose of this test is as an assessment of mutagenic potential and details of its conduct and of its use in the evaluation of pyrethroids in this context have been described in the mutagenicity section. However, the data presented from this test can also provide information for the assessment of male fertility as measured by the number of successfully mating males and the number of pregnant females. Using this test, fenvalerate at 25, 50 and 100 mg/kg/day and permethrin at 15, 48 and 150 mg/kg/day were administered orally to male mice with no consequent effect upon their fertility (FAO, 1980). Cypermethrin was the subject of two studies, one with single oral doses of 6·25, 12·5 or 25 mg/kg and the other with daily oral doses of 2·5, 5, 7·5 or 10 mg/kg/day for five days (FAO, 1980). Together they showed no evidence for an effect on male fertility. In a separate study, male mice were dosed 2·5, 5, 7·5 or 10 mg/kg/day cypermethrin for five days and selected animals examined histopathologically one or seven days after the last dose. There were no treatment-related changes in either the testis or epididymus as a result of cypermethrin administration. Deltamethrin was subjected to the test using either single oral doses of 6 or 15 mg/kg or daily doses of 3 mg/kg/day for seven days. The highest dose used was equivalent to one-half of the LD50 and gave rise to clinical signs of toxicity and some mortality. However, there was no effect on fertility in the male mice at any dose level and there was no evidence for treatment-related abnormalities in the testis of animals selected for histopathological examination (FAO, 1981).

Teratology studies

The pyrethroids have been evaluated for their teratogenic potential in the rat, mouse or rabbit in standard teratology studies. In these studies the pyrethroid was dosed orally to groups of at least 20 pregnant rats or mice or at least 15 pregnant rabbits over the period of organogenesis, usually, day 6 to 16 of pregnancy for rats or mice and day 6 to 18 of pregnancy for rabbits. The pups were obtained by Caesarian section just prior to the termination of the gestation period and given a visceral and skeletal examination. Table 3.16 shows the species tested for each pyrethroid and the dose levels used in each study. In some studies a proportion of the pregnant females were allowed to

Table 3.16. Teratology studies on the pyrethroids

Pyrethroid	Species	Dose levels (mg/kg/day)	Teratogenic potential	Source
Allethrin	Rabbit	Up to 215	None	Miyamoto (1976)
	Mouse[a]	Up to 150	None	Miyamoto (1976)
Cypermethrin	Rat	17·5, 35, 70	None	FAO (1980)
	Rabbit	3, 10, 30	None	FAO (1980)
Deltamethrin	Rat and Mouse	0·1, 1, 10	None	FAO (1981, 1982)
	Rabbit	1, 4, 16	None	FAO (1981, 1982)
	Rat	1·25, 2·5, 5	None	Kavlock et al (1979)
	Mouse	3, 6, 12	None	Kavlock et al (1979)
Fenvalerate	Mouse	5, 15, 50	None	FAO (1980)
	Rabbit	Up to 50	None	FAO (1980)
Permethrin	Rat	22·5, 71, 225	None	FAO (1980)
		10, 20, 50	None	FAO (1980)
		4, 41, 83	None	Metker et al. (1978)
	Mouse	15, 50, 150	None	Miyamoto (1976)
Phenothrin	Mouse[a]	30, 300, 3000	None	Miyamoto (1976)
	Rabbit[a]	10, 100, 1000	None	Miyamoto (1976)
	Rabbit	3, 10, 30	None	FAO (1981)
Resmethrin	Rat	10, 20, 50	None	Miyamoto (1976)
	Rabbit	10, 30, 50	None	Miyamoto (1976)

[a]Tested with non-racemic isomer mixtures.

deliver and rear their pups which were examined for growth and behavioural characteristics.

A survey of the results from these studies showed that there was some evidence for a toxic effect in the top-dose dams in most cases. There was no evidence for a teratogenic effect in any of the studies shown in Table 3.16. In those cases where the pups were delivered normally there was no evidence for an effect on their subsequent growth pattern or behavioural characteristics. In addition to the studies shown in Table 3.16, resmethrin has been reported not to be teratogenic in the rat at doses up to 80 mg/kg/day or the rabbit at doses up to 100 mg/kg/day (Federal Register, 1983). A limited assessment in the rabbit has also been undertaken with the pyrethrins at a dose level of 90 mg/kg/day with no effects, as described by Williams (1973). Another aspect was studied by Spencer & Berhane (1982) who measured the protein and glycogen content of the foetuses delivered from pregnant rats fed permethrin over the period of organogenesis. There was some decrease in the concentration of both parameters at high levels of 2500–4000 p.p.m. permethrin, but the response was not strictly dose-related or marked.

Multigeneration studies

The multigeneration reproduction study is the most complex of the studies described in this section, involving the assessment of the influence of a chemical over three generations in the rat. The study is designed to provide general information on the effects of a compound on gonadal function, oestrous cycles, mating performance, conception, parturition, lactation, weaning and the growth and development of the offspring. The study does not necessarily determine cause and effect in all these cases. Five pyrethroids have been evaluated in studies following the same design (FAO, 1980, 1981). Groups of female and male rats were administered the pyrethroid in diet for a period of up to 12 weeks after which time the animals were mated either on a two female to one male or a one female to one male basis. The females were allowed to deliver and rear their pups for the first litter of the first generation and then they were re-mated to give a second litter. The parents for the next generation were usually selected from the pups in this second litter. The pre-mating, mating and rearing sequence was then repeated for the production of two more generations with two litters per generation. Pyrethroid administration continued throughout the study.

Permethrin was evaluated in two studies, one at dose levels of 20 or 100 p.p.m. and the other at 500, 1000 or 2500 p.p.m. (FAO, 1980). In the second study clinical signs of toxicity (tremoring) were seen at the top dose of 2500 p.p.m. and similar effects were also noted in pups at the same dose level. There was no evidence for an effect on any of the indices of reproductive performance in either study. A similar pattern of results was obtained for the four other pyrethroids, cypermethrin at dose levels of 10, 100 or 500 p.p.m. deltamethrin at 2, 20 or 50 p.p.m., fenvalerate at 1, 5, 25 or 250 p.p.m. and phenothrin at 200, 600 or 2000 p.p.m. Toxic effects, such as bodyweight gain decreases were seen at the top dose in the studies with cypermethrin, deltamethrin and fenvalerate, but in all cases there was no effect on the reproductive performance of the test animals over three generations (FAO, 1980, 1981). A three-generation rat reproduction study has been reported on resmethrin in which there was a slight increase in the number of pups born dead and a decrease in pup weight at a dose of 500 p.p.m. (Federal Register, 1983). The pyrethrins were subjected to a one generation two-litter study of limited design at a dietary level of 5000 p.p.m. (Williams, 1973). There was growth suppression in the pups of both litters but the reproductive performance of the parent animals did not appear to be affected on the evidence of the indices measured.

Summary

The lack of reproductive toxic potential of the pyrethroids is self-evident from the survey of the studies described above. Overall there was little evidence for an effect on reproductive performance from the different studies undertaken on several pyrethroids, even when doses sufficient to elicit toxic effects in the parent animals were used.

Human experience

The inherent purpose of laboratory animal studies is to identify the toxicological properties of a compound, to determine its mechanism of action and to put its toxicity into perspective with that of other materials. The studies described previously in this chapter have adequately addressed these points for the pyrethroids. An ultimate objective of these studies is to provide an assessment of the likely effect of a compound in man. With newly developed materials such as the synthetic pyrethroids their toxicological assessment in man relies heavily if not totally upon the results from the animal studies, since experience with man builds up only slowly over a long period of time. However, some experience has arisen in their usage and this section reviews the available information. It must be noted that there has been a long history of the preparation and usage of pyrethrum and its extracts and that this provides relevant information for this survey. Reviews of the effects of pyrethrum have been given by FAO/WHO (1967), Barthel (1973) and Hayes (1982) and the reader is referred to these for more detail. Human experience derives from the results of accidental or intentional poisoning, manufacturing practice and the normal use of the materials.

Accidental poisoning

It is a notable fact that during the long period of preparation and use of pyrethrum extracts there appear to have been only two recorded serious poisonings, one fatal, and they were in the nineteenth century. A similar picture is shown for the synthetic pyrethroids during their shorter but intensive development and usage in recent years. Only one fatal poisoning, due to an accidental usage of cypermethrin formulation as cooking ingredient, has been documented in the literature (Poulos et al., 1982).

Manufacturing practice

In a perceptive report McCord et al. (1921) described the skin reaction of 85 workers engaged in pyrethrum processing. The workers, 75 females and 10 males, were employed in grinding the pyrethrum flower and filling, weighing and sealing bags of the ground material. Three types of dermatitis were observed in a proportion of the workers, one of which could be definitely ascribed to pyrethrum. This was described as a vesicular dermatitis which developed within a 48-hour period but which disappeared 48 hours after removal from the source of irritation. Dermatitis due to contact with pyrethrum has been recorded by other investigators as reviewed by FAO/WHO (1967). Dermal effects on workers have been reported from deltamethrin manufacture prior to the installation of a new plant (FAO, 1982). This was observed as a pricking sensation initially and later developed into a blotchy erythema. No other symptoms have been recorded during the manufacture of deltamethrin nor for other synthetic pyrethroids (FAO, 1982).

Recommended uses

The most direct human experience derives from the therapeutic uses of pyrethrum extracts. They have been applied extensively for the control of body lice, particularly during World War II. They have also been administered orally to man as an anthelminthic. There have been few recorded cases of skin effects and none for any other adverse reaction as a result of these uses (Hayes, 1982). Pyrethrum formulations have also been applied for insecticidal purposes for indoor use. There have been occasional reports of dermatitic type responses to this usage.

Reports are beginning to accumulate on human experience with the use of synthetic pyrethroids in a variety of situations. Two trials were undertaken with 5% deltamethrin for indoor pesticide application with checks on the clinical reaction of the applicators and support operators. No adverse reactions were reported from the first trial but in the second some of the personnel complained of heat around the eyes or on the face (Hayes, 1982). The effects disappeared a few hours after work finished. A similar trial was undertaken with permethrin and no effects were seen. A few cases of skin irritation have been reported from spray applicators applying deltamethrin for orchard pesticide treatment and in confined spaces in greenhouses (FAO, 1982). Cypermethrin was the subject of study during cotton spray trials on the Ivory Coast. A cypermethrin formulation was used in six spray sessions at two-weekly intervals for one season of spraying. The seven operators wore their normal clothing. These personnel were given a comprehensive medical, clinical and neurological examination before and after the trials and no abnormal findings were encountered, including blood biochemical and peripheral nerve function parameters (FAO, 1982).

Other investigators have recorded the reactions of personnel using pyrethroid formulations in other circumstances. Kolmodin-Hedman et al. (1982) surveyed the subjective symptoms arising from 139 workers planting conifer seedlings treated with fenvalerate or permethrin formulations. A proportion of workers handling conifers treated with either formulation complained of skin irritation. Additionally some of those handling the fenvalerate-treated conifers only also noted a tingling sensation of the face and hands, a reaction which disappeared overnight. A small number of planters reported other symptoms such as respiratory tract or eye irritation, however it is difficult to place these findings in perspective since there was no comparison with an unexposed group of people. LeQuesne et al. (1980) examined 23 laboratory workers involved in field trials (2), formulation (8) and general laboratory work (13) with pyrethroids, notably cypermethrin, fenvalerate, permethrin and fenproparthrin. The study involved interviews, to ascertain subjective symptoms, and electrophysiological monitoring. The most frequently reported symptom was a facial sensation described as "tingling", "burning", "nettle rash" by workers who had experienced it on one or more occasions. This sensation usually occurred about 30 min to 3 h after exposure and lasted for about 30 min to 8 h. Apparently this did not occur when permethrin only was involved. Neurological examination was undertaken on all the workers and no abnormal findings were recorded. The electrophysiological measurements from these workers were compared with an age-matched control group of subjects and no difference in response could be discerned between the two groups.

In the most recently reported study to date, Tucker & Flannigan (1983) describe symptoms of what they call cutaneous sensation on aerial applicators, loaders and farmers applying fenvalerate during field operations. The effects and their time course were similar to those outlined above.

Commentary on human effects

At this time the only recorded adverse reactions of pyrethrum and the pyrethroids in man appear to be on the skin. No systemic toxicological effects have been noted except for the handful of accidental poisonings, two involving pyrethrum and one a synthetic pyrethroid. The skin reaction from pyrethrum has been described as a dermatitis and appears to be quickly reversible. Investigations have revealed that the pyrethrum flower and crude extracts are the main cause of this dermatitis, and that refined extracts and the pyrethrins as such have little or no effect upon the skin (Williams, 1973). With the synthetic pyrethroids the most commonly observed symptom has been described as a transient facial sensory symptom the onset, duration and reversal of which has been detailed by LeQuesne et al. (1980). Volunteer studies to investigate these effects have not been very helpful. No irritant or allergic responses were observed from 7 volunteers given 10 daily topical applications of 0·5 ml barthrin and later challenged by 0·1 ml of the pyrethroid (Ambrose et al., 1963). Similarly, fenvalerate (in two studies) applied either as the technical material or as an EC formulation did not cause any dermal effects on volunteers (FAO, 1980). Phenothrin (Sumithrin) also gave no adverse dermal reaction when administered to male volunteers over a three-day period (FAO, 1981). The relevance of the skin effects on man is discussed in the following concluding review of the overall assessment of pyrethroid toxicology in man.

Overall assessment for man

The overall assessment is based upon the knowledge of events in man as described above and upon the extrapolation from animal toxicology. The human data, so far, indicate that man is unlikely to be affected systemically by the small amounts of pyrethroids he is exposed to in normal circumstances. This view is supported by information from the few studies monitoring the excretion of pyrethroid metabolites from humans after exposure to the insecticides. The methods used have been based upon the measurement of the cyclopropanecarboxylic acid moiety which is excreted as the glucuronide conjugate in urine from animals and man and is more appropriate than measuring metabolites of the 3-phenoxybenzyl moiety which shows species specificity in its conjugation (Eadsforth & Baldwin, 1983). Kolmodin-Hedman et al. (1982) using this type of method could detect only a trace amount of urinary permethrin metabolite (0.26 µg/ml) from 1 person out of 17 who were engaged on dipping, packing and planting conifer seedlings treated with permethrin formulation. Air analysis showed that permethrin concentrations in the breathing zone of the dipper and packers varied between 0·011 and 0·085 mg/m^3 and they were lower for the planters. The measurement of the permethrin metabolite from sprayers in a Nigerian field trial led to an estimate that less than 2 mg of permethrin per person was absorbed in any 12 h period of spraying (FAO, 1982). Similarly the amount of cypermethrin absorbed per person was calculated to be less than 1 mg per spray session during spraying of the pyrethroid formulation in the Ivory Coast (FAO, 1982). There is very strong evidence therefore that man is unlikely to experience any acute or subacute effect due to normal exposure to pyrethroids. The amounts he is exposed to are small and are substantially lower than levels which elicit even a mild toxicological response in animals in oral, dermal and inhalation studies. The same lack of effect in man can be predicted for subchronic and chronic exposure since the animal toxicology has shown that chronic administration does not lead to irreversible damage even at high-dose levels. The assessment of mutagenic, carcinogenic and reproductive effects in man is always difficult due to the latency or subtlety of the changes against a background of spontaneous occurrence. However, the evaluation of the range of animal studies show the potential in these respects to be very low or non-existent for the pyrethroids. It should be remembered also that the long period of usage of pyrethrum, and hence the pyrethrins, in a variety of applications has not led to any detectable systemic effect in man.

Despite the favourable toxicological profile shown by the pyrethroids there is the continuing need to build up data on man by careful recording of observations during the manufacture and the use of the compounds in field trials and by taking the opportunity to monitor developed products for possible adverse effects.

The only established effect of pyrethrum and the pyrethroids in man is on the skin. The dermatitic response from pyrethrum and its crude extracts have been known for many years but it is worth noting that constituents other than the insecticidal pyrethrins were responsible for this effect (Mitchell et al., 1972). Frequent exposure does not appear to have led to any chronic manifestation of these effects. Some of the synthetic pyrethroids produce a temporary facial sensation in certain exposure situations. It has been surmised that this facial sensation could be due to spontaneous firing of the sensory nerve fibres or nerve endings in response to local concentrations of the pyrethroid on the skin (LeQuesne et al., 1980). These effects in man could not have been predicted from conventional dermal irritancy and sensitisation tests in animals which had shown a generally low potential for adverse skin reaction. There does seem to be good reason for designing an animal test which can assess this particular type of response. It could be used also to establish criteria with the existing pyrethroids by which newly developed compounds could be evaluated. This kind of development and the regular monitoring during use advocated previously should ensure that the extending use of the synthetic pyrethroids continues with the good safety record shown by the current compounds.

References

Ames, B. N., McCann, J. & Yamasaki, E., 1975, Methods for detecting carcinogens and mutagens with the *Salmonella* mammalian – microsome mutagenicity test. *Mutat. Res.*, **31**, 347–64.

Ambrose, A. M., Miller, C. V. & O'Dell, F. A. 1963, Toxicologic studies on pyrethrin-type esters of chrysanthemumic acid. *Toxicol. Appld. Pharmacol.*, **5**, 414–26.

Ambrose, A. M., Miller, C. V., O'Dell, F. A. & Pepper, J. L., 1964, Toxicologic studies on pyrethrin-type esters of chrysanthemumic acid. *Toxicol. Appld. Pharmacol.*, **6**, 112–20.

Ashwood-Smith, M. J., Trevino, J. & Ring, R., 1972, Mutagenicity of Dichlorvos. *Nature, Lond.*, **240**, 418–20.

Barthel, W. F., 1973, Toxicity of pyrethrum and its constituents to mammals. In *Pyrethrum: the Natural Insecticide*, edited by J. E. Casida (Academic Press: New York & London), pp. 123–42.

Bartsch, H., Malaveille, C., Camus, A. M., Martel-Planche, G., Brun, G., Hautefeuille, A., Sabadie, N. & Barbin, A., 1980, Validation and comparative studies on 180 chemicals with *S. typhimurium* strains and V79 Chinese hamster cells in the presence of various metabolising enzymes. *Mutat. Res.*, **76**, 1–50.

Bond, H., Mauger, K. & DeFoe, J. J., 1973, Interactions in the toxicity of pyrethrum, synergists and other chemicals to mammals. In *Pyrethrum: the Natural Insecticide*, edited by J. E. Casida (Academic Press: New York & London), pp. 177–94.

Bradbury, J. E., Forshaw, P. J., Gray, A. J. & Ray, D. E., 1983, The action of mephenesin and other agents on the effects produced by two neurotoxic pyrethroids in the intact and spinal rat. *Neuropharmacol.*, **22**, 907–14.

Bridges, B. A. 1972, Simple bacterial systems for detecting mutagenic agents. *Lab. Pract.*, **21**, 413.

Brodie, M. E., & Aldridge, W. N., 1982, Elevated cerebellar cyclic GMP levels during the deltamethrin-induced motor syndrome. *Neurobehav. Toxicol. Teratol.*, **4**, 109–13.

Carlson, G. P. & Schoenig, G. P., 1980, Induction of liver microsomal NADPH cytochrome *C* reductase and cytochrome *P*-450 by some new synthetic pyrethroids. *Toxicol. Appld. Pharmacol.*, **52**, 507–12.

Carpenter, C. P., Weil, C. S., Pozzani, U. C. & Smyth, H. F., 1950, Comparative acute and subacute toxicities of allethrin and pyrethrins. *Arch. Ind. Hyg. Occup. Med.*, **2**, 420–32.

Casida, J. E., Kimmel, E. C., Elliot M. & Janes, N. F., 1971, Oxidative metabolism of pyrethrins in mammals. *Nature, Lond.*, **230**, 326–7.

Cole, L. M., Ruzo, L. O., Wood, E. J. & Casida J. E., 1982, Pyrethroid metabolism: comparative fate in rats of tralomethrin, tralocythrin, deltamethrin and (*IR* α-*S*)-*cis*-cypermethrin. *J. Agric. Food Chem.*, **30**, 631–6.

Crampton, R. F., Gray, T. J. B., Grasso, P. & Parke, D. V., 1977, Long term studies on chemically induced liver enlargement in the rat. *Toxicology*, **7**, 289–306.

Crawford, M. J., Croucher, A. & Hutson, D. H., 1981a, Metabolism of *cis* and *trans* cypermethrin in rats. Balance and tissue retention study. *J. Agric. Food Chem.*, **27**, 130–5.

Crawford, M. J., Croucher, A. & Hutson, D. H., 1981b, The metabolism of the pyrethroid insecticide cypermethrin in rats; excreted metabolites. *Pestic. Sci.*, **12**, 399–411.

Eadsforth, C. V. & Baldwin, M. K., 1983, Human dose–excretion studies with the pyrethroid insecticide cypermethrin. *Xenobiotica*, **13**, 67–72.

Elliott, M., Janes, N. F., Kimmel, E. C. & Casida, J. E., 1972, Metabolic fate of pyrethrin I, pyrethrin II and allethrin administered orally to rats. *J. Agric. Food Chem.*, **20**, 300–13.

FAO, 1977. Evaluations of some pesticide residues in food. Result of the deliberations of the Joint Meeting of the FAO Panel of Experts on pesticide residues and the environment and the WHO Expert Group on pesticide residues, Rome 22–30 November 1976. Food and Agriculture Organisation of the United Nations, Rome.

FAO, 1980, Pesticide residues in food – 1979 evaluations. FAO plant production and protection paper 20 Suppl. Food and Agricultural Organisation of the United Nations, Rome.

FAO, 1981, Pesticide residues in food – 1980 evaluations. FAO plant production and protection paper 26 Suppl. Food and Agricultural Organisation of the United Nations, Rome.

FAO, 1982, Pesticide residues in food – 1981 evaluations. FAO plant production and protection paper 42. Food and Agricultural Organisation of the United Nations, Rome.

FAO, 1983, Pesticide residues in food – 1982 report. FAO plant production and protection paper 46. Food and Agricultural Organisation of the United Nations, Rome.

FAO/WHO, 1967, Evaluation of some pesticide residues in food. Result of the deliberations of the Joint Meeting of the FAO Working Party and the WHO Expert Committee on pesticide residues, Geneva 14–12 November 1966. Food and

Agricultural Organisation of the United Nations, Rome and World Health Organisation, Geneva.

Federal Register, 1982, Tolerances and exemptions from tolerances for pesticide chemicals in or on raw agricultural commodities; Permethrin. *Federal Register*, **47**, 45008–10.

Federal Register, 1983, Tolerances for pesticides in food administered by Environmental Protection Agency; Resmethrin. *Federal Register*, **48**, 36246–7.

Feuer, G., Golberg, L. & Le Pelley, J. R., 1965, Liver response tests. *Food Cosmet. Toxicol.*, **3**, 235–49.

Fitzhugh, O. G., 1966, Problems related to the use of pesticides. *Canad. Med. Ass. J.*, **94**, 598–604.

Gammon, D. W., Brown, M. A. & Casida, J. E., 1981, Two classes of pyrethroid action in the cockroach. *Pestic. Biochem. Physiol.*, **15**, 181–91.

Gammon, D. W., Lawrence, L. J. & Casida, J. E., 1982, Pyrethroid toxicology: protective effects of diazepam and phenobarbital in the mouse and the cockroach. *Toxicol. Appld. Pharmacol.*, **66**, 290–6.

Gaughan, L. C., Unai, T. & Casida, J. E. 1977, Permethrin metabolism in rats. *J. Agric. Food Chem.*, **25**, 9–17.

Gilbert, D. & Golberg, L., 1965, Liver response tests. *Food Cosmet. Toxicol.*, **3**, 417–32.

Glickman, A. H. & Lech, J. J., 1982, Differential toxicity of *trans*-permethrin in rainbow trout and mice. *Toxicol. Appld. Pharmacol.*, **66**, 162–71.

Glomot, R., 1982, Toxicity of deltamethrin to higher vertebrates. In *Deltamethrin* (Roussel-Uclaf), pp. 109–37.

Gray, P. K., 1981, Advisory opinion on the oncogenic potential of permethrin. Federal Insecticide, Fungicide and Rodenticide Act (FIFRA). Scientific Advisory Panel, March 20 1981.

Gray, A. J., Connors, T. A. Hoellinger, H. & Nguyen-Hoang-Nam, 1980, The relationship between the pharmokinetics of intravenous cismethrin and bioresmethrin and their mammalian toxicity. *Pestic. Biochem. Physiol.*, **13**, 281–93.

Gray, A. J. & Rickard, J., 1982a, Toxicity of pyrethroids to rats after direct injection into the central nervous system. *Neurotox.*, **3**, 25–35.

Gray, A. J. & Rickard, J. 1982b, The toxicokinetics of deltamethrin in rats after intravenous administration of a toxic dose. *Pestic. Biochem. Physiol.*, **18**, 205–15.

Hayes, W. J., 1982, Pesticides derived from plants and other organisms. In *Pesticides Studied in Man*, edited by W. J. Hayes (Baltimore & London: Williams & Wilkins), pp. 75-111.

Hutson, D. H., 1979, The metabolic fate of synthetic pyrethroid insecticides in mammals. In *Progress in Drug Metabolism*, edited by J. W. Bridges & L. F. Chasseaud, Vol. 3 (Chichester: John Wiley), pp. 215–52.

Hutson, D. H. Gaughan, L. C. & Casida, J. E., 1981, Metabolism of the *cis* and *trans* isomers of cypermethrin in mice. *Pestic. Sci.*, **12**, 385–98.

IARC, 1980. The evaluation of the carcinogenic risk of chemicals to humans. *Monographs of the International Agency for Research on Cancer*, **22**, 18–19.

James, J. A., 1980, The toxicity of synthetic pyrethroids to mammals. In *Developments in Animal and Veterinary Sciences*. **6**, *Trends in Veterinary Pharmacology and Toxicology*, edited by A. S. J. P. A. M. van Miert, J. Frens & F. W. van der Kreek (Amsterdam: Elsevier), pp. 249–55.

Joy, R. M., 1982, The chlorinated hydrocarbon insecticides. In *Pesticides and Neurological Diseases*, edited by D. Ecobichon & R. M. Joy (Boca Raton, Fl: CRC Press), pp. 91–150.

Kadry, A. M. & Abu-Hadeed, A. H., 1982, Genotoxic activities of the synthetic pyrethroid insecticide fenproparthrin (S-3206) in rat bone marrow and *S. typhimurium* TA 98. *Pharmacologist*, **24**, 151.

Kavlock, R., Chernoff, N., Baron, R., Linder, R., Rogers, E., Carver, B., Dilley, J. & Simmon, V., 1979, Toxicity studies with decamethrin synthetic pyrethroid insecticide. *J. Environ. Pathol. Toxicol.*, **2**, 751–65.

Kimbrough, R. D., Gains, T. B. & Hayes, W. J., 1968, Combined effect of DDT, pyrethrum and piperonyl butoxide on rat liver. *Arch. Environ. Health.*, **16**, 333–41.

Kimmel, E. C., Casida, J. E. & Ruzo, L. O. 1982, Identification of mutagenic photoproducts of the pyrethroids allethrin and terallethrin. *J. Agric. Food Chem.*, **30**, 623–6.

Kolmodin-Hedman, B., Swensson, A. & Akerblom, M., 1982, Occupational exposure to some synthetic pyrethroids (permethrin and fenvalerate). *Arch. Toxicol.*, **50**, 27–33.

Lawrence, L. J. & Casida, J. E., 1982, Pyrethroid toxicology: mouse intracerebral structure–toxicity relationships. *Pestic. Biochem. Physiol.*, **18**, 9–14.

LeQuesne, P. M., Maxwell, I. C. & Butterworth, S. T. G., 1980, Transient facial sensory symptoms following exposure to synthetic pyrethroids: a clinical and electrophysiological assessment. *Neurotoxicol.*, **2**, 1–11.

Litchfield, M. H., 1983, Characterisation of the principal mammalian toxicological and biological actions of synthetic pyrethroids. In *Pesticide Chemistry: Human Welfare and the Environment*, edited by J. Miyamoto & P. C. Kearney, Vol. 2 (Oxford: Pergamon Press), pp. 207–11.

Lock, E. A. & Berry, L. N., 1981, Biochemical changes in the rat cerebellum following cypermethrin administration. *Toxicol. Appld. Pharmacol.*, **59**, 508–14.

Malone, J. C. & Brown, N. C., 1968, Toxicity of various grades of pyrethrum to laboratory animals. *Pyrethrum Post*, **9**, 3–8.

Masri, M. S., Hendrickson, A. P., Cox, A. J. & DeEds, F., 1964, Subacute toxicity of two chrysanthemumic acid esters: barthrin and dimethrin. *Toxicol. Appld. Pharmacol.*, **6**, 716–25.

Matsuoka, A., Hayashi, M. & Ishidate, M., 1979, Chromosomal aberration tests on 29 chemicals combined with S9 mix *in vitro*. *Mutat. Res.*, **66**, 277-90.

McCord, C. P., Kilker, C. M. & Minster, D. K., 1921, Pyrethrum dermatitis. *J. Amer. Med. Assn.*, **77**, 448–9.

Metker, L., Angerhofer, R. A., Pope, C. R. & Swentzel, K. C., 1978, Toxicological evaluation of 3-(phenoxyphenyl) methyl (±) *cis*, *trans*-3-(2,2-dichloroethenyl)-2,2-dimethylcyclopropanecarboxylate (permethrin). Report No. 51-0831-78, US Army Environmental Hygiene Agency.

Metker, L. W., 1980, Subchronic inhalation toxicity of 3-(phenoxyphenyl) methyl (± *cis*, *trans*-3-(2,2 dichloroethenyl)-2,2-dimethylcyclopropane carboxylate (Permethrin). Report No 75-51-0026-80. US Army Environmental Hygiene Agency.

Mitchell, J. C., Dupuis, G. & Towers, G. H. N., 1972, Allergic contact dermatitis from pyrethrum (chrysanthemum SPP). *Brit. J. Derm.*, **86**, 568–73.

Miyamoto, J., 1976, Degradation, metabolism and toxicity of synthetic pyrethroids. *Environ. Health Perspec.*, **14**, 15–28.

Moriya, M., Ohta, T., Watanabe, K., Miyazawa, T., Kato, K. & Shirasu, Y., 1983, Further mutagenicity studies on pesticides in bacterial reversion assay systems. *Mutat. Res.*, **116**, 185–216.

Paldy, A., 1982, Examination of the mutagenic effect of synthetic pyrethroids on mouse bone marrow cells. *Chem. Abs.*, **96**, 212242–3.

Parker, C. M., McCullough, C. B., Gellatly, J. B. M. & Johnston, C. D., 1983, Toxicologic and carcinogenic evaluation of fenvalerate in the B6C3F1 mouse. *Fund. Appld. Toxicol.*, **3**, 114–20.

Parkin, P. J. & LeQuesne, P. M., 1982, Effect of a synthetic pyrethroid deltamethrin on excitability changes following a nerve impulse. *J. Neurol. Neurosurg. Psychiatry*, **45**, 337–42.

Polakova, H. & Vargova, M., 1983, Evaluation of the mutagenic effects of decamethrin: cytogenic analysis of bone marrow. *Mutat. Res.*, **120**, 167–71.

Poulos, L., Athanaselis, S. & Coutsenlinis, A., 1982, Acute intoxication with cypermethrin (NRDC 149). *J. Toxicol. Clin. Toxicol*, **19**, 519–20.

Purchase, I. F. H., Longstaff, E., Ashby, J., Styles, J. A., Anderson, D., Lefevre, P. A. & Westwood, F. R., 1978, An evaluation of six short term tests for detecting organic chemical carcinogens. *Brit. J. Cancer*, **37**, 873–959.

Ray, D. E. & Cremer, J. E., 1979, The action of decamethrin (a synthetic pyrethroid) on the rat. *Pestic. Biochem. Physiol.*, **10**, 333–40.

Rose, G. P. & Dewar, A. J., 1983, Intoxication with four synthetic pyrethroids fails to show any correlation between neuromuscular dysfunction and neurobiochemical abnormalities in rats. *Arch. Toxicol.*, **53**, 297–316.

Ruzo, L. O., Unai, T. & Casida, J. E., 1978, Decamethrin metabolism in rats. *J. Agric. Food Chem.*, **26**, 918–25.

Ruzo, L. O., Engel, J. L. & Casida, J. E., 1979, Decamethrin metabolites from oxidative, hydrolytic and conjugative reactions in mice. *J. Agric. Food Chem.*, **27**, 725–31.

Sher, S. P., 1974, Tumours in control mice: literature tabulation. *Toxicol. Appld. Pharmacol.*, **30**, 337–59.

Sher, S. P. 1982, Tumours in control hamsters, rats and mice: literature tabulation. In *CRC Critical Reviews in Toxicology*, Vol. 10, edited by L. Golberg (Boca Raton, Fl: CRC Press), pp. 49–79.

Shono, T., Ohsawa, K. & Casida, J. E., 1979, Metabolism of *trans* and *cis*-permethrin, *trans* and *cis*-cypermethrin and decamethrin by microsomal enzymes. *J. Agric. Food Chem.*, **27**, 316–25.

Soderlund, D. M. & Casida, J. E., 1977, Substrate specificity of mouse liver microsomal enzymes in pyrethroid metabolism. In *Synthetic Pyrethroids*, edited by M. Elliott, ACS Symposium Series 42 (Washington, DC: American Chemical Society), pp. 162–72.

Spencer, F. & Berhane, Z., 1982, Uterine and foetal characteristics in rats following a post-implantational exposure to permethrin. *Bull. Environ. Contam. Toxicol.*, **29**, 84–8.

Springfield, A. C., Carlson, G. P. & DeFeo, J. J., 1973, Liver enlargement and modification of hepatic microsomal drug metabolism in rats by pyrethrum. *Toxicol. Appld. Pharmacol.*, **24**, 298–308.

Staatz, C. G., Bloom, A. S. & Lech, J. J., 1982, A pharmacological study of pyrethroid neurotoxicity in mice. *Pestic. Biochem. Physiol.*, **17**, 287–92.

Swentzel, K. C., Angerhofer, R. A., Haight, E. A., McCreesh, A. H. & Weeks, M. H., 1978, Safety evaluation of the synthetic pyrethroid insecticide, resmethrin, as a clothing impregnant. *Toxicol. Appld. Pharmacol.*, **45**, 243.

Takahashi, M. & LeQuesne, P. M., 1982, The effects of the pyrethroids deltamethrin and cismethrin on nerve excitability in rats. *J. Neurol. Neurosurg. Psychiatry*, **45**, 1005–11.

Tucker, S. B. & Flannigan, S. A., 1983, Cutaneous effects from occupational exposure to fenvalerate. *Arch. Toxicol.*, **54**, 195–202.

Ueda, K., Gaughan, L. C. & Casida, J. E., 1975, Metabolism of (+) *trans* and (+) *cis* resmethrin in rats. *J. Agric. Food Chem.*, **23**, 106–15.

Verschoyle, R. D. & Aldridge, W. N., 1980, Structure activity relationships of some pyrethroids in rats. *Arch. Toxicol.*, **45**, 325–9.

Verschoyle, R. D. & Barnes, J. M. 1972, Toxicity of natural and synthetic pyrethrins to rats. *Pestic. Biochem. Physiol.*, **2**, 308–11.

Vijverberg, H. P. M. & van den Bercken, J., 1982, Action of pyrethroid insecticides on the vertebrate nervous system. *Neuropath. Appld. Neurobiol.*, **8**, 421–40.

Vijverberg, H. P. M., Ruigt, G. S. F. & van den Bercken, J., 1982, Structure related effects of pyrethroid insecticides on the lateral-line sense organ and on peripheral nerves of the clawed frog *Xenopus laevis*. *Pestic. Biochem. Physiol.*, **18**, 315–24.

Vijverberg, H. P. M., van der Zalm, J. M. & van den Bercken, J., 1982, Similar mode of action of pyrethroids and DDT on sodium channel gating in myelinated nerves. *Nature, Lond.*, **295**, 601–3.

Vijverberg, H. P. M., van der Zalm, J. M., van Kleef, R. G. D. M. & van den Bercken, J., 1983, Temperature and structure dependent interaction of pyrethroids with the sodium channels in frog node of Ranvier. *Biochim. Biophys. Acta*, **728**, 73–82.

Waters, M. D., Sandhu, S. S., Simmon, V. F., Mortelmans, K. E., Mitchell, A. D., Jorgenson, T. A., Jones, D. C. L., Valencia, R. & Garrett, N. E., 1982, Study of pesticide genotoxicity. In *Genetic Toxicology: an Agricultural Perspective*, edited by R. A. Fleck & A. Hollaender (New York: Plenum Press), pp. 275–324.

Weisburger, J. H. & Williams, G. M. 1980, Chemical carcinogens. In *Toxicology. the Basic Science of Poisons*, edited by J. Doull, C. D. Klaasen & M. O. Amdur (New York: Macmillan Publishing Co.). pp. 84–138.

WHO, 1973, Evaluations of some pesticide residues in food. Prepared by the Joint Meeting of the FAO Working Party of Experts on pesticide residues and the WHO Expert Committee on pesticide residues, Rome. 20–28 November 1972, World Health Organisation, Geneva.

Williams, C. H., 1973, Tests for possible teratogenic, carcinogenic, mutagenic and allergenic effects of pyrethrum. In *Pyrethrum: the Natural Insectide*, edited by J. E. Casida (New York & London: Academic Press), pp. 167–76.

Yamamoto, I., Elliott, M. & Casida, J. E., 1971, The metabolic fate of pyrethrin I. pyrethrin II and allethrin. *Bull. World Health Org.*, **44**, 347.

4. Effects on non-target organisms in terrestrial and aquatic environments

I. R. Hill

Introduction

Methods of pesticide application are determined by available technology, by economic considerations and by the spatial and temporal distribution of the pest organism, be it microbial, invertebrate, vertebrate or plant. The naturally occurring and synthetic pyrethroids are highly active broad-spectrum, non-systemic insecticides, which are effective following contact or ingestion. Therefore, their use in the control of agricultural, horticultural, forestry and certain public health pests is most often by "boom-spraying" or "mist-blowing", to maximise the degree of contact between chemical and target organism. Inevitably a high proportion of the applied material misses the target insect, much being deposited on plant surfaces but some entering the soil or even aquatic environments. A wide range of non-target organisms are therefore exposed to pyrethroid insecticides, both during and following spraying.

As a result of our lack of knowledge of the structure–activity relationships of pesticides on diverse groups of organisms, and because of the varying degrees to which these chemicals can enter, redistribute and transform in the environment, it is impossible to predict with confidence their effects on non-target biota. Consequently, practical studies need to be undertaken to evaluate the environmental hazards posed by each new pesticide. A programme of studies is designed taking into account the intended uses of the product, its use rates, physicochemical characteristics and speed and mechanism of dissipation. Many of the studies carried out by industry are now legally enforced requirements of government-controlled pesticide registration agencies. Others are done on the initiative of the manufacturer or by researchers in academic institutions and government research stations. The earliest investigations of the effects of a pesticide are usually carried out on a spectrum of organisms under laboratory conditions. Later studies in the field are more complex, expensive and often difficult to interpret, and are therefore generally only done where laboratory data has indicated a potential hazard.

The natural product pyrethrum (active ingredient, pyrethrins) has been commercially used as an insecticide for over a century, and the photolabile synthetic pyrethroids for 20–30 years. Although the instability of these products in sunlight limits their usefulness for many horticultural and agricultural purposes they have proved effective for use in animal husbandry and in the home. More recently a number of analogues have been developed with sufficient stability in sunlight for use on outdoor crops. This review covers both published and unpublished work on pyrethrins and synthetic pyrethroids, and discusses their effects on a wide and representative range of the non-target organisms that might be at risk.

Where studies have been done with formulated products data have been assumed to be calculated as the formulation, unless stated otherwise. Organisms are mostly referred to in the text by their scientific names. Lists of the scientific and common names of the terrestrial and aquatic invertebrates and fish mentioned in this review are given at the beginning of the appropriate sections. Published references are indicated by author names in the text and by a reference number in the majority of tables.

Effects on birds and mammals

In the natural environment birds and mammals may be exposed to pesticides during application, and subsequently by eating treated seed or sprayed plant material and by contacting or "licking" water droplets containing the pesticide. Furthermore, insectivorous birds may consume insects exposed to the pesticide.

The dermal toxicity of pyrethroids to mammals is very low, greater than approximately 1000 mg ai/kg, over a 24 hour period (Table 4.1). This amount is equivalent to a 20 g animal adsorbing on to the skin all the chemical applied at 100 g ai/ha to a 2 square metre area. For larger animals and/or lower application rates, the area is proportionally greater. In practice most of the pyrethroid with which an animal comes into contact will be adsorbed to the fur and not absorbed by the skin. The dermal route thus poses no hazard either during spraying or in the short period while the spray is still wet on the plant surface.

The acute oral toxicities (LD50) of the photostable pyrethroids to mammals are approximately proportional to the level of insecticidal activity and therefore also to application rate, permethrin being least toxic at 400 mg ai/kg (Table 4.1) and deltamethrin most toxic at $\geqslant$ 20 mg ai/kg. The respective application rates of permethrin and deltamethrin are approximately 200 and 10 g ai/ha. To consume a dose equivalent to the LD50, a 20 g mammal would need to eat the sprayed plant material in a 0·5 square metre area, whilst the parent pyrethroid is still all available for absorption in the gut.

Photodecomposition together with plant uptake and metabolism continually

Table 4.1. Toxicity of pyrethroids to mammals and birds[a]

Pyrethroid	Mammal		Bird[b]	
	Acute oral toxicity to rat and/or mouse (LD50, mg ai/kg)	Acute dermal toxicity to rat and/or rabbit (24-h LD50, mg ai/kg)	Acute oral toxicity (LD50, mg ai/kg)	Subacute (8-day) oral tocicity (LC50, mg ai/kg/day)
Natural pyrethrum	200–2500	>1500	>10 000 (1,2)	>5000 (1,2,4)
Photolabile pyrethroids				
allethrin	500–2500			
resmethrin	700–>5000			
Photostable pyrethroids				
permethrin	400–>5000	>200–>2000	>13 000 (1–4)	>27 000 (1–4)
cypermethrin	100–>4000	>2000–>5000	>10 000 (2)	>20 000 (2,5)
fenvalerate	200–3000	>2500–5000	>4000 (1,2,5)	>15 000 (5)
deltamethrin	20–150	>800–>3000	>4000 (2)	

[a] Approximate values shown are taken from published (see Chapter 3; Hill et al., 1975; Elliott et al., 1978; Bradbury & Coats, 1982) and unpublished data (ICI Plant Protection Division).
[b] Species tested shown in parentheses: 1 = *Coturnix coturnix japonica*, Japanese quail; 2 = *Anas platyrhyncos*, mallard duck; 3 = *Sturnus vulgaris*, starling; 4 = *Phasianus colchicus*, ring necked pheasant; 5 = *Colinus virgianus*, bob-white quail.

reduce the residues of the parent pyrethroid (see Chapter 5), and the degradation products have been shown to be much less toxic than the parent. Therefore, there is a large margin of safety to mammals under field conditions. Further studies have shown that at the levels that could be consumed by these organisms in the field, there is no effect on reproductive processes (see Chapter 3). The toxicity profile to mammals is such that some pyrethroids have been registered for use in dipping, spraying and tagging pest control techniques with farm animals.

Birds have been shown to be even less sensitive to pyrethroids than are mammals. In acute and subacute (approximately 8 day) oral toxicity studies individuals in the treated groups have never consumed and retained sufficient chemical to provide an LD50 or LC50 figure (Table 4.1). David (1981) observed a strong repellant action of technical and formulated permethrin administered to quail in feed although repellancy diminished with repeated exposure. The reported acute and subacute values are >4000–>13 000 mg ai/kg and >15 000–>27 000 mg ai/kg/day, respectively. To obtain a dose of 5000 mg ai/kg from a pyrethroid sprayed at a rate of 100 g ai/ha, a small bird of 20 g weight would have to consume all of the chemical in an area of approximately 10 square metres, whether it be deposited on crop, insect or soil. The same sized bird eating half its own weight in treated insects (such as aphids) during a day could consume no more than about 1 mg of pyrethroid. Bradbury & Coats (1982) found residues of approximately 0·5 mg/kg in dead and moribund aphids following fenvalerate applications to cotton, thus in practice amounts consumed are likely to be even lower. Pyrethroids are not used for seed dressing, thus birds are not exposed to the high-point source levels used in such applications. The pyrethroid insecticides therefore present no acute, subacute or dietary hazard to wild bird populations.

Reproductive studies with permethrin and cypermethrin at dietary concentrations of 25 and 50 mg ai/kg, respectively, greatly in excess of that to which wild birds could be continuously exposed, had no effect on egg production or weight (unpublished data, ICI Plant Protection Division). Nor was there any difference between treated and control groups regarding fertility, embryonic mortality, hatchability or abnormalities in offspring. Indeed, the studies, continued to 14 days after hatching, showed no effect on reproductive success. Bradbury & Coats (1982) suggested that the higher "tolerance" of birds to pyrethroids was due to efficient metabolism and excretion.

It has therefore been unnecessary to carry out field studies of the effects of pyrethroids on wild populations of mammals or birds. However, during riverine forest studies by Takken et al. (1978) and Smies et al. (1980), and as part of forest studies by Pillmore (1973) and Kingsbury and co-workers (see Table 4.30), the effects of pyrethrin and synthetic pyrethroids on birds and small mammals was investigated. The application rates, methods and areas treated are described in Tables 4.29 and 4.30.

Pillmore (1973) studied the bird population in a 16 ha area in the centre of 64 ha of pyrethrum-treated forest. Bird "trend" counts, reflecting day-to-day activity (but which only show reasonably conspicuous effects) were carried out before, on the day of and on the day after, spraying. In addition visits were made to the same area one year later. From observations of about 30 species it was concluded that there were no changes in species composition or relative abundance. Some birds marked in the year of spraying were observed the following year. A breeding census of 6 species showed no effects due to the treatment in the study period of up to 2 weeks after application.

In the riverine forest studies with permethrin, cypermethrin and deltamethrin, Smies et al. (1980) carried out regular morning counts of birds seen along 1 km stretches of the river bank. At least 20 species were recorded, including insectivores, herbivores and raptors. No change in abundance of species was noted. The observers searched for affected or dead birds and mammals, but found none. In a similar study by Takken et al. (1978), following deltamethrin applications, up to 35 species of birds were seen in regular transect counts carried out until 2 weeks after application. Once again no effects were reported. Kingsbury & McLeod (1979) also studied the bird populations along transects, from 5 days before to 5 days after a $2 \times 17\cdot5$ g ai/ha application of permethrin. Searches for sick or dead birds were made on the day of treatment. Observations of over 40 species, and territory mapping using singing and sighted birds showed no increase in mortality, no sign of "pesticide stress" and no effect on nesting territories.

In three coniferous forest trials in Canada (Kingsbury & McLeod, 1979; Kingsbury & Kreutzweiser, 1980b; Kreutzweiser, 1982) traps were used to study small mammal populations before and after applications of permethrin (400–900 ha plots; 17·5 g ai/ha and 2 × 17·5 g ai/ha with a 6 day interval). Specimens were identified, aged and sexed, and adult females were dissected to determine breeding condition. In one study (Kingsbury & McLeod, 1979, two applications) 41 animals from 8 species were caught in the treated plot and 87 animals from 5 species in the control during a period of several days approximately 3 weeks after the final application. In both "blocks" 60% of the females were in breeding condition. Total "young of the year" constituted 22% of the mammals from the treated area and 60% from the control area, but within the shrew population (*Sorex araneus* and *Blarina brevicauda*) 60% were "young of the year" in both areas. In the two other studies, although numbers trapped were fairly low, fecundity was high in treated and control groups. The three studies led the authors to conclude that small mammal populations were unaffected by permethrin at rates recommended for spruce budworm control.

From extensive laboratory toxicological studies and from a few large scale field studies it can be concluded that pyrethrins and the synthetic pyrethroids currently marketed present no hazard to wild populations of birds or mammals.

Effects on terrestrial arthropods

Pyrethrum and the synthetic pyrethroids are broad-spectrum insecticides active to varying degrees on an extensive range of arthropod species and used widely against insect pests in agriculture and horticulture and also in veterinary and public health outlets (Elliott et al., 1978; Lhoste & L'Hotellier, 1982; Chapter 6; and Table 4.2). The action of these compounds is usually seen as an initial and rapid knockdown, followed by recovery or death. At sub-lethal concentrations with some organisms an anti-feeding response or repellancy may occur (Hall, 1979; Moore, 1980; Penman et al., 1981; Iftner, 1982).

It is inevitable that any broad-spectrum insecticide will also have an adverse effect on a wide range of non-target insect species in laboratory tests. However, under field conditions a number of factors relating to the crop and climate, and to the distribution and behaviour of the arthropod populations determine the relative amounts of chemical that come into contact with target and non-target organisms. The importance of the majority of non-target organisms to the integrity of the ecosystem is poorly understood and thus the significance of any pesticide-induced changes to the ecological balance is often unclear. However, some arthropods such as bees are important pollinators of wild and cultivated plants. Other arthropods are known to be natural enemies of the pests of agricultural and horticultural crops (Bartlett, 1964; Neuenschwander et al., 1975; Cameron et al., 1980). For example, some members of the Coccinellidae

Table 4.2. Representative range of target pests of the synthetic pyrethroid, cypermethrin

Pest organism[a]		Target crop, etc.[b]
Scientific name	Common name	
Lepidoptera		
(E,L,A)	*Agrotis ipsilon*	Black cutworm	Corn
	Choristoneura fumiferana	Spruce budworm	Conifers
	Heliothis zea	Cotton bollworm	Cotton
	Phalonia hospes	Sunflower moth	Sunflowers
	Pieris brassicae	Cabbage white	Cabbage
	Plutella xylostella	Diamondback moth	Cabbage
	Spodoptera frugiperda	Fall armyworm	Cotton
	Trichoplusia ni	Cabbage looper	Lettuce
	Tineola bisselliella	Clothes moth	Carpets
Coleoptera		
(L,A)	*Anthomonus grandis*	Boll weevil	Cotton
	Epilachna varivestis	Mexican bean beetle	Beans
	Hypera brunneipennis	Egyptian alfalfa weevil	Alfalfa
	Leptinotarsa decemlineata	Colorado potato beetle	Potatoes
	Attagenus spp.	Carpet beetle	Carpets
Hemiptera/		
Heteroptera	*Dysdercus fasciatus*	Cotton stainer	Cotton
(N,A)	*Lygus lineolaris*	Tarnished plant bug	Apples
	Nezara viridula	Green stinkbug	Cotton
	Nomius pygmaeus	Stinkbug	Peaches
Hemiptera/		
Homoptera	*Aphis fabae*	Black bean aphid	Beans
(N,A)	*Macrosiphum euphorbiae*	Potato aphid	Potatoes
	Macrosiphoniella sanborni	Chrysanthemum aphid	Chrysanthemum
	Trialeurodes abutilonea	Banded-wing whitefly	Cotton
Diptera		
(L,A)	*Drosophila melanogaster*	Fruit fly	Apples
	Hylemya antiqua	Onion maggot/fly	Onions
	Musca spp.	Flies	⎱ Public health
	Simulium spp.	Blackflies	⎰ and veterinary
	Tabanus spp.	Blood-sucking flies	
Thysanoptera		
(N,A)	*Frankliniella* spp.	Thrips	Cotton

[a] Insecticide often targeted at specific stages in life cycle, but shown to be active against stages in life cycle shown in parentheses: E = egg; L = larvae; N = nymph; A = adult.
[b] Individual pests are often active on a range of crops.

(ladybirds) are predatory both in their larval and adult stages, feeding on aphids, scale insects, mealybugs, thrips, mites, etc. Populations of insects are also important food sources for some species of bird and their young.

Consequently, studies have been carried out in the laboratory and the field of the effects of pyrethroids on a number of pollinator insects and a range of predator arthropods of agronomically important pests. Soil microarthropod populations exposed to pyrethroids have also been studied under field conditions. Table 4.3 lists the scientific and common names of the terrestrial arthropods discussed below.

Bees

The group of insects collectively known as bees, members of the order Hymenoptera, includes both social and solitary members. Bees are vegetarians feeding on the nectar and pollen of flowers and in the process often pollinate the plants they visit. Indeed, some plants such as red clover depend on bees exclusively to carry pollen, as the blossoms are so constructed that smaller and weaker insects cannot get to the nectar. During feeding, bees may be exposed to residues of pesticides, whether on the target crop or from spray drifted on to other plants in the vicinity. Non-foraging members of a colony of social bees and the young of social or solitary bees may also be subjected to pesticides from residues present in the pollen or nectar collected by the foragers. Atkins et al. (1978) reported that significant honey bee colony losses in California resulted from pesticides affecting the brood.

Bees, and in particular honey bees, are important pollinators of a number of crops, many of these crops also being susceptible to pests that can be controlled with pyrethroid insecticides. For example, synthetic pyrethroids are active against fruit tree pests such as the winter moth (*Operophtera brumata*) and codling moth (*Cydia pomonella*) and against the seed weevil (*Ceutorhyncus assimilis*) and pollen beetle (*Meligethes aeneus*) that damage oil seed rape (*Brassica napus*). The synthetic pyrethroids being "broad-spectrum" insecticides therefore have the potential to put bees at risk, especially when applications coincide with flowering.

Honey bees (*Apis mellifera*), social organisms in colonies maintained by man, are not only the principal bee of agronomic importance in many countries but are also a highly suitable test species for studying pesticide toxicity. Honey bees are readily available for laboratory toxicity tests, field observations can be effectively carried out and samples of bees, pollen and wax can be collected for analysis either during test programmes or on the occasion of a suspected poisoning incident.

Laboratory tests

Laboratory studies of acute oral and contact toxicity have shown the honey bee to be highly sensitive to synthetic pyrethroids under the test conditions (Tables 4.4 and 4.5).

In acute oral toxicity tests, honey bees are fed (individually or in groups) with sugar or honey into which has been mixed a pesticide. Direct contact toxicity techniques vary widely with, for example, topical applications in an organic solvent (usually acetone; to "wet" the surface of the bee) or water, direct

Table 4.3. Scientific and common names of terrestrial invertebrates referred to in text

---

ARTHROPODA, INSECTA

Hymenoptera (bees, wasps, ants)	Aculeata	
	Apis mellifera	Honey bee
	Megachile pacifica	Alfalfa leaf cutting bee
	Nomia melanderi	Alkali bee
	Parasitica	
	Apanteles ornigis	Braconid
	Campoletis sonorensis	⎫
	Venturia canescens	⎬ Ichneumonid
	Diaeretiella rapae	
	Tetrastichus julis	Eulophid
	Trichogramma pretiosum	
	Formicidae	Ants
Coleoptera (beetles)	Carabidae	Ground beetles
	Agonum dorsale	
	Amara spp.	
	Harpalus affinis	
	Loricera pilicornis	
	Nebria brevicollis	
	Pterostichus spp.	
	Chrysomelidae	
	Diabrotica longicornis	Wheat bulb fly
	Oulema melanopus	Leaf beetle
	Coccinellidae	Ladybirds, ladybeetles
	Adonia variegata	
	Coccinella spp.	
	Stethorus punctum	
	Staphylinidae	Rove beetles
	Lathrobium fulvipenne	
	Tachyporus hypnorum	
	Others	
	Lasioderma serricorne	
	Tribolium castaneum	
Hemiptera	Heteroptera	
	Cyrtorhinus lividipennis	Capsid bug
	Deraeocoris punctatus	
	Geocoris spp.	Big-eyed bugs
	Hyaloides vitripennis	
	Jalysus spinosus	
	Microvelia atrolineata	Water cricket
	Nabis spp.	Damsel bugs
	Orius spp.	Pirate bugs
	Xylocoris flavipes	
	Homoptera	
	Acyrthosiphon spp.	⎫
	Aphis spp.	⎬ Aphids
	Nilaparvata lugens	Brown plant hopper
Neuroptera (lacewings)	*Chrysopa* spp.	Green lacewings
	Austromicromus sp.	Brown lacewing

Table 4.3 (cont.)

Lepidoptera (butterflies, moths)	*Ephestia kuhniella* *Heliothis virescens* *Phyllonorycter* sp. *Plodia interpunctella*	Flour moth Tobacco budworm Leafminer
Diptera	Syrphidae *Aphidolites* sp.	Hoverflies Midge
Dermaptera (earwigs)		
Thysanoptera (thrips)		

ARTHROPODA, ARACHNIDA

Araneae (spiders)	*Araneus* spp. *Erigone atra* *Lycosa pseudoannulata* *Tetragnatha* spp.	

Acari	Phytoseiidae
	  Amblyseius spp.
	  Phytoseiulus spp.
	  Typhlodromus spp.
		Phytoseiid mites

	Other
	  Tetranychus spp.
	  Pronematus spp.
		Red spider mite

spraying in water droplets and vacuum dusting (Atkins et al., 1954). A classification of pesticide toxicity using 24 hour LD50 values from such laboratory tests, proposed by the International Commission for Bee Botany (1980, 1982) is as follows:

> 100 µg ai/bee	virtually non-toxic
10–100 µg ai/bee	slightly toxic
1–10 µg ai/bee	moderately toxic
0·1–1·0 µg ai/bee	highly toxic
<0·1 µg ai/bee	extremely toxic

Based on laboratory acute toxicity data, the synthetic pyrethroids are considered to be highly or extremely toxic to honey bees (Table 4.4). The majority of the reported data shows LD50 values for the photostable synthetic pyrethroids permethrin, cypermethrin, fenvalerate and deltamethrin in the range 0·02–0·3 µg/bee. The LD50 values for those photolabile synthetic pyrethroids tested were from 0·02–10 µg/bee, reflecting their wider variation in insecticidal potency. Studies with cypermethrin showed that neither temperature in the range 12–32 °C nor bee age between 2–6 and 12–18 day-old groups influenced direct contact toxicity by more than a factor of 3-fold (unpublished data, ICI SOPRA).

I. R. Hill

Table 4.4. Laboratory acute toxicity of pyrethroids to *Apis mellifera* (honeybee)[a]

| | | | 24h LD 50 (μg ai/bee) | |
| | | | Direct contact | |
Pyrethroid	Test material[b]	Oral	Topical, acetone	Vacuum dusting
Permethrin	Technical	0·19–0·28	0·05–0·10	0·16
	E.C.	1·7–2·4	0·017[c]	0·16
	Cis-technical			0·25
Cypermethrin	Technical	0·18–0·26	0·013–0·056	0·55
	E.C.	0·13–>3	0·043–0·18	
Fenvalerate	Technical	0·29	0·077–0·34	0·41
	E.C.	~3		
Deltamethrin	Technical	0·079	0·02–0·051	0·067
	E.C.	>0·4		
Cyhalothrin	Technical		0·027	
Flucythrinate	Technical		0·27	
Allethrin	Technical	4·6–9·1	1·6–3·4	>0·31
Bioallethrin	Technical		0·1	
Resmethrin	Technical	0·069	0·015	
Bioresmethrin	Technical	0·055	0·006–0·05	
Pyrethrum	Technical	0·15	0·13–0·29	>0·63

[a] Data from published (Ref. nos. 5, 13, 15, 52, 94, 129, 153, 239, 252, 253) and unpublished (ICI Plant Protection Division, ICI Americas Inc., ICI SOPRA and ICI Holland) sources.
[b] E.C. = emulsifiable concentrate formulation.
[c] Topical applications in water and spray application in water droplets of a size similar to those in conventional agricultural sprays both gave a value of 0·022.

In an attempt to represent the field exposure of honey bees to plant surface residues in the laboratory, a number of authors have treated filter paper or plant surfaces at known rates. The plants were mostly sprayed under field conditions, harvested, and sometimes cut up into small pieces and placed in small cages such that the bees had to forage amongst the plant material to reach a food supply (see, for example, the method of Lagier et al., 1974). Table 4.5 summarises the available data. Although such experiments are unlikely to provide a measure of field toxicity, the results do show that even under the severe conditions of the test the toxicity of pyrethroid residues declines rapidly with time. Except at rates considerably higher than those recommended for field uses, "aging" the pyrethroid residues for a few hours to a few days on plant surfaces resulted in considerably reduced honey bee mortality. Gerig (1979, 1981) also reported an initial "knockdown" of the bees, from which many had recovered within 12 hours.

Table 4.5. Laboratory surface contact toxicity of E.C. formulations of pyrethroids to *Apis mellifera* (honeybee)[a]

Pyrethroid	Filter paper			Plant surfaces			
	Application rate (g ai/ha)	Age of residues (h)	Mortality after 24 h (%)	Plant	Application rate (g ai/ha)[b]	Age of residues (h)	Mortality after 24 h (%)
Permethrin	"field"	"dried"	0–2	*Phacelia*	(0·005%)	1, 14	1, 6[c]
	125, 250	"fresh"	100	*Solidago*	(0·005%)	1, 14	10, 6[c]
				Alfalfa	35–56	2–24	100–16
				Alfalfa	70–140	2–24	100–61
				Alfalfa	224	2–24	100
				Alfalfa	224	6–56	47·4
Cypermethrin	200, 400	"fresh"	100	*Phacelia*	(0·015%)	1, 14	6, 13[c]
				Solidago	(0·015%)	1, 14	82·2, 46[c]
Fenvalerate	100, 200	"fresh"	80	*Phacelia*	(0·015%)	1, 14	6, 7[c]
				Solidago	(0·015%)	1, 14	30, 2[c]
				Alfalfa	112	3, 8	57, 17
				Alfalfa	448	3, 8	100, 97
Deltamethrin	25, 50	"fresh"	"high"	*Phacelia*	(0·002%)	1, 14	1, 2[c]
				Solidago	(0·002%)	1, 14	2, 1[c]
				Alfalfa	11	6–56	44·0
				Alfalfa	17	6–56	100·3
				Apple	(0·0012%)	4	63
AC222,705				*Solidago*	(0·003%)	1	0·7[c]
Bioresmethrin	100, 200	"fresh"	100				
		3	0				

[a] Data from published (Ref. nos. 11, 15, 83, 84, 94, 124) and unpublished (ICI Americas Inc.) sources.
[b] Values in parentheses are percentage active ingredient in spray; if pyrethroid applied at 300 l/ha (typical ground application rate), 0·001% = 3 g ai/ha.
[c] Bees exposed to residues for 2 h only.

Table 4.6. Effect of permethrin (E.C.) residues on alfalfa to alfalfa leaf cutting bees, alkali bees and honeybees

| Application rate (g ai/ha)[b] | Bee mortality (%) after 24 hours exposure to permethrin residues on alfalfa aged for: | | | | | | | | |
| | 2–3 hours | | | 8 hours | | | 24 hours | | |
	ALB[a]	AB[a]	HB[a]	ALB	AB	HB	ALB	AB	HB
0	10	10	2	6	7	4			
35	79	63	86	24	25	44		Not done	
71	93	83	99	64	59	65			
140	100	90	100	88	78	93			
0	9	8	7	7	8	4	6	10	8
56	100	73	100	89	36	79	76	12	16
112	100	86	100	100	51	87	98	43	61
224	100	97	100	100	74	100	100	56	100

[a] Alfalfa leaf cutting bee (ALB), *Megachile pacifica*; alkali bee (AB), *Nomia melanderi*; honey bee (HB), *Apis mellifera*.
[b] 0–140 kg ai/ha, Johansen and Mayer (1976); 0–224 kg ai/ha, (unpublished data, ICI Americas Inc.)

   The toxicity of permethrin residues on alfalfa plants, to the alfalfa leaf cutting bee (*Megachile pacifica*) and the alkali bee (*Nomia melanderi*) in comparison to the honey bee has been reported by Johansen and co-workers (Johansen & Mayer, 1976; and unpublished data, ICI Americas Inc.). Using small cages filled with chopped plant tissue the results showed that permethrin was toxic to all three species of bee under the severe conditions used, but that toxicity decreased with increasing age of residues. The alkali bee was least affected by the permethrin residues, and the alfalfa leaf cutting bee exhibited similar or slightly greater sensitivity than the honey bee (Table 4.6). The toxicity of fenvalerate to megachilid bees pollinating lucerne has also been reported (Agrishell, 1982). Direct spraying at rates of 50 and 100 g ai/ha caused a rapid knockdown of approximately 60% of the bees (within 2 hours). Many of these bees recovered and mortality after 24 hours was 25%. Treated surface contact toxicity tests were carried out on paper "impregnated" with rates equivalent to 50 and 100 g ai/ha. Exposure of the megachilid bees to freshly treated paper resulted in deaths at 24 hours of less than 10% of the males but approximately 40% of the females, with numbers still increasing for the latter.

Field cage tests

Where pesticides are shown under laboratory conditions to be sufficiently toxic to honey bees to present a potential hazard under agricultural conditions, it is common practice to carry out field studies. Whilst the most relevant test is to expose honey bee colonies during large-scale field trials, the high cost of such studies has stimulated researchers to look for cheaper (and consequently smaller scale) techniques. Field cages, usually about 5–40 m^2 × 2 m high, are favoured by workers in a number of European countries (for example, France, Germany and Switzerland). A colony of bees is placed in the cage and exposed to treated plants (see, for example, Gerig, 1979; Debray, 1981). Greenhouse studies have also been done. Whilst supporters of these methods claim more control over exposure levels and conditions there is also considerable criticism on the basis of the restricted foraging habit imposed by the enclosures. Emulsifiable concentrate formulations have been used in all cage studies reported with pyrethroids.

In studies by Gerig (1979), permethrin and fenvalerate were sprayed on to *Phacelia* plants in cages at "field" rates (0·005% and 0·015% ai, respectively) during bee flight. The only effect reported was repellancy of the bees from the treated crop. With either compound, the honey bees stopped visiting flowers following spraying but foraging gradually returned to normal over a period of 2–3 days. In further studies (Gerig, 1981) with deltamethrin (0·001% ai) and cypermethrin (0·005% ai), a post-application decrease in foraging was again observed. After bee visits to the crop had returned to normal, water sprayed on to treated plants to "freshen-up" the residues caused some further repellancy by deltamethrin, although less strongly than initially observed. No such return of repellancy was noted when water was similarly sprayed on to permethrin residues (Gerig, 1979).

A number of workers have shown that whilst deaths may occur of bees exposed during spraying, the toxicity of pyrethroid residues on the crop declines rapidly with time. Atkins et al. (1977a) exposed caged bees during application of a "high" rate (224 g ai/ha) of permethrin to alfalfa and to the residues 2 hours after treatment. Bee mortality during the 24 hours following spraying was 100% where bees were present during spraying but less than 14% when they were introduced 2 hours later (controls had 7% mortality). Cage studies with fenvalerate (Cazenave et al., 1980; Debray, 1981; Agrishell, 1982), in which *Phacelia* or mustard was treated at 50 g ai/ha and honey bees exposed to the residues 2 or 20 hours after spraying, showed mortality similar to that in the control population. In studies with deltamethrin (unpublished data, Roussel-Uclaf) in which *Phacelia* and *Solidago* were sprayed in the presence of honey bees, and oil seed rape was treated prior to exposure of the bees, there was no greater mortality on treated plots than on controls. Repellancy was noted in the case of *Phacelia* and *Solidago*.

Kindt & Stark (1981) exposed honey bees to fenvalerate-treated oil seed rape in a greenhouse during the winter. No effects were observed when the plants were treated at up to three times normal field rate prior to introduction of the bees, but when application at normal field rate was made in the presence of foraging bees, flying activity soon decreased and a marked rise in mortality was observed. However, the authors urge caution in the interpretation of the data, in view of reported physiological differences between bees in the summer and winter (Maurizio, 1961; Zherebkina & Schagun, 1971), and the unnatural environment.

Megachilid bees (lucerne pollinating) in field cages were exposed to fenvalerate "on the day of treatment" of clover sprayed with 50 g ai/ha (male bees only) or 15 days after treatment (male and female bees) (Agrishell, 1982). The male bees (40 per cage), allowed to forage from the day of treatment, showed approximately 25% mortality within 1 day; with some knockdown and recovery prior to this. No further losses occurred over the 10 day observation period. Reduced foraging activity was noted during the first hour following introduction of the bees. When male and female bees were introduced 15 days after application, mortality during the following 5 days was less than 10%.

Field cage trials clearly demonstrate that the synthetic pyrethroid insecticides are unlikely to express the toxic potential to bees indicated by laboratory acute oral and contact tests. However, such tests may not adequately represent the conditions of pesticide exposure and bee behaviour that occur during agricultural use. Consequently it has been necessary to carry out further tests in large-scale field trials.

Field trials

In the majority of honey bee field studies, formulated pesticides (emulsifiable concentrates in the case of pyrethroids) are applied to plots of 2 ha or greater. However, some European protocols use replicated small-plot trials, sometimes with sequentially sown plots and sequential applications (see, for example, Agrishell, 1982). Bees are most at risk when applications of pesticides are made to flowering crops during the period of foraging, and with some non-pyrethroid products colony losses may be considerable. Large reductions in mortality have resulted from spraying many pesticides during the evening, night or early morning when few if any bees are active in the crop (Atkins et al., 1978). Whilst such practices may be agronomically desirable, studies should also include treatments made whilst bees are foraging, to provide "worst case" data. Consequently, trials with pyrethroids have often been carried out during periods of bee activity. Observations made with hives in or around the test area have included mortality at the hive and in the field, foraging activity, brood development and behavioural observations. In some studies pyrethroid residues in the bee, pollen or wax have been measured. Untreated fields and/or fields treated with toxic or non-toxic standards (best sited at least 1 km from the pyrethroid treatment) are necessary in order that data may be adequately interpreted.

The considerable number of studies carried out with permethrin, cypermethrin, fenvalerate and deltamethrin are summarised in Table 4.7. The overall data demonstrate that pyrethroids may be safely applied to flowering crops being foraged by bees.

In a number of trials with permethrin in orchard crops (apples and pears) no effects at all were seen on bees entering the crop as little as 1 hour after application (Wilkinson & Bull, 1984). Following applications of permethrin to forest plantations, Kingsbury & McLeod (1979) reported only minor increases in bee mortality.

Trials with permethrin, cypermethrin, fenvalerate and deltamethrin applied during flowering to winter and spring oil seed rape, alfalfa, mustard and sweet corn in different countries under a wide range of climatic conditions have produced remarkably consistent results. Observations of hive or foraging mortality showed no or very little effect from any of the pyrethroids tested (Table 4.7). Rapid "knockdown" of bees followed by recovery, reported in field cage studies, was also observed in large-scale field studies by Bocquet et al. (1980) when deltamethrin spraying was carried out whilst the bees were visiting mustard flowers and by Gerig (1981) following permethrin applications to spring rape. In a considerable number of field studies a change in foraging behaviour was noticed following spraying. Bocquet et al. (1980) and Shires & Debray (1982) reported that bees flew away rapidly immediately after spraying at the highest rate, whilst other investigators (Atkins et al., 1977a; Debray, 1981; Pike et al., 1982) noticed that bee visits to the treated crop were unusually short and that the organisms alighted on foliage only briefly. Reductions in numbers of bees active in the crop ranged from 30–100% and persisted from a few hours to a few days (Table 4.7). This repellancy effect of pyrethroids was paralleled by a fall in the amount of pollen collected by the bees (Gerig, 1981; Agrishell, 1982). However, following a return to normal foraging, which occurred within a few days, pollen collection was similar to that on control plots. Furthermore, no toxicity was detected once foraging re-commenced, consistent with aging of residues in laboratory tests.

Mixtures of pesticides are often applied to crops. Extensive studies in California (Atkins et al., 1978) have shown that such combinations are less hazardous to bee colonies than are the same pesticides when applied separately. The beneficial effect of mixtures resulted from reducing the frequency of pesticide applications whilst not increasing the bee kill above that occurring from an application of any one of the pesticides in the mixture. Atkins et al. (1977a) investigated the possibility that the repellency property of pyrethroids might be used to reduce the field bee toxicity of other pesticides. They sprayed a mixture of chlorpyrifos and methamidophos with and without permethrin (as a 24% E.C., 224 g ai/ha) and permethrin alone to 6 ha plots of alfalfa, at 00·30 hours and 05·30 hours. Bee colonies were placed in the spray path together with caged bees. It was established that the presence of permethrin on all occasions reduced the extent to which the honey bees foraged. However, the early morning applications containing permethrin suppressed crop visits for 5 days, whereas the night treatments reduced

Table 4.7. Field tests of toxicity of pyrethroids to *Apis mellifera* (honey bee)

Pyrethroid (as E.C.s)	Application rate (g ai/ha)	Crop (area, ha)	Time of application and stage[a]	Bee entry (time after spraying)[b]	Effects[c] Colony mortality	Foraging mortality	Foraging suppression	Ref. no.
Permethrin	50	Apples (3–5)	GP	10–18 days	–	–		
	50	Pears (2)	GP	14 days	–	–		
	50	Apples (2)	06·00 h F	4 h	–			
	17, 50[d]	Apples (4)	07·45 h F 20·45 h F	3 h <1 h	–			281
	70[e]	Spring rape (13)	10·00 h F	NC	–	–		85
	70[e]	Spring rape (19)	evening F	NC	–	–		13
	70[e]	Spring rape (16)	07·30 h F	2 h	–	–	–	15
	(0·005%)	Spring rape	11·00 h F	NC	–		80%(1 day)	
	224[e]	Alfalfa (6)	02·00 h F	NC	+		35%(2 days)	
	224[e]	Alfalfa (6)	00·55 h F	NC	+		60%(2 days)	204
		Alfalfa (6)	05·40 h F	NC			60%(5 days)	
	220 × 6[e]	Sweet corn (0-4)	07·30 h F	NC	–			
	220 × 4[e]	Sweet corn (0-4)	F	NC	–			138
	17·5 × 2(oil)[e]	Forest	Early a.m. F	NC	+		>90%(<4 days)	
Cypermethrin	50 × 2	Winter rape (1,2)	20·15 h F	NC	+		*f*	†
	25[e]	Winter rape (5) Spring rape (19)	Early p.m. F Early p.m. F	NC NC	+ +		90%(1 day) 90%(1 day)	236
Deltamethrin	11·2	Alfalfa (6)	02·00 h F	NC	–		35%(1½ days)	13
	16·8	Alfalfa (6)	02·00 h F	NC	–		65%(2 days)	
	5	Mustard (0·005)[g]	mid-day F	NC	–	–	100%(1-2 h)	
	7·5	Mustard (0·005)[g]	mid-day F	NC	–	–	100%(1-2 h)	24
	10	Mustard (0·005)[g]	mid-day F	NC	–	+	100%(2 h)	

Fenvalerate							
50	Mustard (0.002)[g]	10.00 h F	NC	—	—	10% (2 h)	} 36
50	Mustard	17.00 h F	NC	—	—	—	
50[e]	Winter rape (5)	16.00 h F	NC	+		50% (1 h)	} 2
50[e]	Spring rape (5)	13.15 h F	NC	—		30% (2 h)	
110	Alfalfa (12)	24.00 h F	NC	—		70% (1 day)	} 175
110	Alfalfa (12)	11.00 h F	NC	—		70% (1 day)	
400[e]	Alfalfa (1.6)	07.20 h F	NC	—			

[a] GP = green cluster/pink bud stage; F = crop in flower.

[b] NC = hives not closed during spraying.

[c] Colony/foraging mortality: — = no effect; + = slight increase in mortality. Foraging suppression; percentage reduction in foraging and (period of reduction).

[d] Same site sprayed on successive days.

[e] Aerial application.

[f] Foraging was reduced but may have been due to cold, windy conditions.

[g] Small plots, replicated in space and time, to provide different climatic conditions.

† Unpublished data (ICI SOPRA).

foraging for 2 days only. Slight increases in hive mortality were observed following all-night sprayings and also after early morning applications without permethrin. Where synthetic pyrethroid applications are necessary and their use is cleared by registration authorities it would seem sensible to apply them in combination with compatible pesticides also required in the spray programme. However, much more work is required before pyrethroids should even be considered for inclusion in other spray mixtures solely on the basis of their repellency properties.

Where brood behaviour was observed (Gerig, 1979; Agrishell, 1982; Moffett et al., 1982; Pike et al., 1982; Shires & Debray, 1982) or the longer term behaviour and development of the hive monitored (Kingsbury & McLeod, 1979; Agrishell, 1982), no adverse effects of the synthetic pyrethroids were reported.

Analysis of cypermethrin residues in pollen collected by honey bees (Shires & Debray, 1982) showed a decline from 0·9 μg/g shortly after spraying to 0·01 μg/g six days later, paralleling the declining residues in the flowers of the oil seed rape crop. On removing the frames from the hives at the end of the study there were no detectable residues in the wax and only very small amounts (0·002 μg/g) in the honey. Analytical determination of cypermethrin in dead bees collected at the hive following two applications of 50 g cypermethrin ai/ha to winter rape (unpublished data, ICI SOPRA) showed residues equivalent to one-tenth to one-hundredth of the laboratory value for acute toxicity.

Monitoring

In the United Kingdom, beekeepers suspecting bees of being poisoned send samples of the organisms to the National Bee Unit of ADAS (Agricultural Development Advisory Service; Ministry of Agriculture, Fisheries and Food). Bees are analysed and where poisoning is attributable to normal use of a pesticide the data submitted for review to the Environmental Committee of the Pesticides Safety Precautions Scheme (PSPS), the organisation responsible for registering pesticides in the United Kingdom. Results of the analyses of these bees are published in the Rothamsted Experimental Station Annual Report. No incident of poisoning attributable to pyrethroid insecticides has been reported to date (Annual Reports 1977–81).

Conclusions

The large number (over 30) of field studies on the effects of synthetic pyrethroids on honey bees, and the comparative toxicity studies with other species of bee, have demonstrated that these insecticides pose little or no hazard to bee populations following agricultural applications. Some workers commend the added precaution of spraying on to flowering crops whilst bee activity is minimal (during the evening, night or very early morning) or closing the hives during treatment. The absence of toxic effects under field conditions when laboratory acute toxicity tests classify these compounds as "highly" or "extremely" toxic is partly due to the low field application rates, partly to the repellancy effect, and also because the residues on plant surfaces show a rapid fall in toxic effect with time due to adsorption and/or absorption by the plant.

Predators and parasites

Whilst many of the predators and parasites of agronomically important pests are only present as natural populations, considerable efforts have been made over recent years to develop biological control procedures whereby the predator or parasite is introduced into the crop by man. As most crops are susceptible to a wide range of pests of which only some may be suitable for biological control, it is mostly necessary to institute a programme of pest management using both biological and chemical techniques. Any pesticide used in situations where biological control is important should preferably have either no effect on the beneficial predators or parasites or a selective effect, being more harmful to the pest. Pyrethroid insecticides have therefore been studied under both laboratory and field conditions for their effects upon agronomically important predators and parasites. Croft & Whalon (1982) have previously reviewed much of the available data.

Phytoseiid mite predators

Phytoseiid mites (e.g. *Amblyseius*, *Phytoseiulus* and *Typhlodromus* [*Meta-seiulus*] spp.) have over recent years been extensively used as predators of spider mites (pests) in horticultural crops, whilst other pests in the same crops are chemically controlled. The pyrethroid insecticides are highly active against many of the insect pests present in these crops. Comprehensive investigations under laboratory conditions have shown the pyrethroid insecticides to be between one and three orders of magnitude more toxic to phytoseiid predator mites than to the pest spider mites (Table 4.8). Field studies have confirmed these observations (Table 4.9), although at the lower rates of application (often approximating field rates) the differential effect on prey and predator was less marked and even absent in some trials. Where predator numbers were reduced, recovery of the populations was usually slow. This type of effect is not specific to pyrethroids, but is common to a number of insecticides; and acaricides are therefore often included for mite control in orchard spray programmes.

Increased fecundity of the pest spider mites at sublethal concentrations of pyrethroids has been suggested by Hoyt et al. (1978) although other workers

Table 4.8. Comparative toxicity of pyrethroid insecticides to mite pests and their predators under laboratory conditions[a]

Mite pest[c]	Phytoseiid mite (predator)[c]	"Ratio" of toxicity (LC50 or LD50) of pest : predator[b]						Ref. no.
		Permethrin	Cypermethrin	Fenvalerate	Fluvalinate	SD77706	NCI85913	
Tetranychus urticae	*Typhlodromus occidentalis*	2						223
T. urticae	*T. occidentalis*		3	2/3				285
Tetranychus pacificus	*T. occidentalis*	2						112 }
Pronematus anconai	*T. occidentalis*	1						
Tetranychus urticae	*Amblyseius fallacis*	2	3	3				285
T. urticae ("OP resistant")	*A. fallacis* ("OP resistant")			2				222
T. urticae ("OP resistant")	*A. fallacis* ("P + OP susceptible")				2	≥2	≥3	43 }
T. urticae ("OP resistant")	*A. fallacis* ("P + OP resistant")				2	≥−1	≥3	
T. urticae ("OP susceptible")	*A. fallacis* (P + OP susceptible")				1	3	3	
T. urticae ("OP susceptible")	*A. fallacis* ("P + OP resistant")				1	1	3	
T. urticae	*Typhlodromus pyri*		3					285

[a] Organisms attached to a glass slide or a leaf and dipped into a solution of the pesticide.

[b] "Ratio" of pest : predator toxicity: 1:1–1:10 = −1; 2:1–10:1 = 1; 11:1–100:1 = 2; 101:1–1000:1 = 3. Values ≥1 indicate insecticide more toxic to predator than pest.

[c] OP = organophosphate; P = permethrin.

Table 4.9. Effects of pyrethroid insecticides on mite pests and their predators on apple trees under field conditions[a]

Pyrethroid	Application rate (g ai/ha)[b]	Effect on mite pest[c,e]			Effect on predators[d,e]			
		T.u.	P.u.	A.f.	Ty. spp.	Ph.m	H.v.	S.p.
Permethrin	14–56	+	0/+	0				0
	70–140	+	0/+	−	−	+	0/−	0/−
	~450		+	−	−		−	−
Fenvalerate	14–56	(0)	0/+	0/−			(0)	0/−
	224		+	−				−

[a] Data from published (Ref. nos. 114, 282) and unpublished (ICI Americas Inc.) sources. Additional data from studies giving application rates only as percentage active ingredient in spray can be found in Ref. nos. 6, 112, 113 and 220.
[b] Applied on up to 12 occasions, at 1–2 week intervals.
[c] T.u. = *Tetranychus urticae*; P.u. = *Panonychus ulmi*.
[d] A.f. = *Amblyseius fallacis*; Ty. spp. = *Typhlodromus* spp.; Ph.m. = *Phytoseius macropilus*; H.v. = *Hyaliodes vitripennis*; S.p. = *Stethorus punctum*.
[e] 0 = no effect; + = increase in numbers; − = decrease in numbers; (0) = low numbers or variable results.

have reported a suppression of egg laying (Riedl & Hoying, 1980; Penman et al., 1981).

The predatory mites have long been known to develop resistance to organochlorine and organophosphate insecticides (Smith et al., 1963; Moto-yama et al., 1970; Croft & Meyer, 1973; Croft, 1976; Schulten & van de Klashorst, 1977) and to pyrethrins (Schulten et al., 1976). Recent studies with the synthetic pyrethroid insecticides have shown increased levels of phytoseiid mite resistance following selection under laboratory conditions. Strickler & Croft (1982) described strains of *Amblyseius fallacis* in which the level of permethrin resistance had been elevated by up to 500-fold, and which was stable over 25 generations (Croft & Whalon, 1983). Croft et al. (1982) noted varying degrees of cross-resistance to pyrethrins and several synthetic pyrethroids in a permethrin-selected strain of *A. fallacis*. Under field conditions two "permethrin-resistant" strains survived permethrin applications of 48 and 190 g ai/ha (sequential) equally well, but one of the strains was reduced in numbers by a subsequent application of 145 g ai/ha fenvalerate (Whalon et al., 1982). However, unless pyrethroid applications are maintained throughout the growing season, immigration of susceptible biotypes can reduce resistance by hybridisation (Croft & Whalon, 1983). Hoy et al. (1982) have described methods for the large-scale production of organophosphate- and pyrethroid-resistant phytoseiid mites.

Residues of pyrethroids dried on leaf surfaces have been shown to cause repellancy of the phytophagous spider mites (Hall, 1979; Hoy et al., 1979; Riedl & Hoying, 1980; Penman et al., 1981; Iftner, 1982) and their predators (Penman et al., 1981). Under field conditions the ability of organisms to move to residue-free areas will depend upon the efficiency of spray coverage. Whilst most studies have shown pyrethroid residues dried on surfaces to remain toxic to insects for several days (Plapp & Vinson, 1977; Plapp & Bull, 1978; Hoy et al., 1979), investigations using the phytoseiid mite predators *Amblyseius longispinus* and *Phytoseiulus persimilis* showed pyrethrins to have almost no residual activity (Shinkaji, 1976; Shinkaji & Adachi, 1978), probably as a result of photodegradation. Under field conditions the age and concentration of residues as well as the type of surface and the sensitivity of the organism may all influence pest and predator sensitivity.

A hemipteran mirid (*Hyaliodes vitripennis*) and the coleopteran "ladybird" *Stethorus punctum*, both predators of spider mites, have been shown to be less susceptible to pyrethroids than are the phytoseiid mites (Table 4.9). Hull & Starner (1983) have suggested that combinations of organophosphate insecticides and low rates of pyrethroids may be most effective in maintaining phytophagous mite pest control whilst allowing the predator *S. punctum* to survive and reproduce.

Other predators and parasites

A number of predators and/or parasites of pests other than mites have been studied under laboratory conditions. Where comparative data for pest and predator or parasite are available the results are shown in Table 4.10. The relative toxicity values vary with the different association. For example, the hymenopterous parasite *Venturia canescens* was more susceptible to a number of pyrethroids than its host *Ephestia kuhniella* (flour moth, Lepidoptera; Elliott et al., 1983). Conversely, a predaceous hemipteran (*Xylocoris flavipes*) exhibited a greater tolerance to pyrethrin and permethrin than did the three prey species of stored-product beetle tested (Press et al., 1978). Similarly, *Chrysopa carnea* (green lacewing, larval predator of *Heliothis virescens*, tobacco budworm), *Apanteles ornigis* (braconid parasite of *Phyllonorycter blencardella*, a leafminer pest) and *Austomicromus tasmaniae* and *Coccinella undecimpunctata* (brown lacewing and a coccinellid respectively, predators of two aphid pests) were between one and four orders of magnitude less sensitive than their respective pests (Table 4.10). The relative sensitivity of the members of the prey–predator or prey–parasite association may also vary with different pyrethroids as was seen with *Campoletis sonorensis*, an ichneumonid parasite of *Heliothis virescens* (Plapp & Vinson, 1977; Rajakulendran & Plapp, 1982a) and with a number of organisms studied in a cotton agroecosystem (Sukhoruchenko et al., 1982).

Sub-lethal effects other than repellancy have not been observed. Syrett & Penman (1980) showed that sub-lethal doses of fenvalerate did not affect the numbers of eggs laid by a lacewing predator and that there was no change in the hatching period or larval development time. Furthermore, the level of contact toxicity decreases with time. Sukhoruchenko et al. (1982) showed that whilst all test organisms died after exposure to leaves treated 1 day previously, only 20% mortality was observed 4 weeks later.

The tests described above were carried out under laboratory or similar conditions using mainly topical and surface contact on glass or paper. Such techniques do not necessarily represent field exposure (where organisms may be exposed to direct spray or to residues on plants, soil or food), nor need they accurately reflect the relative toxicities to organisms differing widely in size. Some authors have used different methods for pest and predator or parasite (Pree, 1979; Kalushkov, 1982). Test methods should either be comparable or reflect likely exposure routes. Where the differential toxicity between the pest and its predator or parasite is unequivocal and established by good methodology, observations of possible ecological or agronomic significance should be substantiated by field studies. However, deciding upon the significance of a laboratory result is often made difficult by a fundamental lack of understanding of the relationship between pests and potentially "beneficial" organisms under field conditions.

Some authors have not included pest species in their studies but have compared toxicity of pyrethroids to the different life stages of predators or parasites (Stenseth, 1979; Wilkinson et al., 1979; Warner & Croft, 1982; Bull & House, 1983). Others have investigated the relative effects of a range of insecticides, including pyrethrins and pyrethroids (Shinkaji & Adachi, 1978;

Table 4.10. Comparative toxicity of pyrethroid insecticides to pests and their predators or parasites under laboratory conditions[a]

Pest	Predator (Pr) or parasite (Pa)	"Ratio" of toxicity (LC50 or LD50) of pest : predator or parasite[b]									Ref. no.
		Pyrethrum	Bioallethrin	Permethrin	Cypermethrin	Fenvalerate	Deltamethrin	Fluvalinate	Cyhalothrin	AC222,705	
Ephestia kuhniella A	*Venturia canescens* Pa, A		2	2[c]		1	2		2		74,75
Tribolium castaneum A	*Xylocoris flavipes* Pr, A	−1		−1							211
Lasioderma serricorne A	*X. flavipes*	−1		−1							
Plodia interpunctella A	*X. flavipes*	−1		−1							
Heliothis viriscens L	*Campoletis sonorensis* Pa, A	3		2	1			1		−1	212
H. viriscens	*C. sonorensis*	1				−1	−1				208
H. viriscens	*Chrysopa carnea* Pr, L			−2		−1	−2				207
H. viriscens	*C. carnea*			<−3[d]	−1	−4[d]					49,237
H. viriscens	*C. carnea*				−1			<−2		<−1	212
Phyllonorycter blancardella A	*Apanteles ornigis* Pa, A			−2	−1	−1					96
Acyrthosiphon kondoi	*Austromicromus tasmaniae* Pr, A					<−2					
Acyrthosiphon pisum	*A. tasmaniae*					<−3					262
Acyrthosiphon kondoi	*Coccinella undecimpunctata* Pr					−1					
Acyrthosiphon pisum	*C. undecimpunctata*					−2					
Aphis gossypii	*Adonia variegata* Pr, L			−1	−1	1	−3				
A. gossypii	*Nabis palifer* Pr, L			−1	−1		−2				
A. gossypii	*Orius niger* Pr, L			−1	−1		1				
A. gossypii	*Deraeocoris punctatus* Pr, L			−1	−1		−1				258
A. gossypii	*Chrysopa carnea* Pr, L			−1	−1	−1	−1				
Acyrthosiphon gossypii	*Aphidolites aphidimyza* Pr, L			−1	−1	−3	−4				
Ac. gossypii	*Adonia variegata*			1	1	2	1				
Ac. gossypii	*Nabis palifer*			2	2		2				
Ac. gossypii	*Orius niger*			2	2		4				
Ac. gossypii	*Deraeocoris punctatus*			2	2		3				

Acyrthosiphon gossypii	Chrysopa carnea	2	1	2	3	
Ac. gossypii	Aphidolites aphidimyza	2	1	−1	−1	258
Heliothis armigera L	Adonia variegata	−2	−1	2	−2	
H. armigera	Nabis palifer	−1	−1		1	
H. armigera	Orius niger	−1	−1		2	
H. armigera	Deraeocoris punctatus	1	1		2	
H. armigera	Chrysopa carnea	−1	−1	−1	1	
H. armigera	Aphidolites aphidimyza	−1	−1	−2	−2	
Diabrotica longicornis L	Coccinellidae (8 spp.) Pr, A	−1/1	−1/2	−2/1		39
Oulema melanopus	Coccinellidae (8 spp.) Pr, A	−1/1	−1/2	−1/2		
O. melanopus	Testrastichus julis Pr, A	3	3	3		

[a] Topical application, surface residue exposure or sprayed.

[b] "Ratio" of pest : predator or parasite toxicity: <1:1000 = −4; 1:101–1:1000 = −3; 1:11–1:100 = −2; 1:1–1:10 = −1; 2:1–10:1 = 1; 11:1–100:1 = 2; 101:1–1000:1 = 3; >1000:1 = 4. Values ≥ 1 indicate insecticide more toxic to predator or parasite than to pest.

[c] (1R)-cis-permethrin tested.

[d] Includes pupation.

Waddill, 1978; Warner & Croft, 1982; Surulivelu & Menon, 1982; Rajakulendran & Plapp, 1982a) and the potential for synergistic or antagonistic effects of pesticide mixtures (Rajakulendran & Plapp, 1982b).

In association with the International Organisation for Biological Control (IOBC), a number of Research Groups are attempting to develop laboratory tests that realistically evaluate the toxicity of pesticides to predatory arthropods. Edwards & Wilkinson (1983), as part of this programme, have studied the effects of a number of products, including cypermethrin, on four species of predatory ground beetles (Carabidae; *Agonum dorsale*, *Pterostichus cupreus*, *P. melanarius* and *Nebria brevicollis*). The beetles were exposed to spray deposits on soil, soil in which barley was growing (with the organisms present at spraying), direct spraying (although only diurnal species would normally be at risk by this route) and dried spray deposit on glass. Cypermethrin sprayed at field rate (25 g ai/ha) on to a glass surface caused 100% mortality of all species. However, when applied directly to beetles, to soil or to soil/barley, no effect was observed except with *P. cupreus* and *A. dorsale* where some knockdown (and recovery) occurred. At 125 g ai/ha some knockdown was seen with all except *N. brevicollis*, but mortality remained low. Dunning et al. (1982) in a similar comparison of laboratory test methods reported 80% mortality of *P. melanarius* following exposure to soil treated with 12 g ai/ha deltamethrin. Treated food had no effect. The researchers in both studies suggested that soil tests best represented the field situation for carabids. However, the most appropriate laboratory test method for any organism will depend upon its behavioural patterns, which influence the route by which it may be exposed to pesticides.

A number of field studies of the effects of permethrin, fenvalerate and deltamethrin on predators and parasites have been reported. These are summarised in Table 4.11. Populations were sampled using a wide range of techniques including suction sampling, soil traps, cuvette traps and direct observation and counting. However, it is not always easy to draw conclusions from such data, for the numbers of predators may be influenced not only by a direct effect of the chemical but also indirectly by a reduction in the numbers of prey following the pesticide treatment. Furthermore, under field conditions the relative effect on prey and predator may depend upon crop variety. Reissig et al. (1982), using three rice varieties with different degrees of susceptibility to the brown planthopper pest (*Nilaparvata lugens*), found that whilst the use of deltamethrin at 30 g ai/ha resulted in a suppression of predators and a resurgence of the pest, only the latter was greatly influenced by the degree of plant resistance. Field results can also vary from trial to trial and in different years as a result of climatic and crop factors, population densities, etc. Finally, the reported studies differ widely in quality. Hagley et al. (1980), for example, collected beetles in pitfall traps assessing live and dead ("pesticide affected") at collection time. The deaths were assumed by the author to result from pesticide treatment. However the 3–4-day period between collections could also have resulted in natural deaths or death by cannibalism. Furthermore, whereas their pyrethroid treatment was to an apple orchard, the control was an area of unsprayed peach trees. For all the above reasons it is usually unwise to draw conclusions from any one study.

Effects were recorded on a number of non-target organisms in many of the studies (Table 4.11). However, on almost all occasions where observations were continued for several weeks after spraying, full or partial recovery of affected groups or species was also reported. Initial population declines, seen in most cases, were followed by recovery within three weeks. The consequences of the observations must be assessed on a crop-to-crop basis and with knowledge of the contribution of the organisms studied to the ecosystem. In general the significance of many predator and parasite populations is poorly understood. In cereal crops, for example, the role of particular groups of natural enemies in controlling pest aphids is currently in some dispute. There is a strong belief that aphid-specific predators and parasites are of little importance in preventing initial outbreaks of aphid infestation as they do not exert a controlling influence until the crop threshold for aphid damage has been exceeded. The polyphagous predators are presently thought of as being more important, as they are likely to be present in the crop when the aphids arrive and therefore available to feed on them. Groups commonly considered important, although still the subjects of much debate, are carabid and staphylinid beetles, araneaeid spiders and possibly the Dermaptera (earwigs).

Studies on the cereal ecosystem have been carried out over a number of years by ICI Plant Protection Division (unpublished data). In one study with winter wheat, permethrin was applied in mid-May 1979, and non-target arthropods assessed by pitfall trapping and suction sampling until harvest (Figure 4.1). Two hundred and forty taxa were recognised; but only abundant, consistent or "important" groups were considered in detail. Approximately 20% of the populations showed post-treatment changes, of which the majority were decreases. However, the populations recovered, some very rapidly, and numbers of most taxa were not significantly different from the controls six weeks after treatment. It was also concluded that the permethrin treatment would not seriously affect the supply to insect food for game birds such as the partridge, as populations had recovered by the time of the young hatching. Subsequent studies with cypermethrin have been carried out in spring and autumn cereal crops (unpublished data) but are not yet fully evaluated. Initial consideration of the data shows similar levels of effect to those described above; with some population reductions after spraying, but followed by rapid recovery. There is no evidence that subsequent spring populations on non-target arthropods are affected by autumn sprays of cypermethrin.

Some studies have suggested that controlling the time of spraying or the use of combinations of pesticides or pesticides and pheromones might in certain circumstances be desirable. Bull & House (1983) demonstrated that there was a high rate of emergence of the parasite *Trichogramma pretiosum* from the eggs

Table 4.11. Effects of pyrethroid insecticides on predators and parasites under field conditions

Pyrethroid and application rate (g ai/ha)	Crop	Organisms (predator = Pr, parasite = Pa)	Effect[a]	Ref. no.[b]
Permethrin				
3 × 52, 160, 210	Apples	*Pterostichus melanarius* Pr	(−)	
		Harpalus affinis Pr	(+)	95
		Amara spp. Pr	(+)	
50	Winter wheat	Carabidae Pr	(−)	
		Staphylinidae Pr	(−)	40,†
		Araneae Pr	(+)/R	
50, 100	Cotton	Coccinellidae Pr	(+)	
		Staphylinidae Pr	+	
		Orius sp. Pr	+	3
		Chrysopa spp. Pr	(+)	
		Aphelinidae Pa	−, (+)	
		Encyrtidae Pa	−	
Fenvalerate				
50	Cereals	Staphylinidae (on soil) Pr	−	
		Entomophagous arthropods on soil Pr	−	2
		Entomophagous arthropods on plants Pr	−	
		Phytophagous arthropods on plants	+/−	
1, 3, 5 × 110	Cotton	Carabidae Pr	(+)	
2 × 56, 110		Coccinellidae Pr	+	
		Geocoris spp. Pr	+	
		Orius insidiosus Pr	+	
		Nabis spp. Pr	+	221
		Jalysus spinosus Pr	(+)	
		Chrysopa spp. Pr	(−)	
		Hymenoptera (ants) Pr	(+)	
		Arachnida Pr	+	

150	Maize	Coleoptera	Carabidae Pr	–	
		Coleoptera	Coccinellidae Pr	+/R	
		Neuroptera	Chrysopidae (eggs) Pr	–	2
		Hymenoptera	Aphidiidae Pa		
		Diptera	Syrphidae (larvae Pr; pupae)	+/r	
Deltamethrin					
7·5	Winter rape	Coleoptera	*Loricera pilicornis* Pr	–	
		Coleoptera	*Tachyporus hypnorum* Pr	+/R	225
		Coleoptera	*Lathrobium fulvipenne* Pr	+/R	
		Arachnida	(*Erigone atra* dominant) Pr	+/–	
				10; 40	
3 × 10, 40	Rice	Hemiptera	*Microvelia atrolineata* Pr	–; +/R	
		Hemiptera	*Cyrtorhinus lividipennis* Pr	+/R; +/R	105
		Arachnida	*Lycosa pseudoannulata* Pr		
		Arachnida	*Tetragnatha* sp. Pr	–; +/R	
		Arachnida	*Araneus* spp. Pr		

[a] – = no effect; + = decrease in numbers; /R = recovery; /r = partial recovery; /– = no recovery; ( ) = low numbers, few samplings or variable results.

[b] Additional data from studies giving application rates only as percentage active ingredient in spray can be found in Ref. nos. 96, 238 and 258.

† Unpublished data (ICI Plant Protection Division).

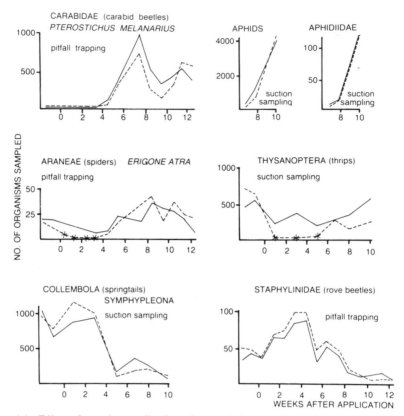

Figure 4.1. Effect of a spring application of permethrin on arthropods in a cereal ecosystem. Winter wheat plots (duplicate 4.5 ha) sprayed with 50 g ai/ha permethrin in mid-May (unpublished data, ICI Plant Protection Division). (---), permethrin-treated; (—), control; *, significantly different from control ($P = 5\%$)

of its host *Heliothis viriscens*, but only when permethrin was sprayed on to eggs after their exposure to the parasite. Spraying before parasitisation considerably reduced emergence. Legner & Medved (1981), studying the use of female sex pheromone in the control of *Pectinophera gossypiella*, pink bollworm of cotton, found that a pyrethroid (permethrin) plus pheromone regime was much more effective that either releases of hymenopteran parasites or use of the pheromone alone.

Conclusions

Where programmes of pest management include both biological and chemical control it is essential that any pesticides used either do not adversely affect the biological agent or affect it substantially less than the pest and allow its recovery. Similar constraints apply where natural populations of predators or parasites are known to be an important and irreplaceable component of pest control. Laboratory tests or small field plot studies carried out to identify potential "problem pesticides" should include both pest and predator or parasite, and the testing procedure should reflect the likely routes of exposure. Where the results suggest that a pesticide might be agronomically or ecologically unsuitable for specific uses this should be confirmed by field studies. To allow for temporal and spatial variations, field trials are best done over several seasons, but should at least be over a period of time sufficient to monitor the rate of recovery of affected populations.

Spider mite pests in orchards have been extensively studied and shown to be much less susceptible to pyrethroids than the pytoseiid mite predators that are often "inoculated" into the crop. The use of pyrethroid-resistant strains of mite predators, currently under investigation, may eventually enable pyrethroids to be used on a wide scale in programmes of integrated control, although synthetic pyrethoid "pressure" must be maintained to prevent excessive dilution of the resistant population.

In the majority of other systems investigated predators have either exhibited pyrethroid sensitivities similar to or less than the pest. In these cases and even where the reverse was also apparent, predator populations generally recovered rapidly from the effects of the pyrethroids. However, in the majority of agricultural and horticultural uses, natural predator or parasite control of pests is considered to be of little significance compared to the use of pesticides. Where biological control is practised and pyrethroids (or any other pesticide) are undesirably toxic to the predator or parasite, selectivity of the pesticide against the pest may still be developed by exploitation of any spatial or temporal asynchrony in the prey–predator/parasite relationship. Where releases of cultured predators or parasites are made, development of resistant strains may be of value.

Arthropod populations

Considerable numbers of the organisms in the environment have no noticeable or direct role in controlling pests of crops. Neither is their importance in the ecosystem understood. Few studies are carried out on these organisms because of the difficulty of interpreting any effects observed. However, the effect of permethrin on the soil microarthropod population has been investigated in 36 square metre plots of uncropped, cultivated soil (unpublished data, ICI Plant Protection Division). Rates of 500 and 5000 g ai/ha, 2·5 and 25 times maximum recommended field rates, respectively, were applied and 15 cm soil cores analysed 3 days before, and 22 and 91 days after, treatment. There were very few effects on the arthropod populations. At 500 g ai/ha no significant effects were observed. At the higher rate a significant reduction in the gamasid and one genus of astigmatid mites was recorded. A few significant increases in numbers were observed, notably amongst the Collembola at the high rate. Representative results of a similar study with cypermethrin at 100 and 1000 g ai/ha, sampled on 7 occasions over 3 years, are shown in Table 4.12. Again very few effects were observed.

On a wider and non-agricultural field scale, observations of the effect on terrestrial insects during forest applications of synthetic pyrethroids (for spruce budworm control) have been described by Kingsbury and colleagues (see Table 4.30) and in riverine forest spray programmes (for blackfly control) by Takken et al. (1978), Smies et al. (1980) and Everts et al. (1983). Terrestrial insect knockdown or "fallout" was observed but lasted no more than 48 hours in the Canadian forests and up to 5 days in the African riverine forests. A wide variety of groups of insects were represented. Where sampling was sufficiently detailed (see, for example, Kingsbury, 1982) post-treatment numbers of arboreal and/or flying insects indicated either rapid recolonisation or strong residual populations following application.

Effects on earthworms

Little data has been published from studies of the effects of pyrethroids on earthworms. Lofs-Holmin (1982) treated fields of winter wheat with a pesticide programme of benomyl, triadimefon, fenvalerate and chlorothalonil. The soil was sampled on only one occasion, two months after the final application, and *Lumbricus terrestris*, *Allolobophora caliginosa* and *A. rosea* identified, counted and weighed. Total numbers were unaffected by the pesticide treatments but there was a significant reduction in the weight of juveniles, which were the dominant part of the population.

Annual applications of permethrin and cypermethrin have been made to a number of soils, using 36 square metre plots with a grass sward (unpublished data, ICI Plant Protection Division). Rates of up to 5000 g ai permethrin/ha and 1000 g ai cypermethrin/ha were applied and earthworms identified, counted and weighed (see Edwards & Brown, 1982), following sampling by a formalin-expellent method similar to that described by Raw (1959). Species of *Lumbricus* and *Allolobophora* were present in all trials. There were no significant effects on total numbers at the lower application rates of permethrin or cypermethrin (representative of field rate), and only occasional significant reductions at the high rates (Figure 4.2). Total earthworm weight gave a very similar picture, mostly showing no effect from pyrethroid treatments. Numbers for any one species collected varied quite widely, but there was no consistent effect on the adults or juveniles of the worm species present.

Lhoste & L'Hotellier (1982) reported that there were negligible deaths of *Allolobophora chlorotica* in soils treated at field rate (12·5 g ai/ha), and that concentrations five times this were required to cause a large mortality after 28 days.

Table 4.12. Effect of cypermethrin on microarthropod populations in soil[a]

Organism/group			Predominant behaviour		No. of genera/species identified	No. of organisms (×100)/m² [c]							
			Habitat[b]	Feeding habit		Pre-application		Post-application					
								3 wk		30 wk		3 yr	
						T	C	T	C	T	C	T	C
Arachnida	Acari	Mesostigmata (Gamasina)	S	Predators	13/19	6	7	4	22	5[d]	18	10	20
		Cryptostigmata	D	Predators	3/3	37	24	27	25	18	28	25	36
		Astigmata	S	Detritus	7/10	35	52	28	37	63	72	148	106
			S	Rotting vegetation	2/–	144	190	75	129	53	124	111	115
	Aranea	Prostigmata	Various	Various	>3 families	32	33	17	13	7	11	39	30
			S	Predators	ND	2	1	0	1	1	1	1	1
Insecta	Collembola	Neanuridae	S	Fungi	2/2	18	14	9	14	3	10	12	2
		Onychiuridae	D	Fungi	2/–	78	57	67	57	59	97	64	73
		Isotomidae	S	Fungi	4/5	10	6	9	6	7	7	32	24
		Entomobryidae	D	Fungi	2/2	42	27	32	21	16	16	25	19
			S	Fungi	4/5	0	2	4	2	7	7	9	6
	Hemiptera		S	Various	ND	0	1	0	1	1	1	6	2
Myriapoda			Various	Various	ND	1	2	3	0	0	0	2	2

[a] Unpublished data (ICI Plant Protection Division).
[b] S = Soil surface or near surface; D = "deep" soil.
[c] T = 1000 g ai/ha; C = control (no treatment).
[d] Significantly different from control (P = 6%).
ND = not differentiated.

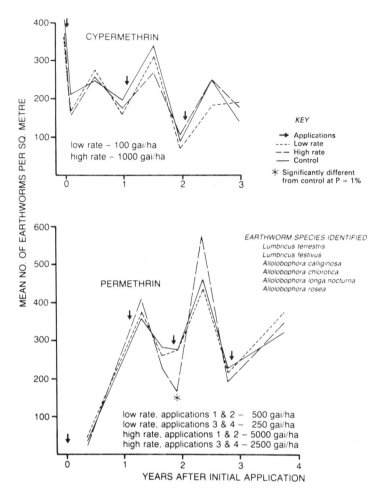

Figure 4.2. Total number of earthworms in a sandy-loam soil treated with annual applications of synthetic pyrethroids

The available data indicate that synthetic pyrethroids at recommended field rates will not adversely affect earthworm populations.

Effects on terrestrial molluscs (slugs and snails)

Effects of pyrethroids on terrestrial molluscs have been reported by Crowell (1977) as part of a series of investigations to identify effective molluscicides. More than 170 preparations were tested over a period of several years, including permethrin, fenvalerate and fenpropathrin. It is not clear as to how the molluscs were exposed to the pyrethroids although a method of preparing a wheat bran bait containing 2% product is described. Nor were the test species stated, but in the overall programme *Helix aspersa* (European brown snail, common snail), *Arion ater* (European black slug) and *Limax flavus* (tawny slug) were all used. Experimental animals were starved for at least 48 hours and exposed to bait and observed for a period of 10 days. All of the pyrethroids tested were classified as "poor" molluscicides, indicating no or negligible activity.

In a laboratory test with *Agriolimax agriolimax* (field slug) fed delta-methrin-treated lettuce leaves, mortality was only 8% at four times the normal field application rate (Lhoste & L'Hotellier, 1982).

Laboratory and field studies in aquatic environments have also shown aquatic molluscs to be unaffected by concentrations of pyrethroids much higher than will occur in the environment following normal use patterns (see p. 195ff.).

Effects on microorganisms and algae

Microorganisms play a major role in the maintenance of soil fertility (Lynch & Poole, 1979) and, in particular, in the functioning of the carbon and nitrogen cycles. Because pyrethroids come into contact with the soil organisms the effect of these chemicals on the activities of microbes has been determined.

In vivo studies have shown that pyrethrins and pyrethroids can affect microbial processes. However, the rates at which effects have been observed were often grossly in excess of those which occur under "in use" conditions. Furthermore, the experimental conditions did not always enable the extrapolation of possible environmental effects.

Draughon & Ayres (1978, 1980) studied the inhibition of fungal mycotoxins by allethrin and pyrethrins in an attempt to find a pesticide capable of controlling both insect pests and fungi in stored food products. In liquid culture pyrethroids at 100 µg ai/l caused a small reduction in mycelial biomass and a large (60–90%) inhibition of the formation of the toxins citrinin (from *Penicillium citrinum*) and patulin (from *Penicillium* and *Aspergillus* spp.). "In use" studies were not undertaken.

Laboratory tests were carried out by Tu (1982) on the effects of five synthetic pyrethroid insecticides on the plant pathogenic fungus *Rhizoctonia solani*. *In vitro* tests in an agar medium gave EC50 values of >60 000 µg ai/l for permethrin, cypermethrin, fenvalerate and deltamethrin. None of the pyrethroids tested gave complete control of the fungal disease on soybean (*Glycine max*) seedlings germinating in a fungus-inoculated soilless compost drenched with the insecticides at concentrations up to 500 mg ai/kg deltamethrin, 2000 mg ai/kg fenpropanate and 3000 mg ai/kg permethrin, cypermethrin and fenvalerate.

In a series of investigations, Stratton & Corke (1982a,b) have described the effects of permethrin and 10 of its degradation products on 10 fungi, 2 algae and 3 blue-green algae (cyanobacteria). The studies, carried out in liquid culture, showed permethrin to be less toxic to the organisms than were most of the degradation products. Different mixtures of the parent and products caused synergistic, additive or antagonistic responses. However, the EC50 values for fungal growth, algal growth, photosynthesis and nitrogenase activity were several orders of magnitude higher that "in use" applications could generate. The EC50 values for permethrin and its metabolites were mostly >2000 µg ai/l, whereas the maximum expected concentrations in shallow (15 cm deep) bodies of water contaminated by spray drift or run-off following recommended application rates would be ≤ 20 µg ai/l permethrin or total degradation products. Walsh & Alexander (1980) incubated the marine diatom *Skeletonema costatum* in laboratory culture with technical permethrin and determined the 96 h EC50 value for growth as 70 µg/l.

In the one published study of the effects of pyrethroids on soil organisms *in vivo*, Tu (1980) treated a sandy loam with 0·5 and 5 mg ai/kg permethrin, cypermethrin, fenvalerate, deltamethrin and fenpropathrin. Numbers of viable bacteria and fungi were determined, and nitrification, non-symbiotic nitrogen fixation, dehydrogenase activity and urease activity measured. Numbers of organisms and nitrogen fixation were depressed following treatment but recovered rapidly, within the 4 week period of study (Table 4.13). On occasions both numbers and activity were greater in treated soils than in the control. The author considered the effects "short lived and minor in nature".

Studies of the effects of permethrin and cypermethrin on microbial populations in soil have also been carried out by ICI Plant Protection Division (unpublished data), and are shown in Table 4.13. The only effects seen in a wide range of tests were transient and probably not treatment-related, even at excessively high rates (up to 10 000 g ai/ha) or with repeated applications of 50 g ai/ha.

Nitrogen fixation by the symbiotic *Rhizobium* spp. associated with legumes is of importance to many agricultural crops and to the fertility of soil. Studies of the effects of permethrin and cypermethrin in soil at approximately field rate and 10 × field rate have been carried out using soybeans under greenhouse conditions (unpublished data, ICI Plant Protection Division). The appearance of the plants, their aerial height, fresh or dry weights (foliage and roots separately) and numbers or weight of pods at intervals up to harvest were all unaffected by the pyrethroids. Neither was there any effect upon the numbers of nodules, nodule colour or nitrogen-fixing ability (acetylene reduction test).

There is still considerable debate over how to measure and interpret the effects of pesticides on soil microorganisms or their activities. However, the results of the test described above suggest that the synthetic pyrethroids even at

Table 4.13. The effects of pyrethroids applied to soil on microorganisms and their activities

Test	Permethrin	Cypermethrin	Fenvalerate	Deltamethrin	Fenpropathrin	Reference and experimental conditions
	Effects[a]					
Number of viable bacteria	+	+	+	+	+	
Number of viable fungi	+	+	+	+	+	
Carbon turnover (respiration)	+	+	+	+	+	Tu (1980); 0·5 and 5·0 μg
Nitrification; soil organic matter	(+)	(+)	−	−	(+)	ai/g sandy loam soil; study
Nitrogen fixation; non-symbiotic	+	+	+	+	+	period: 4 weeks
Dehydrogenase activity	−	+	+	+	+	
Urease activity	(+)	(+)	(+)	(+)	(+)	
Number of viable bacteria	−					
Number of viable fungi	−					
Number of viable actinomycetes	−					
Total microbial propagules	−	−				
Total algal cells	−					Unpublished data (ICI Plant
Microbial biomass (ATP)	−	−				Protection Division)[g]
Carbon turnover (respiration)						
soil organic matter	(+)	−[f]				Permethrin 500 and 10 000 g
[14C]glucose[b]	(+)	(+)				ai/ha; coarse sand and
plant material[b,c]	−	−				sandy loam
miscellaneous substrates[b,d]	−/(+)					
Nitrification						Cypermethrin 50 and 500 g
plant material[b,e]	(+)	−				ai/ha; loam and loamy sand
ammonium sulphate[b]	(+)	−				
Dehydrogenase activity (+ glucose)	(+)					
Phosphatase activity	(+)					

[a] − no effect; + transient effect; ( ) minor or variable effect, probably not treatment related.
[b] Soil amendments.
[c] Permethrin, 14C-labelled wheat straw; cypermethrin, 14C-labelled maize.
[d] 14C-labelled sucrose, urea, starch, phenol; unlabelled pectin, cellulose, vanillin, tripalmitin.
[e] Permethrin, lucerne + wheat straw; cypermethrin, lucerne.
[f] Also as 5 applications (5 × 50 g ai/ha; 1 × 500 and 4 × 50 g ai/ha) at approximately weekly intervals; soil allowed to air dry for 12 days before final application.
[g] Study periods from 1 week (enzymic activity) to 5–8 weeks (carbon and nitrogen turnover).

10 × field rate will not adversely affect either the carbon and nitrogen transformations occurring in soil or the size of the microbial community.

Two studies have been carried out on the effects of permethrin on the microbial populations of natural aquatic environments (unpublished data, ICI Plant Protection Division). In a short (8 day) laboratory study of the adsorption and degradation of permethrin in a river water-sediment system, measurement of microbial biomass (ATP), total numbers of propagules and algal cells and viable bacteria showed no treatment-related effects in either water or sediment. A more extensive study was carried out in field ponds sprayed with permethrin at approximately field rate (see p. 216). Water and sediment were sampled before and for 4 weeks after treatment. Assessment of the microbial populations by the techniques described above for laboratory studies showed a slight increase in numbers of organisms after treatment. However, this might have been due to an increased food supply following the deaths of some invertebrates or to changes in environmental conditions. It was considered that permethrin had no detrimental effect on the microbial populations.

Effects on plants

Studies under greenhouse conditions (unpublished data, ICI Plant Protection Division) have shown cypermethrin (as an E.C.) applied as a post-emergence spray at 70 or 350 g ai/ha to have little or no effect (growth rate or phytotoxicity) on a wide range of monocotyledonous and dicotyledonous plants (Table 4.14). Neither were plant growth regulatory effects seen nor was there any influence on the photosynthetic activity (oxygen evolution) of macerated fronds of *Asparagus officinalis*. Greenhouse studies have also shown there to be no or negligible phytotoxic effects from permethrin or deltamethrin, even at excessively high rates. With fenvalerate, stunting and chlorosis were commonly observed, the latter being severe and causing some leaf drop at very high rates. However, when applications ceased new growth appeared normal.

Substantial numbers of field observations for the effects of formulated pyrethroids have been made during efficacy trials and following "in use" applications. There have been very few reports of phytotoxic or adverse yield effects in the field (Dowson & Garvie, 1979; Ruscoe, 1979; Weaver et al., 1979; Cheng, 1980), except with fenvalerate (Hattori, 1977; Mowlam et al., 1977; Hargreaves & Cooper, 1979) which can cause noticeable damage in some crops to seedlings and the soft, young leaves of older plants. However, Hattori (1977) reported that plants generally grew away from the damage without apparent long-term effect. Table 4.15 shows yield data from trials with permethrin and cypermethrin in which pest "pressure" was naturally low, enabling valid comparison of yields from treated and control plots. On no occasion did the pyrethroids tested reduce the yield of cotton, brassicae, potatoes or apples.

No effects of pyrethroids on aquatic plants has been recorded (see p. 213 ff.) except by Rawn et al. (1981) who applied permethrin at 28 g ai/ha to experimental ponds and found that the pyrethroid was sorbed by the duckweed (*Lemna* sp.) on the surface, causing growth inhibition. The concentration of permethrin in the plant was 30–55 µg ai/g dry weight after 24 h, decreasing to less than 0·1 µg ai/g after 2–3 weeks (Rawn et al., 1982).

Table 4.14. Phytotoxic effects of cypermethrin under glasshouse conditions[a]

Test plant	Application rate (g ai/ha)	Application stage(s)[b]	Phytotoxic effect
Tomato	1000,4000[c] (2 × 0·005% ai)[d]	— 3 wk old	} no effect
Maize	1000,4000[c] 70,350	— pre; 2·5L	} No effect
Cotton	200,20000, 400,800 70,350	25 cm 4L pre; 2L	} No effect
Soybean	70,350	pre; 1·5Tr	No effect
Sugarbeet, oil seed rape, winter wheat, rice	70,350	pre; 2·5–4·5L	No effect
Impomea purpurea, Amaranthus retroflexus, Polygonium aviculare, Chenopodium album, Xanthium spinosum, Abutilon theoprasti, Avena fatua, Digitaria sanguinalis, Poa annua, Echinchloa crusgallii, Sorghum halepense, Setaria viridis, Agropyron repens	70,350	pre; 3–5·5L	No effect
Cyperus rotundus	70,350	pre; 13 cm	No effect
Senecio vulgaris	70,350	pre	} Slight
Portulaca oleracea	70,350	pre; 6L	} stunting

[a] Data from published (Ref. no. 100) and unpublished (ICI Plant Protection Division) sources.
[b] L = leaf stage; Tr = trifoliate stage; pre = pre-emergence application.
[c] Technical cypermethrin; remainder E.C.'s.
[d] Spray concentration.

Effects on aquatic invertebrates and fish

Pyrethroid entry into aquatic environments

The entry of pesticides into aquatic environments can either occur intentionally or unintentionally. The available routes of entry, together with mechanisms of dissipation have been described by Hill & Wright (1978).

The pyrethroids are marketed principally as agricultural and horticultural insecticides. From such applications small amounts of a pesticide can enter bodies of water indirectly, either as a result of spray drift, or from soil surface erosion by wind and water. Direct entry will only result from spraying over drainage or irrigation channels, and where misuse or spillage occurs.

The possible use of pyrethroids as replacements for DDT in *Simulium* (blackfly; Mohsen & Mulla, 1981; Muirhead-Thomson, 1981a) and mosquito (Mulla et al. 1973, 1975, 1978a, 1980; Haskins et al., 1974; Mount & Pierce, 1975; Thompson & Meisch, 1977; Darwazeh et al., 1978; Rettich, 1979, 1980) larvae control, and their potential as a spruce budworm insecticide in forestry, would result in some overspraying of aquatic environments. Furthermore, there is also current interest in the use of these insecticides in rice paddies. Whereas such proposals would involve application to large areas of water, the use of pyrethroids for the control of invertebrates either destructively grazing on the marine alga *Chondrus crispus* in tank culture (Shacklock & Croft, 1981) or presenting a nuisance value in piped mains water (Abram et al., 1980) will only result in their limited and very localised entry into natural aquatic environments. Mulla et al. (1978b) have also suggested that synthetic pyrethroids might provide excellent substitutes for rotenone (see Marking & Bills, 1976) or isobornyl thiocyanoacetate (see Burress et al., 1976) for the control of undesirable fish.

The degradation, adsorption and leaching of pyrethroids in soil has been extensively studied (see Chapter 5). The data clearly demonstrate that negligible if any parent chemical will reach aquatic environments via leaching and entry into ground water. As regards the potential for soil surface movement (run-off), it is important to appreciate that the synthetic pyrethroids are rapidly degraded in most soil types, under both aerobic and anaerobic conditions and are strongly adsorbed to soil. For run-off to occur there would need to be heavy rain shortly after spraying on a site adjacent to a body of water and where the topography is conducive to surface run-off of water containing soil particles. Not all sprayed areas will be found alongside ponds, streams etc., neither will all bodies of water be bordered by a sprayed field, thus in practice only a very small proportion of all the pyrethroid applied will run-off into water. The rapid and strong adsorption of synthetic pyrethroids to soil surfaces suggests that when such aquatic contamination does occur it will most likely take the form of erosion of soil particles containing the strongly adsorbed insecticide. Published evidence shows that for a compound having the adsorptive characteristics of permethrin, less than 1% of the total amount of compound applied to a watershed is likely to find its way into water (Caro & Taylor, 1971; Harvey & Pease, 1973; Pionke & Chesters, 1973; Willis & Hamilton, 1973; Caro et al., 1974; Willis et al., 1975; Wauchope, 1978, 1980). In a study of the Big Creek watershed in Ontario, an area of intensive use of synthetic pyrethroids, no residues were detected in water and bottom sediments when monitored during the summer of 1979 (personal communication, C. R. Harris, Research Institute, London, Canada).

Modelling of cypermethrin run-off has been done on the "Pesticide Runoff Simulation" computer programme SWRRB. Using application rates of 15 × 100 g ai/ha applied at 7 day intervals to "river basin scenarios" in the programme (Coshocton Basins 115 and 130, Ohio; Tifton Basin, Georgia; Watkinsville Basin, Georgia; all cropping with corn), the run-off of cypermethrin was modelled for a total of 71 years climatic data. In most instances the annual run-off was equal to, or less than, 0·1% of that applied (occasionally 0·2%, and never greater than 0·4%), equivalent to approximately 1 g ai/ha.

Table 4.15. Effect of cypermethrin and permethrin on crop yields in field studies with low pest "pressure"[a]

Pyrethroid	Crop (and yield parameter)	Pest(s) (and infestation parameter)	Application nos. and rate (g ai/ha)	Pest infestation[b] H.v.	Pest infestation[b] P.g.	Yield[c]
Cypermethrin	Cotton (seed cotton; tonnes/ha)	*Heliothis virescens* (*H.v.*) *Pectinophora gossypella* (*P.g.*) (% infested squares)	8 × 112	8	0	4·6
			8 × 56	7	1	4·9
			8 × 28	4	1	4·5
			0	12	4	4·6
	Cabbage (wt of head kg)	*Trichoplusia ni* (no./10 plants)	5 × 35	2·1		0·52
			5 × 17·5	4·1		0·66
			5 × 8·8	5·1		0·56
			0	6·6		0·43
	Apple (fruit/tree)	*Cydia pomonella* (no./tree)	2/3 × 0·006%[d]	0		301
			2/3 × 0·003%	1		294
			2/3 × 0·0015%	2		322
			0	47		266
Permethrin	Broccoli (kg/plot)	e	6 × 224	e		8·9
			6 × 112			8·6
			6 × 56			8·8
			0			7·6
	Potato (kg/plot)	*Leptinotarsa decemlineata* (no./6 metre row)	1 × 224	63		30
			1 × 112	49		24
			1 × 56	55		31
			0	237		26

[a] Unpublished data (ICI Americas Inc.).
[b] Mean of assessments at intervals throughout growing season.
[c] Yields on treated and control (untreated) plots not significantly different.
[d] Two replicate trials; application rate as percentage active ingredient in spray.
[e] *Bacillus thuringiensis* applied to control plots to minimise infestation.

Table 4.16. Scientific and common names of aquatic invertebrates and amphibia referred to in text

---

MOLLUSCA

Gastropoda	*Limnaea stagnalis*	Pond snail
Bivalvia	*Indoplanorbis exustus*	
	Cipangopaludina malleata	
	Crassostrea gigas	Pacific oyster
	Crassostrea virginica	Eastern oyster

ANNELIDA, HIRUDINEA

	Haemopsis sanguisuga	} Leech
	Helobolella stagnalis	

ARTHROPODA, CRUSTACEA

Cladocera	*Daphnia* spp.	
	Ceriodaphnia sp.	
	Moina sp.	} Water flea
	Simocephalus serrulatus	
	Alona spp	
Copepoda	*Nitocra spinites*	
Ostracoda	*Cyprois* sp.	
Amphipoda	*Gammarus lacustris*	Scud
	Gammarus pseudolimnaeus	
	Gammarus pulex	Freshwater shrimp
Isopoda	*Asellus aquaticus*	Water louse
	Caridina africana	Shrimp
Decapoda	*Crangon septemspinosa*	Grass shrimp
	Homarus americanus	Lobster
	Macrobrachium raridus	Shrimp
	Macrobrachium vollenhovenii	Shrimp
	Orconectes sp.	Crayfish
	Pennaeus aztecus	Brown shrimp
	Pennaeus duoarum	Pink shrimp
	Procambarus clarkii	Red swamp crayfish
	Procambarus blandingi	Crayfish
	Uca pugilator	Fiddler crab
Paracarida	*Mysidopsis bahia*	Mysid shrimp

ARTHROPODA, INSECTA

Ephemeroptera	*Baetis* spp.	
	Cloeon dipterum	
	Ephemerella spp.	} Mayfly
	Hexagenia bilineata	
	Hexagenia rigida	Burrowing mayfly
Plecoptera	*Pteronarcys dorsata*	Stonefly
Odonata	*Agrion splendens*	Dragonfly
Hemiptera	*Corixa punctata*	Lesser water boatman
Coleoptera	*Gyrinus natator*	Whirligig beetle
Trichoptera	*Brachycentrus americanus*	
	Brachycentrus subnulis	
	Hydropsyche pellucidula	} Caddis fly
	Hydropsyche californica	
	Rhyacophila dorsalis	

Diptera	*Aedes* spp.	Mosquito
	Atherix sp.	Rhagionid fly
	Chaoborus crystallinus	Phantom midge
	Chironomus thumni	Midge
	Culex spp.	
	Culiseta incidens	Mosquito
	Psorophora columbiae	
	Simulium spp.	Blackfly
ARTHROPODA, ARACHNIDA		
Hydracarina	*Piona carnea*	Water mite
VERTEBRATA, AMPHIBIA		
Anura	*Bufo bufo*	Toad
	Rana catesbriana	Bullfrog
Urodela	*Triturus* sp.	Newt

The inadvertent aerial drift of pesticide droplets during application has long been an area of concern. Numerous studies have quantified the losses via this route by sampling for target area residues and by measuring airborne particles outside the application zone. However, relatively few workers have examined the amount of pesticide deposited on non-target crops and on to the ground or aquatic environments at a range of distances from the target crop. The factors affecting drift and droplet deposition (for example, climatic and microclimatic conditions, topography and application equipment, Yates & Akesson, 1973; Yates et al. 1978; Lawson, 1979) have inevitably resulted in considerable variation in the data from field trials. Where well designed agricultural studies have been carried out using ground and aerial (fixed wing and helicopter) spraying, surface deposits 100 m from the area of application have almost always been less than 1% of the spray rate whilst at 1000 m the deposition rate is less than 0·1% (Akesson & Yates, 1964; Ware et al. 1969a, b; Frost & Ware, 1970; Yates et al., 1978, 1981; Currier et al., 1982). Such levels of deposition are only likely to occur with aerial spraying, tractor "boom" spray units usually resulting in approximately 10-fold lower rates. In forestry applications the height of spraying can result in higher deposition rates in surrounding areas. Studies with fenitrothion applied aerially to Canadian forests showed that a 400 m "no-spray" buffer zone around a small lake reduced spray deposit on to the water by 80% and 90% on two successive occasions (Ernst et al., 1980). Where the terrain is extremely variable, for example, within canyons in some forestry applications, drift distances and fall-out can be somewhat higher (Ghassemi et al., 1982).

The wide spectrum of biological activity of pyrethroids to target pests indicates that these insecticides are likely to be highly toxic to a range of aquatic invertebrates. Fish are also present in almost all naturally occurring bodies of fresh and salt water. They are harvested for food and caught for sport. In order

Table 4.17. Scientific and common names of fish (Teleosteii) referred to in text

Alburnus alburnus	Bleak
Astyanax fasciatus fasciatus	Banded astyanax
Barbus spurelli	Barb
Carassius auratus	Goldfish
Cottus cognatus	Slimy sculpin
Culaea inconstans	Brook stickleback
Cyprinodon macularius	Desert pupfish
Cyprinodon variegatus	Sheepshead minnow
Cyprinus carpio	Mirror carp
Elassoma zonatum	Banded pygmy sunfish
Epiplatys sp.	Epiplatyd
Esox lucius	Northern pike
Etheostoma spp.	Darters
Fundulus notti	Starhead topminnow
Gambusia affinis	Mosquitofish
Hemichromis bimaculatus	Red cichlid/jewel fish
Ictalurus melas	Black bullhead
Ictalurus nebulosus	Brown bullhead
Ictalurus punctatus	Channel catfish
Labeo parvus	
Lebistes reticulatus	Guppy
Lepomis cyanellus	Green sunfish
Lepomis gibbosus	Pumpkinseed sunfish
Lepomis gulosus	Goggle-eye
Lepomis macrochirus	Bluegill sunfish
Lota lota	Ling, Burbot
Menidia menidia	Atlantic silverside
Micropterus dolomieu	Smallmouth bass
Micropterus salmoides	Largemouth bass
Misgurnus anguillicaudatus	Japanese weather fish
Morone americana	White perch
Mugil cephalus	Striped (grey) mullet
Notemigonus crysoleucas	Golden shiner
Oncorhynchus kisutch	Coho salmon
Opsanus beta	Gulf toadfish
Oryzias latipes	Killifish, rice fish
Perca flavescens	Yellow perch
Pimephales promelas	Fathead minnow
Poecilia reticulata	Guppy
Salmo fario	Brown trout
Salmo gairdneri	Rainbow (steelhead) trout
Salmo salar	Atlantic salmon
Salmo trutta	Brown trout
Salvelinus fontinalis	Brook trout
Salvelinus namaycush	Lake trout
Scardinius erythrophthalmus	Rudd
Semotilus corporalis	
Tilapia mossambica	Mouth-brooder, Mozambique cichlid
Tilapia nilotica	Nile mouth-brooder
Tilapia zilli	
Tinca tinca	Tench
Umbra spp.	Mud minnows

to protect these uses and to ensure the continued integrity of the aquatic environment, governments impose stringent controls over toxic chemicals likely to enter the aquatic environment. Tests to study the effect of a pesticide on aquatic invertebrates and fish are, in the first instance, carried out in the laboratory on a range of organisms representative of those occurring in natural environments. Consideration of the toxicity data, together with predictions of possible aquatic environment exposure levels (from knowledge of entry routes and studies of adsorption, leaching, degradation, etc.), indicate whether or not further studies in simulated or real field conditions are also necessary.

The aquatic organisms described in the text are listed in Tables 4.16 and 4.17, for invertebrates and fish, respectively, together with common names.

Laboratory toxicity and mode of action

Invertebrate toxicity

Synthetic pyrethroids have been extensively tested against a wide spectrum of freshwater, brackish and marine invertebrates. The acute toxicity data are summarised in Table 4.18.

Studies with the molluscs *Limnaea stagnalis* and *Crassostrea gigas* showed permethrin and cypermethrin to exhibit only moderate or low toxicity (48 hours LC50 >5000 µg ai/litre) to this group of organisms. Cypermethrin caused a 50% reduction in *Crassostrea virginica* shell growth rate at approximately 400 µg ai/litre (96 hours) and fenvalerate at >1000 µg ai/litre (24 hours).

As the synthetic pyrethroids are principally used against insect pests, they are also highly toxic to a wide range of aquatic arthropods under laboratory conditions in the absence of sediment or suspended particulate material (Table 4.18). A large number of toxicity tests have been carried out under a wide range of conditions; with static or flowing water, at temperatures of 10–26 °C, for varying periods of time, in waters of different quality, and with assessment of toxicity as either EC50 or LC50. Nevertheless, a clear pattern does emerge. Permethrin and cypermethrin have been studied most extensively. For these two pyrethroids the spectrum of toxic effect, whether by EC50 or LC50 measurements, after 72–96 hours exposure was in the range of 0·02–3 µg ai/litre for permethrin and 0·005–2 µg ai/litre for cypermethrin. Although relatively few tests have been reported for fenvalerate and deltamethrin, a comparison of the available data suggests that the sensitivity of aquatic arthropods to the photostable synthetic pyrethroids is in the order permethrin ≃ fenvalerate < cypermethrin < deltamethin. Not unexpectedly, the activity of these compounds to target terrestrial organisms is in the same order and consequently the recommended application rates are highest for permethrin

Table 4.18. Acute toxicity of pyrethroids to aquatic invertebrates and amphibia[a]

Organism	Organism size/stage	Test method[a]	Assessment	Permethrin	Cypermethrin	Fenvalerate	Deltamethrin	Ref. no.
MOLLUSCA								
Gastropoda *Limnaea stagnalis*	Eggs	S	48 h LC50	>200 000[e]				
	Adults	S	48 h LC50	>10 000[e]				†, ‡
Bivalvia *Crassostrea gigas*[b]	Larvae	F	48 h LC50	>4800	>2300			
	Juveniles	F	48 h LC50	>4800				264
Crassostrea virginica[b]	10 g	F	96 h EC50		370[f]			
	Larvae	S	24 h EC50			>1000[f]		
ARTHROPODA – CRUSTACEA								
Cladocera *Daphnia magna*	First instar	S	48 h EC50	0·6 0·8–1·3[e]	1·3 3·1–22[e]			
	First instar	S	72 h EC50	3·4	0·2–1·6 0·3–0·8[e]			†
	Juv./adults	S	48 h EC50	0·4				
	Ephippia	S	48 h EC50	108				
Copepoda *Nitocra spinites*[b]	0.7 mm	S	96 h LC50	0·15[e]		0·38		150
Amphipoda *Gammarus*	Small juv.	S	96 h LC50			0·05		
Gammarus pseudolimnaeus	Juv./adults	F	96 h LC50			0·03		
Gammarus pulex	5 mm	S	24 h EC/LC50		0·04/0·1			†, 1
	5 mm	S	24 h EC/LC50		0·02/0·2			
Isopoda *Asellus aquaticus*	5 mm	S	72 h EC50	0·09	0·009			250
	10 mm	IF	96 h LC50	~3				

EC/LC50 (µg ai/litre)[c,d]

Group	Species	Size/stage	[a]	Endpoint					References
Decapoda	*Crangon septemspinosa*[b]	1·3 g	R	96 h LC50	0·13	0·01	0·04		‡, 125, 126, 149
	Penaeus aztecus[b]	20 mm	S	96 h LC50	0·34				171
	Penaeus duoarum[b]		F	96 h LC50	0·22	0·04	0·84	1·5	
	Uca pugilator[b]		S	96 h LC50	2·2	0·20		<0·56	
	Homarus americanus[b]	450 g	R	96 h LC50	0·73	0·04	0·14		
	Procambarus blandingi[b]	24 g	F	96 h LC50	0·12				
	Procambarus clarkii	0·05–0·5 g	S	96 h LC50	0·15–0·23[e]				
	Orconectes sp.	2·3 g	F	96 h LC50		0·07			
Paracarida	*Mysidopsis bahia*		F	96 h LC50	0·02	0·005	0·008		‡, 83, 229

ARTHROPODA – INSECTA

Group	Species	Size/stage	[a]	Endpoint					References
Ephemeroptera	*Baetis rhodani*	Early instar	F	96 h EC/LC50		0·006/0·012			†, ‡, 9, 250
	Cloeon dipterum	5 mm	S	24 h EC/LC50		0·07–0·6			
		Early instar	S	72 h EC50	0·03		0·93		
	Hexagenia bilineata		F	96 h LC50	0·10				
	Ephemerella sp.		F	96 h LC50			0·13/>1		
Plecoptera	*Pteronarcys dorsata*		F	72 h EC/LC50	0·15/>0·4				9, 250
Hemiptera	*Corixa punctata*	Adults	S	24 h EC/LC50		0·7/≥5			
Coleoptera	*Gyrinus natator*	Adults	S	24 h EC/LC50		0·07/0·6			
Trichoptera	*Brachycentrus americanus*		F	96 h EC/LC50	0·4/>0·5				
Diptera	*Chaoborus crystallinus*	Larvae	S	24 h EC/LC50		0·03/0·2			9, 191
	Chironomus thumni	Larvae	S	24 h EC/LC50		0·2/≥5			194, 250
	Culex/Aedes spp.	Larvae	S	24 h LC50	0·5–3	0·07–1·0	0·9–2·8	0·02–0·4	
		Larvae	S	24 h LC50	0·7–6		1·2–5·3	0·07–0·6	
		Pupae	F	96 h LC50			0·32		
	Atherix sp.		F						

ARTHROPODA – ARACHNIDA

Group	Species	Size/stage	[a]	Endpoint					References
Hydracarina	*Piona carnea*	Adults	R	96 h LC50		0·04			171

AMPHIBIA – ANURA

Group	Species	Size/stage	[a]	Endpoint					References
		Tadpoles	S	96 h LC50	2670				125, 126

[a] F = continuous flow-through; IF = intermittent flow; S = static; R = 48 hour replacement, static.

[b] Marine or estuarine organisms tested in salt water, all others tested in freshwater.

[c] Technical material (unless stated otherwise) added to water in acetone, ethanol, dimethyl sulphoxide, triethylene glycol or as a solution prepared using "saturation column".

[d] Further data for the above and other pyrethroids can be found in Ref. nos. 1, 26, 174, 191, 194, 199, 205, 229 and 293.

[e] Formulated material (emulsifiable concentrate) added as an aqueous preparation.

[f] Shell growth.

† Unpublished data (ICI Plant Protection Division).

‡ Unpublished data (ICI Americas Inc.).

and fenvalerate and lowest for deltamethrin. The consistently most sensitive organisms tested were the Ephemeroptera and some Crustacea (Amphipoda, Isopoda, Decapoda and Paracarida).

Resting stages or eggs of some organisms can be important mechanisms of surviving adverse environmental conditions for the later re-establishment of populations. The ephippia ("resting eggs") of *Daphnia magna* were found to be approximately two orders of magnitude less sensitive to permethrin than the free-swimming forms (Table 4.18). Eggs "stripped" from adult *Hexagenia rigida* were somewhat less sensitive to permethrin than the nymphs (Friesen et al. 1981), with LC50 values following 9 hours exposure of approximately 2 µg ai/litre and after 24 hours exposure in the range of 0·5–3 µg ai/litre. The eggs at the earliest embryonic stage were least sensitive. The eggs of terrestrial insects have also been shown to exhibit age-related sensitivity (Salkeld & Potter, 1953; Pree & Hagley, 1977).

Organisms in flowing aquatic environments are often only exposed to pesticides for a very short period of time, whilst the chemical flows past, and may therefore tolerate far higher concentrations than are indicated by EC50 data from 24–96 hour exposure studies. Stephenson (1982) measured the EC50 values for ten aquatic invertebrates, representing 7 orders/sub-orders of the Arthropoda, after 2 and 24 hours exposure to cypermethrin. In almost all instances there was less than a three-fold difference between the paired results for any organism. No investigation was made of recovery. Miyamoto (1976) and Nishiuchi (1978, 1979) reported very high LC50 values (2000–>50 000 µg ai/litre) for several cladocerans exposed for only 3 hours to permethrin, pyrethrum and a range of photolabile synthetic pyrethroids. Cladocera exposed for 48 hours are affected at concentrations three to four orders of magnitude lower (see Table 4.18). Friesen et al. (1983) exposed *Hexagenia rigida* nymphs to technical permethrin for 6 hours. The LC50 was approximately 1 µg ai/litre, deaths being observed during a further 8-week period of incubation in uncontaminated water (with sediment present).

Muirhead-Thomson (1977, 1978, 1981a,b) using permethrin and deltamethrin, and Mohsen & Mulla (1981) with deltamethrin and its ± *cis* and chloro analogue NRDC 160, investigated their toxicity to a range of non-target aquatic invertebrates in relation to toxicity to simuliid larvae, the aquatic stage of a dipteran pest (blackfly) for which pyrethroids have been considered as control agents. Under laboratory conditions using a 1 hour exposure followed by observation for mortalities after a further 24 hours in clean water, there was little difference in sensitivity between the *Simulium* species tested and other aquatic arthropods (Table 4.19). Measurement of detachment of simuliids or drift of the non-target organisms, which under field conditions could result in prolonged exposure of the organism to the pyrethroid, was also studied. The results varied markedly for the different organisms. When exposed for 30 minutes (and observed after 30–60 minutes) to concentrations that gave 50% mortality after 1-hour exposure with 24 hours recovery,

Table 4.19. Toxicity of pyrethroids to aquatic invertebrates 24 hours after a 1-hour exposure period

Organism		LC50 (µg ai/litre)[a]		
		Permethrin E.C.[b]	Deltamethrin E.C.[b]	NRDC160 E.C.[c]
Amphipoda	*Gammarus pulex*	0·5	0·02	
Ephemeroptera	*Baetis rhodani*	0·3	<0·005	
	Baetis parvus		0·01[c]	0·10
	Ephemerella ignita		>0·50	
Trichoptera	*Hydropsyche pellucidula*	>50	>0·01, <0·50	
	Hydropsyche californica		0·01[c]	0·07
	Rhyacophila dorsalis		<0·20	
	Brachycentrus subnubilis	0·5		
Diptera	*Simulium* spp.[d]	1·0		
	Simulium equinum	1·0		
	Simulium ornatum		0·10	
	Simulium virgatum		0·02[c]	0·20

[a] Values estimated from tables of percentage mortality in original references.
[b] Studies by Muirhead-Thomson (1977, 1978, 1981a,b) with continuous flow at 18 °C.
[c] Studies by Mohsen & Mulla (1981) with intermittent flow at 20 °C.
[d] Mixture of *Simulium equinum*, *S. erythrocephalum* and *S. ornatum*.

simuliids showed 10% (*Simulium virgatum*) to 50% (*S. equinum*) detachment and *Brachycentrus subnubilis* and *Gammarus pulex* 50% and 30% drift, respectively. At higher concentrations of permethrin, mortality was mostly greater than 90% with variable effects on drifting. Observation of *S. virgatum* exposed to deltamethrin or NRDC 160 for 30 minutes also showed detachment (10–20% 1 hour later) at the "LC50 concentration". At a rate three times higher, equivalent to the LC90 (Mohsen & Mulla, 1981), 40% of the larvae were detached within 1 hour of the 30-minute exposure period. These results suggest that the concentrations required to control the larval stage of the simuliid fly (blackfly) could under field conditions also cause considerable damage to non-target aquatic invertebrates.

The processses of adsorption and degradation will under field conditions rapidly remove pyrethroids from the water phase. However, Anderson (1982) has examined the effect of long exposures of permethrin and fenvalerate at low concentrations to 6 species of aquatic invertebrates: 4 insects, a mollusc and a crustacean. The amphipod *Gammarus pseudolimnaeus* was most sensitive to fenvalerate, showing behavioural changes at 0·022 µg ai/litre within a few hours of exposure starting. After 7 days, approximately 70% of the organisms were dead at this concentration. It was necessary to expose the insects *Ephemerella* sp. and *Atherix* sp. to a similar concentration of fenvalerate for 7 and 14 days, respectively, to cause the same degree of mortality. The mollusc *Helisoma trivolvis* was unaffected by 0·8 µg ai/litre for 28 days.

Mortalities from the equivalent concentrations of permethrin occurred much later during the exposure phase, although behavioural changes were again observed at an early stage. For example, *Brachycentrus americanus* and *Pteronarcys dorsata* stopped feeding and suffered loss of equilibrium. Death of these organisms may have finally resulted from starvation.

The effect of cypermethrin on the survival and reproduction of *Daphnia magna* has been studied over a 23-day period under flow-through conditions (unpublished data, ICI Plant Protection Division). Concentrations of up to 0·02 μg ai/litre had no significant effect on numbers of offspring per parthenogenic female, and had little if any (less than 10%) effect on the size of these females. The concentration that reduced the numbers of young born and surviving by 50% was 0·06 μg ai/litre. At all concentrations of cypermethrin tested (0·005–0·08 μg ai/litre) the surviving young reproduced normally when removed to clean water. In a similar study with permethrin (unpublished data, ICI Plant Protection Division), survival of adults was unaffected at 0·3 μg ai/litre and no significant effect on offspring occurred at 0·1 μg ai/litre.

Muirhead-Thomson (1970) investigated the activity of pyrethrins and pyrethroids in mixtures with the organophosphorus insecticides Dursban and iodophenos on *Simulium ornatum* larvae. After a short (1 hour) exposure period, detachment and mortality were observed for a further 24 hours. At concentrations of organophosphates giving approximately 90% mortality after 24 hours there was less than 10% detachment of the larvae within 1 hour of treatment. Addition of pyrethrins or the photolabile synthetic pyrethroid NRDC 107 at sublethal concentrations to the organophosphates caused 90% of the larvae to detach during the exposure period.

The toxicity of the major degradation products of permethrin (which includes products from cypermethrin, fenvalerate and deltamethrin) to *Daphnia magna* are shown in Table 4.20. The degradation products are considerably more polar than the parent molecule and are of very much lower toxicity. In aquatic environments it is therefore the toxicity of the pyrethroid itself rather than the degradation products which is of potential environmental concern.

Invertebrate mode of action

The mode of action of synthetic pyrethroids on insects is fully discussed elsewhere (see Chapter 2). However, aquatic invertebrates, in particular the crustaceans, have been investigated by a number of workers. Osborne (1980) reported repeated depolarisation, and thus repetitive discharges, of the membrane potential of crayfish slow muscle stretch receptor. The discharges were blocked quickly in the soma but not in the axon, suggesting involvement of the sensory structure rather than the afferent axon. Studies on crayfish giant axons (Wang et al., 1972; Narahashi & Lund, 1980; Lund & Narahashi, 1981a, b) showed effects of allethrin on sodium and potassium activation and

Table 4.20. Acute toxicity of pyrethroid metabolites to *Daphnia magna*[a]

Pyrethroid metabolite	EC50 (µg ai/litre)	
	24 h	48 h
3-(2,2-Dichlorovinyl)-2,2-dimethyl-cyclopropane carboxylic acid (*cis:trans*, 40:60 approximately)[b]	199 000	128 000
3-Phenoxybenzyl alcohol[c]	17 000	10 000
3-Phenoxybenzoic acid[c]	147 000	85 000

[a] Unpublished data (ICI Plant Protection Division).
[b] Ester hydrolysis product of permethrin and cypermethrin.
[c] Ester hydrolysis products of permethrin, cypermethrin, fenvalerate and deltamethrin.

on sodium inactivation currents, altering the amplitude and time course of the action potential in crayfish giant axons. There is currently no proof that these ionic effects, also seen in insects, can cause neurophysiological effects, although hypotheses exist (Narahashi, 1982). Doherty et al. (1981), in studies of *Homarus americanus* axon plasma membranes, reported inhibition by allethrin of Mn^{++} and Ca^{++} ATPases, and discussed the possible relationship to neurotoxicity. Leake (1977) and Leake et al. (1980) have shown, using leeches, that the onset of certain neurophysiological effects parallels the onset of poisoning symptoms.

Fish toxicity

Extensive studies of the acute toxicity to fish of many of the synthetic photolabile and photostable pyrethroid insecticides have been carried out. The synthetic pyrethroids are highly lipophilic compounds that are rapidly and strongly adsorbed to most surfaces, whether biotic or abiotic. Consequently, aquatic toxicity tests carried out under "static" conditions, and without regular replacement of the toxicant preparation, can result in an underestimate of the potential biological effect, due to adsorption of the pyrethroid to the test vessel surface, to the organisms themselves or to faecal matter in the water. Such losses are likely to be higher in fish studies than with aquatic invertebrates as a result of the greater amount of excrement produced. Further losses can result from microbial metabolism of the molecule by ester hydrolysis (see Chapter 5), particularly as the faecal matter and microbial populations increase during the course of the test. Stephenson (1984) reported losses of cypermethrin from water in a "static" test in the order of 30–50% after 24 hours and 60–75% after 96 hours. In order to obtain a true value for the acute toxicity of pyrethroids to fish most workers have used "flow-through" studies, in which concentrations of the toxicant can be maintained throughout the test period.

Studies of pyrethrum and of many of photolabile synthetic pyrethroids, mostly under "static" conditions, have shown this group of chemicals to be highly toxic to fish; 96-hour LC50 values generally being in the range of 0·1–50 µg ai/litre.

Table 4.21 summarises the available data for the acute toxicity of the synthetic photostable pyrethroids to fish using only "flow-through" methods. Values (96-hour LC50) from many species of warm and cold water fish species exposed to technical and formulated (E.C.) material were in the range:

permethrin	0·4 – 30 µg ai/litre
cypermethrin	0·4 – 3·2 µg ai/litre
fenvalerate	0·3 – 5·4 µg ai/litre

Symptoms of intoxication include hyperactivity (including gill movement), loss of balance and the development of darkened areas. Comparable acute toxicity data for deltamethrin have not been published but results from 48 hour "static" and "replacement" tests (Mulla et al., 1978b; Zitko et al., 1979) showed it to be approximately one to two orders of magnitude more toxic to several species of fish than was permethrin or fenvalerate. Thus the photostable pyrethroids under conditions of maintained concentrations are highly toxic to fish.

Although *cis*-permethrin is much more toxic to target insects than is the *trans* isomer, the small amount of data for fish acute toxicity (Tables 4.21 and 4.25) show little difference between the two isomers. The major metabolites of permethrin, cypermethrin and fenvalerate formed by ester hydrolysis in soil and aquatic environments, which also include products from the degradation of deltamethrin, were three or more orders of magnitude less toxic to fish (96-hour LC50 values greater than 3000 µg ai/litre; Table 4.22) than their parent molecule. Therefore, as with aquatic invertebrates (see Tables 4.18 and 4.20), it is the parent pyrethroid and not the metabolites that require particular consideration in aquatic environments.

Most pyrethroids have negative insecticidal temperature coefficients, that is, they are more toxic at lower temperatures. Data from fish studies have, however, been far more variable. In studies of pyrethrins and cypermethrin with *Cyprinus carpio*, by Hashimoto & Nishiuchi (1981) and Stephenson (1982), respectively, no temperature mediated effect was observed. Mauck et al. (1976), investigating the toxicity of pyrethrum and a range of photolabile synthetic pyrethroids to *Lepomis macrochirus*, found that within the range 12–22 °C the 96-hour LC50 values varied by less than one-half an order of magnitude, with most but not all compounds slightly more toxic at the lowest temperature tested (Table 4.23). Kumaraguru & Beamish (1981) observed that permethrin was an order of magnitude more toxic to *Salmo gairdneri* at 5 or 10 °C than at 20 °C (Table 4.23). Conversely, Nishiuchi (1978) found resmethrin more toxic to *C. carpio* at 30 and 35 °C than at 15–25 °C. Neither

Table 4.21. Acute toxicity of pyrethroids to fish in flow-through tests[a]

Pyrethroid	Fish	Temp. (°C)	96-hour LC50 (µg ai/l)[b]		Ref. nos.	Refs. containing further data[c]
			Technical	Formulation		
Permethrin (cis:trans, approx. 40:60)	Salmo gairdneri	10–16	0·4–7	7·6–30	*, 1, 110, 141	38, 92,
	Pimephales promelas	23–25	2·6–16		*, 1, 10	125, 126, 150,
	Lepomis macrochirus	22–23	0·9	2·6–7·1	*	171, 174, 195,
	Cyprinus carpio	23	15			292, 293
	Salvelinus fontinalis	13	4·7			
	Menidia menidia[d]	26	2·2			
	Mugil cephalus[d]	25	5·5		83, 229	
	Cyprinodon variegatus[d]	30	7·8			
Cis-permethrin	Salmo gairdneri	13	20		*	92, 118, 174
Trans-permethrin	Salmo gairdneri	13	8·5			
Cypermethrin	Salmo gairdneri	10–15	0·5–0·9	1·7–3·2	*, 250	38, 171,
	Cyprinus carpio	10–22	0·9–1·6		*	251, 263
	Lepomis macrochirus	22–23	1·8–2·2	1·0–2·1		
	Tilapia nilotica	24–25	2–2·2		250, 251	
	Cyprinodon variegatus[d]	26	1·0		†	
	Salmo trutta	15	1·2		250	
	Scardinius erythrophthalmus	15	0·4			
Fenvalerate	Salmo gairdneri	16	2·1		110	38, 83, 150,
	Pimephales promelas	25	5·4			173, 195, 229
	Menidia menidia[d]	24	0·3			
	Mugil cephalus[d]	26	0·6			
	Cyprinodon variegatus[d]	30	5·0		83, 229	
	Opsanus beta[d]	30	5·4			
Deltamethrin						148, 149, 195, 263, 293

[a] Data for other pyrethroids can be found in Ref. nos. 26, 38, 118, 162, 163, 174, 199, 205, 229, 246 and 292.
[b] Technical material added to water in acetone, ethanol, dimethyl sulphoxide, triethylene glycol or as a solution prepared using a "saturation column"; formulated material (emulsifiable concentrate) added as an aqueous preparation.
[c] From static and replacement tests; and from flow-through studies where a 96-h LC50 value was not determined.
[d] Indicates marine or estuarine fish tested in salt water; all others are freshwater fish.
* Unpublished data (ICI Plant Protection Division)
† Unpublished data (FMC Corporation).

Table 4.22. Acute toxicity of pyrethroid metabolites to fish

Pyrethroid metabolite	Fish	Test method[a]	Temp. (°C)	96-hour LC50 (μg ai/l)	Ref. No.
3-(2,2-Dichlorovinyl)-2,2-dimethyl-cyclopropane carboxylic acid (cis : trans, 40 : 60 approximately)[b]	Cyprinodon variegatus	S	19	3000	†
	Salmo gairdneri	R[24]	15	3100	*
	Salmo gairdneri	R[24]	15	3400	
	Salmo salar	R[48]	10	>500[e]	292
3-Phenoxybenzyl alcohol[c]	Cyprinodon variegatus	S	19	2900	†
	Salmo gairdneri	R[24]	15	6100	*
	Salmo gairdneri	R[24]	15	6600	
	Salmo salar	R[48]	10	>500[e]	292
3-Phenoxybenzoic acid[c]	Salmo gairdneri	R[24]	15	13300	*
	Lepomis macrochirus	R[24]	22	36300	
2-(4-Chlorophenyl)isovaleric acid[d]	Oryzias latipes	—	—	>10000[f]	172
	Oryzias latipes	—	—	>10000[f]	172

[a] S = static; R[24] = 24 hour replacement, etc.
[b] Ester hydrolysis product of permethrin and cypermethrin.
[c] Ester hydrolysis product of permethrin, cypermethrin, fenvalerate and deltamethrin.
[d] Ester hydrolysis product of fenvalerate.
[e] Lethal threshold value given = geometric mean of lowest concentration with and highest concentration without mortality.
[f] 48-hour LC50.
* Unpublished data (ICI Plant Protection Division).
† Unpublished data (FMC Corporation).

Table 4.23. Effect of pH, hardness, temperature, and weight on the acute toxicity of pyrethroids to fish

Test conditions			96-hour LC50 (µg ai/l)						
			Permethrin[a]	Dimethrin[b]	Bioallethrin[b]	S-bioallethrin[b]	Resmethrin[b]	Bioethanomethrin[b]	Pyrethrin[b]
	pH								
Temperature 12 °C	6·5			23	50	>24	2·0	0·27	8·2
Hardness 40–80 mg/l	7·5			29	50	32	2·0	0·23	9·2
Fish wt 0·8 g	9·5			28	54	>24	2·2	0·25	17·4
	pH	*Hardness (mg/l)*							
	6·6	10–13		25	44	38	2·2	0·35	12·4
Temperature 12 °C	7·5	40–48		29	50	32	2·0	0·23	9·2
Fish wt 0·8 g	7·8	160–180		27	44	30	2·2	0·22	10·8
	8·2	280–320		27	38	35	1·9	0·27	9·4
	Temperature (°C)								
	5/–[c]		0·4						
pH 8·0/7·5[c]	10/12		0·4	29	50	32	2·0	0·23	9·2
Hardness 40–80/360[c] mg/l	15/17		3	56	42	27	3·8	0·55	11·8
Fish wt 1·0/0·8[c] g	20/22		6	82	32	—	4·9	0·64	11·6
	Fish wt (g)								
	0·8–1·2		2						
Temperature 15 °C	4·6–5·4		6						
Hardness 360 mg/l	18–23		50						
pH 8·0	47–53		270						
	196–205		300						

[a] Kumaraguru & Beamish, 1981; flow-through test with *Salmo gairdneri*.
[b] Mauck et al., 1976; static test with *Lepomis macrochirus*.
[c] Test conditions for Kumaraguru & Beamish, 1981/Mauck et al., 1976.

pH nor hardness of the water influenced the toxicity to *L. macrochirus* of a number of the pyrethroids studied by Mauck et al. (1976) (Table 4.23). Kumaraguru & Beamish (1981) demonstrated that size of fish had a pronounced effect on permethrin toxicity (Table 4.23), with small fish most sensitive. The majority of data for acute toxicity of the synthetic pyrethroids (Tables 4.21 and 4.22) have been determined with small fish over a wide range of temperatures.

Concern has often been expressed that pesticide mixtures may have greater than additive toxicity (synergism) resulting in an unforeseen environmental hazard. The cost and time of investigating the toxicity of the huge number of possible mixtures renders such an exercise out of the question. However, wherever studies of mixtures have been carried out with either non-pyrethroid pesticides (Ferguson & Bingham, 1966; Mayer et al., 1972; Macek, 1975; Hashimoto & Nishiuchi, 1981) or with pyrethroids (Marking & Mauck, 1975; Tag el-Din et al., 1981), fish toxicity has rarely deviated markedly from additive. Only occasionally was there a synergistic response greater than one order of magnitude above the additive value, a small change in relation to differences likely to occur in the field due to environmental factors.

It has often been reported that salmonid fish are more susceptible to toxicants than are other fish (Wuhrmann, 1952; Herbert, 1965; Edwards & Brown, 1967) and for this reason *Salmo gairdneri* has commonly been proposed as a test species. Mauck et al. (1976) reported that pyrethrum and five synthetic photolabile pyrethroids were more toxic to salmonids (*Oncorhynchus kisutch* and *S. gairdneri*) than to *Ictalurus punctatus*, *Lepomis macrochirus* and *Perca flavescens*. However, the data show this to be not entirely true and furthermore, there was less than one-half an order of magnitude difference between the acute toxicity values for any one chemical with a range of fish. The large amount of data available (Table 4.21) shows little difference in the sensitivities of different fish species to individual photolabile or photostable pyrethroids. Small differences in toxicity can be easily accounted for by variations in the test conditions and the consequent bio-availability of the test chemical.

Dose–response curves (Figure 4.3) of the LC50 concentration of pyrethroid against time, show that during the 96 hours study period the plotted line does not become asymptotic to the time axis. Although the lethal threshold concentration (the concentration beyond which 50% of the population cannot live for an indefinite time: Lloyd & Jordan, 1963; Sprague, 1969) is not reached in many of the studies it can be extrapolated visually or mathematically. However, this serves little practical purpose, for pyrethroid concentrations in the water phase in natural aquatic environments show a rapid decline in concentration with time, initially due to adsorption to bottom sediments, suspended particles, plant material, etc., and in the slightly longer term from degradative processes (see Chapter 5). Mauck et al. (1976) in a long term acute toxicity study of bio-ethanomethrin found that the lethal threshold concentration for *Oncorhynchus kisutch* and *Ictalurus punctatus* using "flow-through" conditions with 9–15 g fish was reached after 35 days and 20 days, respectively, and was approximately one-third of the 96-hour value.

Fish life cycle studies have been carried out with a number of pyrethroids (Table 4.24). Full life cycle studies have not been reported in the literature. However, such a test with permethrin has been carried out using *Pimephales promelas* (unpublished data, FMC Corporation). Fish were exposed to permethrin for 256 days, the study starting with viable eggs and continuing through one mating cycle to include the exposure of second generation fry for 30 days. The "no-effect level" was 0·3 μg ai/litre, fry being the most sensitive stage of the life cycle. Extended life cycle tests were, during the 1960s and early 1970s, considered essential for studying long-term effects of pesticides on fish.

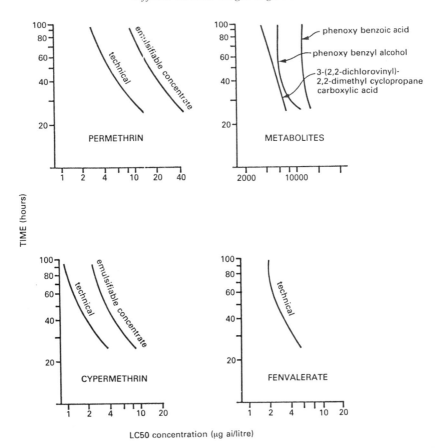

Figure 4.3. Dose–response curves for toxicity of pyrethroids and their metabolites to *Salmo gairdneri* (rainbow trout ≤4 g wt)

More recently, however, many workers (see, for example, McKim et al., 1975; Eaton et al., 1978; McKim, 1977) have shown that partial life cycle studies using the most sensitive parts of the life cycle, in particular the embryo-larval stage, can be used to provide accurate estimations of life cycle maximum acceptable toxicant concentrations. A number of embryo-larval studies have been reported for permethrin, fenvalerate, AC 222,705 (Spehar et al., 1982, 1983; Hansen et al., 1983) and cypermethrin (unpublished data, ICI Americas, Inc.). Exposure was commenced shortly after embryo fertilisation and continued under continuous flow conditions for approximately four weeks. Embryonic development, hatching success and survival and growth of hatched fish were all studied. No-effect levels were ≥0·2 µg ai/l for permethrin, cypermethrin and fenvalerate (Table 4.24), similar to, to no more than one order of magnitude more sensitive than, the 96 hour no-effect levels. AC 222,705 was more toxic than the other pyrethroids tested, with an embryo-larval no-effect level of 0·03 µg ai/l. Survival of newly hatched larvae was the most

Table 4.24. Life cycle and partial life cycle studies of the effects of pyrethroids on fish

Pyrethroid	Fish species	Method (life cycle)[a]	No-effect level (µg ai/l)	Ref. No.
Permethrin	*Pimephales promelas*	full	~0·3	*
	Pimephales promelas	partial	0·7	245
	Cyprinodon variegatus	partial	10	99
Cypermethrin	*Pimephales promelas*	partial	~0·3	†
Fenvalerate	*Pimephales promelas*	partial	0·3	245
	Pimephales promelas	partial	0·2	246
	Cyprinodon variegatus	partial	0·6	246
AC222,705	*Pimephales promelas*	partial	0·03	245
	Cyprinodon variegatus	partial	0·03	99

[a] Partial life cycle = embryo-larval test.
* Unpublished data (FMC Corporation).
† Unpublished data (ICI Americas Inc.).

sensitive stage, but there appeared to be little if any effect on growth of survivors, in contrast to observations with some other products (Jarvinen & Tanner, 1982; Benoit et al., 1982; Holcombe et al., 1982a). However, in natural aquatic environments fish will only be exposed to pyrethroids in the aqueous phase for very short periods of time due to adsorptive and degradative processes.

Because of the rapidity with which pyrethroids are removed from the water phase some workers have exposed fish to relatively high concentrations of deltamethrin for short periods before transfer to clean water and observation for recovery. Juvenile *Salmo gairdneri* exposed to permethrin for 1 hour at concentrations up to 50 µg ai/l showed some effects (jaw spasms and hyperactivity) but recovered completely. Following exposure to 100 µg ai/l for 1 hour, spinal curvature was observed in two out of ten fish after 20 hours, but there were no deaths after 2 days (unpublished data, ICI Plant Protection Division). Whereas deltamethrin has a 48-hour LC50 value of approximately 1 µg ai/l (Mulla et al., 1978b), Francois et al. (1982) found that over 50% of young *Salmo fario* recovered completely from immersion in 100 µg ai/l for 30 minutes, 50 µg ai/l for 1 hour and 25 µg ai/l for 2 hours. *Astyanax fasciatus fasciatus* survived 1 hour in 750 µg ai/l deltamethrin; but within 3–5 hours all fish had died following exposure to >63 µg ai/l (Salibian & Fichera, 1980, 1981). Lhoste et al. (1979) reported that *Salmo trutta* eggs were unaffected by 25 µg deltamethrin/litre for up to 48 hours. Dejoux (1979) using even higher concentrations of deltamethrin with shorter exposure periods, recorded survival of *Tilapia zilli* after 1½ minutes and *Labeo parvus* after 2 minutes at 10 000 µg/l.

Table 4.25. Comparative acute toxicity of permethrin to a mammal and a fish[a]

Route of administration	Isomers	Acute toxicity[b]	
		Mouse	*Salmo gairdneri*
Water	*cis,trans*	—	~0·003
Oral	*cis,trans*	~1000	—
	cis	265	—
	trans	>5000	—
Intraperitoneal (i.p.)	*cis,trans*	514	14
	cis	108	22
	trans	>800	27
Intravenous (i.v.)	*cis,trans*	31	1·8
	cis	17	1·3
	trans	>135	1·2

[a] Glickman et al., 1981, 1982.
[b] Mouse, oral: 24-hour LD50 (mg ai/kg); Mouse and fish, i.p. and i.v.: LD50 (mg ai/kg); Fish, water: 96-hour static LC50 (mg ai/litre).

Fish mode of action

Values for the relatively low acute oral toxicities of many pyrethroids to mammals (Table 4.1) are often contrasted with their very high toxicity to fish (Table 4.21). Fish are consequently reported to be approximately five orders of magnitude more sensitive to pyrethroids than are mammals such as mouse or rat.

The respiratory organs of fishes are the gills. Water is drawn in through the mouth, forced back into the pharynx and then out across the gills. In the action of circulating the water, blood is brought into close contact with water in the thin-walled gills and gas exchange takes place through the thin intervening membrane. During this process any toxicant in the water may partition through the gills into the blood of the fish; and the rate of absorption will in part depend upon the partition coefficient. Pyrethroids are highly lipophilic and readily partition from water into the blood via the gills. Histopathological studies by Kumaraguru et al. (1982), following exposure of *Salmo gairdneri* to permethrin, suggested that the pesticide when present in the diet may also reach the gills, via the circulation. Any comparisons of fish and mammalian toxicity are thus best done using the "body concentrations" required to cause the same degree of toxicity. Glickman et al. (1982) injected mice and *S. gairdneri* with permethrin (Table 4.25). Racemic permethrin was between one and two orders of magnitude more toxic to the fish than to the mouse, a much smaller difference than is obtained by comparing mouse oral and fish "water" toxicity data.

However, the interactions between toxicant and an animal are often very complex. Sun (1968) summarised the factors influencing toxicity into two main categories: (1) penetration, representing a combination of factors such as permeation, absorption, partition, excretion and transportation; and (2) detoxification, resulting from chemical and enzymatic processes, such as decompostion and conjugation. Two further categories: (3) organ and/or tissue and cell sensitivities and (4) indirect effects that inadvertently enhance the overall toxicity to the animal, should also be included.

Holcombe et al. (1982b) proposed that the rapid gill movements of fish exposed to fenvalerate indicated a respiratory effect, although it is difficult to see why such effects might not be secondary. Kumaraguru et al. (1982) observed hyperplasia and fusion of adjacent secondary lamellae in the gills of permethrin-treated *Salmo gairdneri*. The resulting increase in thickness of the epithelial layers could hinder the respiratory, secretory and excretory functions of the gills (Eller, 1975), thus contributing to the overall effect on the fish and possibly resulting in one reason for fish being more sensitive to these chemicals than are mammals.

The synthetic pyrethroids are extensively metabolised in mammals and fish (see Chapter 5) by oxidation, hydrolysis and conjugation. Glickman et al. (1979) observed all three processes *in vitro* when permethrin was added to liver microsomal preparations from *Salmo gairdneri* and *Cyprinus carpio*. However, *trans*-permethrin was principally metabolised by hydrolysis and the *cis*-isomer by oxidation. Anastasi & Bannister (1980) reported *in vitro* inhibition of two mitochondrial enzymes (cytochrome oxidase and succinate dehydrogenase) and stimulation of two extra-mitochondrial enzymes (pyruvate kinase and malate dehydrogenase) in muscle extracts of three fish species.

Comparison of the metabolic activity of extracts of a number of tissues from the mouse and *S. gairdneri* showed the hydrolysis of *trans*-permethrin to be approximately 50–150-fold faster in the mouse than in the fish (Glickman & Lech, 1981). *Cis*-permethrin was hydrolysed relatively slowly in the tissue preparations from both organisms. These observations are consistent with those from two other studies: (1) Little ester hydrolysis occurred *in vivo* in *S. gairdneri* exposed to racemic permethrin in the water (Glickman et al., 1981). Both isomers were rapidly taken up from water at similar rates, distributed to a similar extent in the tissues, and there was little difference in their relative rates of disappearance, following conjugation; (2) *Cis*-permethrin is much more toxic to mice than is the *trans*-isomer (whether by oral administration, intraperitoneal injection or intravenous injection), whereas both isomers are of approximately equal toxicity to *S. gairdneri*, whether entry is through water or by injection (Glickman et al., 1981, 1982; Table 4.25). Studies by other workers have confirmed the differential toxicity of pyrethroid *cis* and *trans* isomers to mammals (permethrin: Soderlund & Casida, 1977; resmethrin: Gray et al., 1980) and their similar toxic levels in fish (Table 4.21).

Glickman et al. (1982), from studies with esterase inhibitors suggest that the relative rates of permethrin isomer metabolism in mammals and fish cannot entirely account for the observed differences in toxicity. Whilst it is difficult to justify such an exact conclusion from the data, studies on target-organ sensitivity (Glickman & Lech, 1982) do indicate that *S. gairdneri* may be physiologically more sensitive to permethrin than the mouse, possibly reflecting different specificities at the site(s) of pyrethroid action.

It is thus probable that the greater sensitivity (approximately 100-fold) to synthetic pyrethroids of fish compared to mammals is the result of differences in both metabolism and physiological response.

Invertebrate and fish toxicity in water–sediment systems

The synthetic pyrethroid insecticides are rapidly and strongly adsorbed in soil and in water–sediment systems (see Chapter 5). Consequently pyrethroids will only be present in the water phase for a relatively short time. Many aquatic organisms may, however, be exposed to bottom or suspended sediments (or to other surfaces) to which the chemical has been adsorbed.

Studies of the uptake of a variety of chemicals from water and water–sediment systems have demonstrated much less bio-availability from the adsorbed phase than from the aqueous phase (Matsumura, 1977; Hamelink, 1979; McLeese et al., 1980a; Lynch & Johnson, 1982). A number of laboratory studies carried out with pyrethroids have also shown a reduction in toxicity to invertebrates and fish in the presence of bottom sediments or particulate material in the water.

Table 4.26 shows the toxicity of permethrin and cypermethrin to *Daphnia magna*, *Cloeon dipterum* and *Asellus aquaticus* in the presence and absence of soil (unpublished data, ICI Plant Protection Division). When permethrin was applied to the water surface (150 ml volume, 7 cm depth of water), a layer of soil (10 g) in the bottom of the vessel reduced toxicity to *D. magna* approximately 5-fold. When permethrin or cypermethrin was added to the soil 4 hours before flooding, the 72-hour EC50 to *D. magna*, *C. dipterum* and *A. aquaticus* was 200- to 400-fold lower than when in water alone. There was no significant difference in toxicity between systems where the treated soil remained as an undisturbed bottom layer and when it was kept in suspension.

First instar *D. magna* and the fish *Lebistes reticulatus* were exposed to a series of concentrations of permethrin for 21 days, in vessels containing a bottom layer of soil (unpublished data, ICI Plant Protection Division). The pyrethroid was added to the water surface, with the organisms present, to give theoretical water concentrations of 0·01–1000 µg ai/litre (300 ml water, depth approximately 10 cm; 25 g sandy loam soil). *D. magna* survived and

Table 4.26. Effect of adsorption on toxicity of pyrethroids to aquatic invertebrates[a]

	72-h EC50 (μg ai/l); where soil present, pyrethroid mixed into soil 4 h before flooding						48-h EC50 (μg ai/l); permethrin applied to water surface	
Organism	Permethrin			Cypermethrin				
		Water + soil			Water + soil			Water + bottom soil
	Water	Bottom soil[b]	Suspend-ed soil[b]	Water	Bottom soil	Suspend-ed soil	Water	
Daphnia magna	3·4	990	810	1·6	310	250	0·5	2·5
Cloeon dipterum	0·027	12	8·4	0·012	2·9	3·3		
Asellus aquaticus	0·085	19	13	0·009	2·8	3·1		

[a] Unpublished data (ICI Plant Protection Division). Values for 'soil' systems are calculated as if the pyrethroid is evenly distributed in the water phase.
[b] Bottom soil = undisturbed layer of soil in vessel; suspended soil = soil kept in suspension throughout test period.

reproduced normally at up to 10 μg/litre, but died at higher concentrations. *D. magna* reintroduced into the 100 and 1000 μg ai/litre vessels after 2 and 8 days, respectively, survived and reproduced. Wherever *D. magna* survived, parthenogenic reproduction was also observed. The fish were unaffected up to 100 μg ai/litre, and also when reintroduced at 1000 μg ai/litre 2 days later. It is noteworthy that direct application of 200 g permethrin ai/ha (maximum recommended field rate) to the whole surface area of a body of water 50 cm deep would only result in a theoretical maximum concentration of 40 μg ai/litre. In agricultural practice aquatic contamination by drift will be much less.

A similar effect was seen when *Salmo gairdneri* was exposed to an emulsifiable concentrate formulation of cypermethrin in microfiltered mains water and in pond water containing 15 mg/litre of suspended solids (unpublished data, Shell Research). In the static test system used, a concentration of 2 μg cypermethrin ai/litre (below the 24 hour flow-through LC50, Table 4.21) distressed fish in the filtered water, but not in the pond water; and after 48 hours one fish had died in the filtered water. At 5 μg ai/litre all the fish in the filtered water showed a gross loss of balance within 3 hours and were dead by 24 hours, whereas fish in the pond water showed no more than an occasional coughing reflex and were all alive after 7 days exposure.

Friesen (1981) and Friesen et al. (1983) studied the long-term effects (up to 70 days) on *Hexagenia rigida* of varying lengths of exposure to permethrin in water–sediment systems. The authors concluded that the nymph stage is affected by permethrin adsorbed to sediment days or even weeks after application. However, careful examination of the methods used, control mortalities and the mathematical and statistical interpretation of the original data (Friesen, 1981) suggests considerable doubt over this conclusion as presented.

A potentially adverse effect of pyrethroids and three other pesticides was claimed by Stratton & Corke (1981) who found that permethrin and cypermethrin induced adhesion to *Daphnia magna* of particulate material (algae, bacteria and silica) added to the water at the start of the exposure phase. However, the particles, principally attached to the setae and setules, did not significantly increase *D. magna* mortality due to permethrin over a 48-hour period (60–80% mortality at 0·5 µg ai/litre). During studies of the toxicity of permethrin and cypermethrin to *D. magna* in the authors laboratory, adhesion of particulate material was no greater for pyrethroid-treated organisms than for controls.

Studies of bio-availability in water–sediment systems are also discussed on pp. 238–246, in relation to accumulation of pyrethroids. A study of the effects of "adsorbed" cypermethrin on aquatic organisms in experimental ponds is discussed on p. 222 ff.

Field studies

Simulated field studies

Where pesticides are found under laboratory conditions to cause adverse effects on aquatic organisms it is accepted practice that further investigation be carried out in natural aquatic environments. Pond studies have been considered by many workers to be of value either prior to or even instead of large-scale field studies (Hurlbert et al., 1970, 1972; Macek et al., 1972; Butcher et al., 1975, 1977). Experiments using natural field ponds mostly have two major shortcomings: (1) replication is not usually possible (although several ponds may be studied), and (2) an adequate control pond is unlikely to be available. Crossland (1982), studying cypermethrin, solved the latter problem by building a concrete partition across a natural pond. However, to avoid these difficulties some workers have attempted to simulate field conditions in laboratory ecosystems, whilst others have used purpose-built experimental ponds.

Investigations of the effects of pyrethroids on communities of aquatic organisms have been carried out in laboratory tanks, in small and large natural ponds and in purpose-built experimental ponds. The results of all these studies are summarised in Table 4.27. Miura & Takahashi (1976) carried out "simulated" field tests with fenvalerate and crustaceans in 25-litre outdoor aquaria, although there was no mention of sediment being present in their tanks. One application equivalent to 53 µg ai/litre was applied. Copepods and ostracods were unaffected but amphipods and cladocerans were eliminated within 48 hours of treatment. The same authors also bioassayed water collected

Table 4.27. Effect of pyrethroids on aquatic organisms following direct applications to simulated, experimental and natural field ponds

Pyrethroid and (Ref. nos.)	Fenvalerate (264)	Fenvalerate (173)	Cypermethrin[a]	Permethrin (190, 193)	Permethrin (242, 244)	Cypermethrin (47)	Permethrin[a]
Pond type and size	Laboratory tanks 5–6 litre, flow-through	Experimental ponds 30 × 6 m	Experimental ponds 5 × 5 m by 1 m deep (see Figure 4d)	Experimental ponds 4 × 7 m by 0.35 m deep	Limnocorrals 25 m² by 4 m deep	2 Natural ponds, 20 × 5 m	3 Natural ponds, 0.02 – 0.4 ha
Application rate	µg ai/l 0.01, 0.1, 1.0, 10	g ai/ha 28, 84	g ai/ha (at 5–7 day intervals) 8 × 1.4, 2 × 1.4, 2 × 14, 2 × 1.4 on soil[c]	g ai/ha 56, 112	µg ai/l 5, 50	g ai/ha 100	g ai/ha 210
Study period (days after final application)	56, 7	6[b]	350 (fish, 14)	7–16	100	14, 112	350
Taxonomic group	Effects on organisms[d]						
Microorganisms	–	–					–
Coelenterata	–						–
Platyhelminthes — Turbellaria	–	–	(–)	–		(–)	–
Rhynchocoela	–		–				–
Rotifera					–		–
Annelida — Oligochaeta			–			(–)	
Hirudinea			–				
Polychaeta	–, ++		–				–

Mollusca	Gastropoda	−	−		−	(−)	−	−	−
	Bivalvia	−	−					++/R	++/R
Arthropoda (Crustacea)	Cladocera	+++/R	++/R		++/R	(−)	++/R	+++/R	−
	Ostracoda	+++/r	−		+++/R		++/R	+++/R	+/R
	Copepoda	++	++/R	+/R	++/R		++/R		
	Amphipoda		+++	++/R	+++/−	++/r	+++/−	+++/−	+++/r
	Isopoda			++/R	++/−	+++/r	+/R	+++/−	
	Cirripedia	−	−						
Arthropoda (Insecta)	Collembola								
	Ephemeroptera	+++	+++/R	+++/R,r	+++/R	+/R	+++/r	+++/r	+++/r
	Plecoptera			++/R	++/R	(−)	+++/r	+++/R	+++/R
	Odonata	++/r	(+++)/R	(+)/R	(+)/R	++/R	(+++)/r	+++/r	+++/R
	Hemiptera	++/r	+++/R	(+)/R	(+)/R	++/R	+++/r	+++/R	+++/r
	Coleoptera	+/r	++/R	(+)/R	(+)/R	++/R	(+++)−	+++/R	++/r
	Trichoptera								+/r
	Diptera	+++/R	++/R	+++/R	++/R	+/R	++/r	+++/R	++/R
	Lepidoptera							+++/R	
Arthropoda (Arachnida)	Hydracarina	+++/R	(−)	(−)	(−)	+/r	+++/r	+++/r	+++/r
Echinodermata		(−)							
Chordata	Cephalochordata	(−,+)							
	Tunicata	+,++							
	Teleostei	−,++	−	−	−		−	−	−,++
	Amphibia		−				−	−	−

[a] Unpublished data (ICI Plant Protection Division).

[b] Samples (approx. 5 litres) of water were collected from the pond and bioassayed in the laboratory.

[c] Mixed with soil for 12 hours before applying to water surface.

[d] −, no adverse effect; +, minor adverse effect; ++, >50% mortality; +++, >95% mortality; /R, complete recovery within 6 months (or test period if shorter); /r, some recovery within 6 months (or test period if shorter); /−, no recovery within 12 months (or test period if shorter).

from fenvalerate-treated experimental ponds (Table 4.27) on each of the 6 days following application. Cladocerans (*Ceriodaphnia* spp. and *Alona* spp.) and ostracods (*Cyprois* spp.) were mostly killed in water taken two hours after application but less than 50% mortality occurred after 1 day and 5% or less by day 5.

Tagatz & Ivey (1981) investigated the effect of fenvalerate on estuarine organisms under laboratory conditions using aquaria colonised "naturally" in the laboratory and field. With concentrations of 0·01–10 µg ai/litre, under flow-through conditions for 56 days, no effects were seen on mixed populations of molluscs, coelenterates, rhynchocoelids and annelids (except for the latter at the highest rate). Severe effects on amphipod populations were noted at all but the lowest rate, where no change in either density of animals or in number of species was observed (see Table 4.27). However, the method of exposure employed (maintained concentrations of fenvalerate for 56 days) is totally atypical of field conditions. The application route used in the study can only represent direct entry into the water phase as could occur from spray-drift, yet in the field only the margins of estuarine environments are likely to be contaminated by this route. Furthermore localised concentrations in estaurine water, although possibly initially within the range used by Tagatz & Ivey (1981), will be rapidly reduced by dilution and by adsorption.

Most other workers have used outdoor ponds, applying pyrethroids at concentrations ranging from field rate (or even higher) to those that more typically result from spray-drift. Such studies are described below.

Extensive studies of the effects of permethrin applied to three natural ponds with between 0·02 and 0·4 ha surface area and 0·3–0·7 m mean depth were carried out in 1976/77 by ICI Plant Protection Division (unpublished data). The pyrethroid was applied by a single spray nozzle traversed back and forth across each pond using a boat (Figure 4.4a, b). The ponds were treated with a single application of 210 g ai permethrin/ha (as an E.C.) during early summer. The effects of permethrin on the organisms in the ponds are summarised in Table 4.27. No control pond was included but the flora and fauna were monitored for up to 2 weeks prior to treatment. Pond samples were examined for 1 month after permethrin was applied and then again 6 months and 1 year after application, samples being taken at several sites in each pond. No effects of the pesticide were seen on microorganisms or protozoa in the water or sediment; from measurements of biomass (estimating ATP using luciferase/luciferin), total number of propagules (by direct microscopy after staining with europium chelate, see Anderson & Slinger, 1975), number of algae (by phase contrast or ultra-violet microscopy) and number of viable aerobic microorganisms that grow on a general laboratory nutrient medium (Oxoid Nutrient Agar). The pond fauna was assessed using pond nets to catch large invertebrates and amphibians, plankton nets for small invertebrates and emergence traps for those insects which leave the water on becoming adult. The permethrin treatment had no adverse effects on Hydra, Turbellaria,

Rotifera, Oligochaeta, Hirudinea, Ostracoda, Gastropoda, Bivalvia, Urodela and Anura tadpoles. There was a major kill of most arthropods, as would be expected with the high rate of permethrin applied, although some groups were recovering within a month of treatment, no doubt aided by a rapid decline in permethrin concentration in the water. Organisms with a short life cycle and those with resistant stages, for example *Daphnia* spp. with ephippia, are most likely to re-establish populations quite rapidly. A year after treatment only one group (*Asellus* sp. initially present only in one pond) had failed to reappear and one other made a poor return. The remainder made partial or complete recoveries. However, the interpretation of such observations in field ponds is complicated by naturally occurring seasonal changes in populations. Fish were studied using caged *Cyprinus carpio* and *Salmo gairdneri* and by seine-netting the native population, mainly *C. carpio*, in one pond (Figure 4.4c). Some deaths of caged populations occurred in two ponds, but only in one case were these attributable to permethrin. After application an increase in activity was noted among the free fish in the very shallow pond, the only pond with an established fish population, and a considerable number died within the following two days. Two months later, this pond dried up due to a drought. The number of live fish rescued suggested than only a small proportion of the population had died due to the pyrethroid. However, with the high rate of application and the shallowness of the pond this was not entirely surprising.

Studies of cypermethrin, applied as an E.C. formulation to small mature, natural ponds at approximately field rate (100 g ai/ha) have been reported by Crossland (1982). In a preliminary trial a $20 \times 5 \times 0.8$ m deep pond was studied for only 2 weeks following treatment. In a later more detailed 16-week study the pyrethroid was applied to one-half of a $20 \times 5 \times 0.7$ m deep pond previously divided with a concrete wall to provide treated and control areas. Residues, fish (a natural population of *Tinca tinca* and *Scardinium erythrophthalmus* in the initial study and introduced *S. erythrophthalmus* in the second study), adult and young amphibia (*Triturus* sp.) and macroinvertebrates were quantified at regular intervals. Concentrations of residues in the water declined rapidly, whilst levels associated with surface vegetation and sediment increased. However, attempted calculation of the mass balance of cypermethrin in all compartments sampled gave less than 50% recovery (as a percentage of applied) after 1 hour and less than 5% after 4 weeks. Submerged vegetation, to which at least some of the cypermethrin would have probably adsorbed, was not sampled. The effects of the treatment on aquatic organisms (Table 4.27) were similar to those described above for field pond studies with permethrin. There were no deaths amongst the fish, although the concentrations of cypermethrin in the water exceeded the laboratory 96 hour EC50 value for *S. erythrophthalmus* (0.4 µg ai/litre) for between 4 hours and 7 days (not analysed between these times). No adverse effects were seen amongst the amphibia, oligochaetes (although numbers were small), turbellarians or gastropods. Arthropods were markedly affected by cypermethrin and,

(a)

(b)

Figure 4.4. Studies of the effects of pyrethroids in natural and experimental ponds. (a) Pond of
0·36 ha sprayed with permethrin; (b) spraying pond by transverse "sweeps"; (c) catching native
fish by seine netting (unpublished study, ICI Plant Protection Division); (d) experimental ponds
for cypermethrin studies (ICI Plant Protection Division)

(*c*)

(*d*)

although the data were rather limited, repopulation of the pond by members of most taxonomic groups was apparent by the end of the study (16 weeks after treatment). No effect was observed on macrophytes but a mat of filamentous algae developed in the treated pond within a few weeks of spraying, probably caused by the absence of herbivorous "grazers" such as the Crustacea and the Ephemeroptera nymphs. Analysis of the overall macroinvertabrate data showed a severe depression in numbers of individuals and species per sample and in the Shannon diversity index (Shannon & Weaver, 1949) for between 4 and 6 weeks post-application. By week 10 the numbers of species and the diversity index had returned to the control pond levels, although total numbers of organisms remained depressed.

Solomon et al. (1980) have carried out studies in limnocorals sited in the centre of a large lake. Permethrin was injected directly into the water column, thus was likely to have a more severe effect than following natural entry routes. Effects on cladocerans and copepods were observed (Smith et al., 1981) but even at the high rates used recovery of populations occurred within 7 weeks (Table 4.27).

Studies in replicated experimental ponds have been carried out with permethrin and deltamethrin (Mulla et al., 1981) and with cypermethrin (Stephenson, 1984; unpublished data, ICI Plant Protection Division), and are discussed below.

Control of mosquito larvae has on occasions been partly achieved using a number of species of predatory fish, for example *Gambusia affinis*, *Cyprinodon macularius* and *Tilapia mossambica*. Should pyrethroids be used to supplement the biological control it is clearly important that the fish are not put at risk by the insecticide. Mulla et al. (1981) studied the combined effect of two species of fish (*G. affinis* and *C. macularius*, commonly used in mosquito larvae control) and permethrin or deltamethrin, on mosquito larval populations, principally *Culex tarsalis*. Concrete tanks approximately $6 \times 8 \times 0.3$ m deep were filled with water and left 2–3 weeks for mosquito larvae and other organisms to become abundant. Neither bottom sediment nor macrophytes were present. Fish were added and after 2–5 days the pyrethroid was applied on 6 to 8 occasions at weekly intervals. E.C. formulations were used at rates of 1.1 and 5.5 g ai/ha for deltamethrin and 28 and 140 g ai/ha for permethrin. No effects on fish were seen with any treatment and no mosquito larvae survived the initial high rate spray of deltamethrin. Although the first applications of "low rate" deltamethrin and both rates of permethrin had a substantial effect on mosquito larval population, 3–6 applications were necessary for a complete kill. On all occasions the pyrethroid and fish removed the larvae quicker than fish alone. Mats of algae (*Cladophora* sp.) were formed in the treated ponds. This in effect reduced cannibalism of young fish and resulted in more fish being recovered from the treated ponds than the untreated.

Using a very similar system of concrete ponds (see Figure 4.4*d*, and Table

4.27) but with sediment, macrophytes and a diverse invertebrate population, researchers at ICI Plant Protection Division (unpublished data) examined the effects of cypermethrin on an established fauna and flora, at a rate more typical of spray drift contamination. Two (replicate) ponds treated in the summer of 1980 were compared with two controls. Applications of 1·4 g ai cypermethrin/ha (approximately 1% of field rate; 36% E.C.) were made on each of 8 occasions at 5-day intervals. A high spray volume (3200 litre/ha) and large droplet size were used, preventing loss by drift and ensuring that a large proportion of the application entered the body of the water and was not lost by rapid adsorption to emergent and surface macrophytes. The theoretical concentration of residues if all the chemical was distributed evenly throughout the water column (0·16 µg ai/litre) was not reached, but immediately after spraying the concentrations at 1, 5 and 70 cm deep were 0·16, 0·09 and 0·08 µg ai/litre, respectively. These levels decreased rapidly to 0·07, 0·05 and 0·06 µg/litre, respectively, after 4 hours and to 0·02 µg ai/litre at all three depths at 21 hours. By the third day the residue level was below the analytical limit of determination (0·01 µg ai/litre). At no time could residues be detected in the sediment or on macrophytes (levels were less than 10 µg ai/kg).

Cyprinus carpio approximately 12 cm long, held in cages in treated and control ponds, remained healthy until removed 2½ weeks after the last application. Invertebrate populations were sampled on 2 occasions before treatment and on 8 occasions in the two months following the first application and 4 more times in the next 8 months. Benthic and surface artificial substrate samplers, zooplankton tubes capable of collecting a vertical column of water 75 × 5 cm in diameter, emergence traps and visual assessments were all used to determine the effects of cypermethrin on the pond invertebrate population. The results are summarised in Table 4.27. There was no treatment effect on Turbellaria or Collembola, nor on the Rotifera. The numbers of Oliogochaeta and Hirudinea and probably Ostracoda (very variable) were increased in the treated ponds. The crustacean Asellidae and Gammaridae were eliminated (absent during the 12-month study period but present again 12 months later) as a result of the cypermethrin applications and the numbers of Hemiptera, Coleoptera and possibly Odonata were reduced. No effect was seen on molluscs (Lymnaeidae, Hydrobidae and Planoribidae were common and Valvatidae and Ancylidae present in small numbers) until after the final application when a decrease in numbers was observed. This was probably not directly due to cypermethrin but more likely resulted from predation by the increased numbers of Hirudinea (principally *Helobdella stagnalis* and *Haemopsis sanguisuga*). Cladocera initially decreased but recovered during treatment and became more abundant in the treated ponds, though with a different species composition. Other groups showing a decline followed by a marked recovery were the Copepoda, Ephemeroptera and Chironomidae. The major reductions of sensitive species occurred after the first spray and multiple applications did not appear to cause cumulative effects. Microbial assessment, over a 20-week period, of numbers of viable bacteria, biomass (ATP) and chlorophyll showed seasonal variation but no effects due to the insecticide. Macrophytes increased in all ponds throughout the trial and, as also reported by Crossland (1982) and Mulla et al. (1981), a mat of filamentous algae developed in the treated ponds, probably due to the reduction in grazing arthropods.

In a further study with cypermethrin E.C., eight experimental ponds were used (unpublished data, ICI Plant Protection Division). Two were sprayed at approximately 10% of field rate ("high" rate; 14 g ai/ha) and two at 1% of field rate ("low" rate; 1·4 g ai/ha), both with volumes typical of agricultural spraying (200 litres/ha) applied through fan nozzles producing a conventional spray droplet size. Cypermethrin was also mixed for 12 hours with the fine fractions of a pond sediment (course sand : fine sand : silt : clay, 0·2 : 29 : 23 : 48; 12% organic matter) and applied to the third pair of ponds at a rate equivalent to the "low" spray rate, using 3·5 kg dry wt sediment per pond. Applications were made twice to each pond, 7 days apart. The remaining two ponds were not treated with cypermethrin but one received two applications of untreated sediment. During spray application all ponds other than the one being treated were kept covered to prevent contamination. Results are shown in Table 4.27 and Figure 4.5.

The "high" rate spray caused a marked decrease in numbers of *Gammarus* and *Asellus* (Malcostraca), Diptera, Emphemeroptera and some Cladocera. Within 1–2 months of treatment all of these groups were again present at numbers comparable to the controls, except for the Malacostraca, although *Asellus* sp. was present in one of the two ponds. Some affected and dying Coleoptera, Hemiptera and Odonata were seen in the ponds soon after treatment, but there were no significant effects on the overall population numbers. No effects were observed on other invertebrates or on caged fish (*Cyprinus carpio*, 1–2 g weight).

Following the first "low" rate spray application, Ephemeroptera and Diptera again decreased in numbers, and after the second application Malacostraca also. All groups showed some recovery within one month and, apart from the Malacostraca, complete recovery within two months. No other effects were noticed.

The treated sediment application caused a marked decrease in the numbers of *Asellus* sp. and *Gammarus* sp. caught after treatment, although the former was present in both ponds and the latter in one following both applications. A small and transient reduction in numbers of Ephemeroptera and Diptera was observed, less than occurred with either spray mate. There were no effects in the control pond to which untreated sediment was applied. The effect of cypermethrin applied with sediment on Malacostraca was intermediate to that of the "high" and "low" rate spray treatments. On all other organisms the treated sediment had less effect than either spray rate.

Pyrethroid insecticides have considerable potential for use in the control of

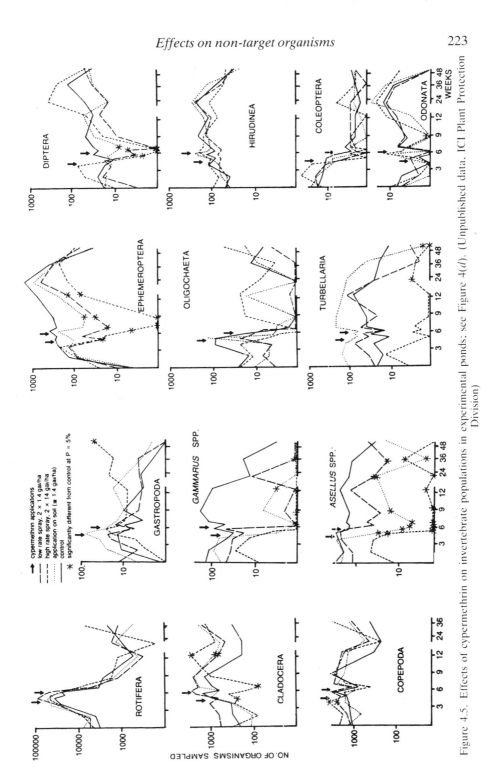

Figure 4.5. Effects of cypermethrin on invertebrate populations in experimental ponds; see Figure 4(*d*). (Unpublished data, ICI Plant Protection Division)

some pests of rice crops. However, fish may be present in rice paddies and are even important sources of food protein in some rice growing areas. For example, in West Malaysia in the early 1970s, fish provided additional income to 35% of rice farmers (Yunus & Soon, 1971), although increasing agricultural developments may have reduced this figure in recent years. Laboratory toxicity data have caused concern over the possible effects on fish in rice paddies, so far preventing pyrethroid development in this market. Field studies with commercial application rates of cypermethrin in experimental rice paddy plots in Korea showed low mortality of caged *Cyprinus carpio* populations (Stephenson, 1984). A number of 6 × 9 m by 0·15 m deep plots (with a 6 × 0·5 by 0·45 m deep trench on one side) were prepared in a paddy field using rigid plastic and earth bunds. Five days after flooding, 40 *C. carpio* of approximately 5 g were placed in cages in each plot (3 plots per treatment), and after 2 days acclimatisation cypermethrin E.C. was applied by knapsack sprayer at 15 and 40 g ai/ha. Control plots were included. At this time the water in the plots contained 66–400 mg suspended solids/litre; however, it is not stated whether this was before or after spraying and whether the spray operator walked through the plot during application. Mortality of *C. carpio* 7 days after spraying was 7% in the control and 13% and 15% in the "low" and "high" rate cypermethrin plots, respectively. In the control and treated plots mortality was lowest in fish cages placed in the deeper trench.

On a larger scale, Stephenson (1984) and Lhoste & L'Hotellier (1982) investigated the effects of application of cypermethrin and deltamethrin, respectively, to fields of rice containing fish. No fish mortalities were reported in either case. Stephenson (1984) aerially applied 25 g ai/l cypermethrin at 2 l/ha to a 250-ha paddy in Spain with 50 cm high crop (giving dense cover) in 5–10 cm water. Specimens of *Cyprinus carpio* (5–10 cm long) were caged in various areas of the paddy, which had a slow intermittent water flow. Fewer details are given for the deltamethrin studies which were carried out in Ivory Coast and the Philippines. In the Ivory Coast trials, 89 fish of various species including *Barbus spurelli*, *Hemichromis bimaculatus* and *Epiplatys* sp. survived in water containing 100 μg ai/l deltamethrin. However, the water flow was such that the "deltamethrin passed rapidly over the fish". At an application rate of 10 g ai/ha the theoretical maximum concentration of deltamethrin in paddy water 10 cm deep would be 10 μg/l, thus the study was at a very high rate. In the Philippine rice study 30-day-old *Tilapia mossambica* "resisted" a concentration of 50 μg/l deltamethrin.

Although further trials are undoubtedly required, the above studies do indicate that the pyrethroid insecticides currently marketed may not have adverse effects on fish populations in rice paddies. However, current concern that there is an inadequate safety margin has led to active interest in the development of new pyrethroids that are several orders of magnitude less toxic to fish than those presently available (see, for example, Anon., 1981; Ide et al., 1983).

Controlled experiments in ponds treated by spray applications to the total surface area have demonstrated that pyrethroids can have a rapid and severe, but transient, effect on the aquatic fauna. However, even at rates up to and above field rate, effects on fish are generally small or absent. Of a wide range of aquatic invertebrates studied only some of the crustaceans and insects are severely affected. The data from a number of studies do also show that these insect populations and some Crustacea recover quite rapidly and that recovery is independent of application rate. This would appear to be due to the rapid fall in the concentration of the pyrethroids due to adsorption, and to the ability of many of the organisms to rapidly recolonise – from the survivors, from resistant stages and from "re-inoculation" of the ponds with eggs layed by the adult winged stages. The Malacostraca are very sensitive to pyrethroids and have little ability to recolonise an enclosed pond system, except from survivors.

Agricultural applications

During the application of pyrethroids to agricultural and horticultural crops small amounts of the products may contaminate aquatic environments by spray-drift and/or run-off. Studies described above in which pyrethroids were applied directly to bodies of water at rates greater than those likely to occur in agricultural practice have shown that whilst the risk to fish is negligible, some aquatic arthropod populations may be severely affected; although the potential for rapid recovery was also noted.

In a number of field trials, crops adjacent to or surrounding bodies of water have been treated with permethrin and cypermethrin (Table 4.28). Sites were deliberately chosen to give a high likelihood of aquatic contamination during spraying by spray-drift and subsequently by soil run-off. In trials by ICI Americas Inc. (unpublished data) permethrin was ground-sprayed, and permethrin and cypermethrin aerially sprayed, up to 17 times in a growing season, on to cotton fields next to 2–3 ha ponds. Crossland et al. (1982) applied double treatments of cypermethrin by tractor boom to potatoes and sugar beet in fields in the UK next to or surrounding 0·02–0·1 ha ponds. Applications were also made to vineyards alongside which ran streams or drainage ditches. The application practices and rates were in all cases consistent with those recommended for normal agricultural use.

The data, summarised in Table 4.28, showed that, as predicted, there was negligible direct hazard to natural fish populations or to caged fish. The observations on the macroinvertebrates were also consistent with those from previous work (see Table 4.27). Some arthropods, in particular surface-living Hemiptera (e.g. Gerridae), Coleoptera (e.g. "whirlygig beetles") and Ephemeroptera were affected at various times during the application programmes. In running water (Crossland et al., 1982) this mostly took the form of a transient increase (for 48 hours after application) in "stream drift", but which had no detectable effect on the numbers or diversity of the benthic populations.

Table 4.28. Effect of pyrethroids on aquatic organisms following applications to adjacent agricultural land

Authors	Crossland et al., 1982		Unpublished data (ICI Americas Inc.)		
Pyrethroid (E.C.)	Cypermethrin	Cypermethrin	Permethrin	Permethrin	Cypermethrin
Experimental site	3 ponds (0·02–0·1 ha, 1–2 m deep) next to sugar beet and potato fields.	Two streams and a drainage ditch, along the edge of vineyards.	A 2-ha cotton field adjacent to a 1·2-ha pond.	17 ha of cotton adjacent to a 3-ha pond (av. depth 2·5 m)	17 ha of cotton adjacent to a 3-ha pond (av. depth 2·5 m)
Application rate and method	70 g ai/ha in 300 l/ha ground application. Two applications 28 days apart.	30 g ai/ha to vineyards by streams and 45 g ai/ha to vineyard by ditch (mist blower 400 l/ha).	224 g ai/ha ground application; 17 times at 5-day intervals.	224 g ai/ha by fixed wing aircraft; 15 applications of 20 l/ha at 5-day intervals.	150 g ai/ha by fixed wing aircraft; 16 applications of 27 l/ha at 5-day intervals.
Study period (days after final application)	7	1–2	50	36	300
Taxonomic group	Effects on organisms[a,b]				
Rotifera			—		—
Annelida Oligochaeta	—	—	—		
Hirudinea	—		—		
Mollusca Gastropoda	—	} −	—		
Bivalvia				(−)	

Arthropoda (Crustacea)	Cladocera	—	—	—	(−)
	Ostracoda	—	—	—	−
	Copepoda	—	—	—	
	Amphipoda	—	+/R	—	+/R
	Isopoda	—	−	—	−
	Decapoda	—	—	C−^d; N−	−
Arthropoda (Insecta)	Ephemeroptera	—	+/R	(++)/r	+/R
	Odonata	−,+^c	—	(++)/r	−
	Hemiptera	—	+/R	(++)/r	+/R
	Coleoptera	−,+^c	+/R	−,−+^e	(+)/R
	Diptera	—	+/R	—	−
Teleostii		N−^g; C−^h	—	N−^f; C−	−

---

[a] −, no adverse effect; +, minor adverse effect; ++, major adverse effect; (including post-treatment absence of organisms); ( ), numbers small or results very variable; /R, complete recovery within 6 months (or test period); /r, some recovery within 6 months (or test period); −, no recovery within 12 months (or test period).

[b] C = caged organisms; N (and all other values) = natural populations.

[c] Immobilisation of specimens of *Notonecta, Gerris* and *Eristalis* ("rat tailed maggot") in one corner of one pond.

[d] *Procambarus clarkii.*

[e] Only surface ("whirligig") beetles affected.

[f] All species of fish in pond were collected and caged (*Tilapia* sp., *C. auratus, N. crysoleucas, I. punctatus, I. melas, M. americana, M. salmoides, L. gulosus, L. cyanellus* [see Table 4.17]).

[g] *L. macrochirus, F. notti, E. zonatum, G. affinis* (see Table 4.17).

[h] *L. macrochirus* only.

[i] Four very small *L. macrochirus* found dying at edge of pond after second application.

Recovery of drifting organisms following transfer to uncontaminated water was reported. However, it is difficult to assess the significance of this to the natural environment, where some of the invertebrates may drift with any residual pyrethroid and where the "drifters" may also be more susceptible to predation. In the UK ponds next to or within fields receiving two cypermethrin applications with a 28-day interval, Crossland et al. (1982) observed few effects that could be attributed to cypermethrin. Where 15–17 pyrethroid applications were made at 5-day intervals to cotton there were some apparent effects, most noticable at the edge of the pond nearest to the sprayed field (Table 4.28).

Whenever effects were observed the aquatic system was also seen to recover during the season of the treatment and mostly very rapidly. The species composition may in some situations have been different before and after a series of pyrethroid applications, but such changes are also a feature of aquatic environments not receiving pesticide treatments. No adverse effects were noted on the macrophyte populations in any of the studies.

Deposition levels were measured on crops and water during spraying, and subsequently residue concentrations were determined in water and sediment. In all of the pond studies it was observed that the levels of pyrethroid deposited on to the water during spraying were several orders of magnitude less than those in the crops. Crossland et al. (1982) reported that surface water samples in ponds contained concentrations of cypermethrin initially in the range <5 μg/l (4 samples) to 6–23 μg/l (10 samples) but these declined rapidly such that after 24 hours the residues were <5 μg/l except in 1 of the 14 surface water samples collected. In water from 20–30 cm depth the concentrations did not exceed 0·07 μg/l and were mostly <0·02 μg/l in the 24 hours following both applications. In flowing water next to vineyards where the pyrethroid was applied by mistblower, initial surface concentrations were much higher than in ponds, at 140–1010 μg/l, but again decreased rapidly, to <20 μg/l after 3 hours. In subsurface water the maximum pyrethroid concentrations were 0·4–1·7 μg/l soon after spraying, but falling to 0·1 μg/l or less within 4 hours of application. In the pond studies of ICI Americas Inc., subsurface water residues were mostly below the limits of determination (0·02–0·05 μg/l).

In none of the studies were detectable amounts of the pyrethroids found in bottom sediments. However, sampling is difficult, for the parent chemical may be adsorbed at the very surface of the sediment, which is easily disturbed. Also sensitivity of the analytical procedure is in the order of 5–10 μg/kg. Residues were not detected in fish by Crossland et al. (1982), nor in fish and crayfish by ICI Americas Inc. (unpublished data).

Where agricultural or horticultural applications of pyrethroids result in spray-drift on to flowing bodies of water, any effects on aquatic invertebrates are likely to be very small and transient. This is a consequence of the intermittent use of the chemical, its rapid dilution and removal by stream flow and adsorption, and the recovery potential of the aquatic organisms by stream-drift and aerial entry. In large enclosed bodies of water only organisms at the margins are likely to be exposed to concentrations of pyrethroid that may have adverse effects, and again the system has the potential for rapid recovery. On occasions the sensitive invertebrates in small ponds may be put more at risk following spray-drift entry of pyrethroids. Here the rate of recovery will depend upon the reinoculation potential. Thus organisms without an aerial stage (for example Crustacea) may, if affected severely, recover very slowly.

Riverine forest applications

Laboratory and field studies have shown the synthetic pyrethroid insecticides to be highly active against pests such as blackfly (Muirhead-Thomson, 1977, 1981a, b; Davies et al., 1982, 1983; Bellec et al., 1983) and tsetse fly (Barlow & Hadaway, 1975; Molyneux et al., 1978; van Wettere et al., 1978; Baldry et al., 1978b, 1981). Blackfly (*Simulium* spp. carrier of onchocerciasis) and tsetse fly (*Glossina* spp., transmitting trypanosomiasis) populations can in part be controlled by applications of insecticides along the forested banks (riverine forests) of African rivers where these vector insects breed by aquatic larval stages. However, at the application rates required for pest control there is a strong possibility that many of the pyrethroid-sensitive non-target species might also be seriously affected. Laboratory studies by Muirhead-Thomson (1979) and experimental pond studies by Mulla et al. (1975) and Mulla & Darwazeh (1976) (see p. 198, p. 217) have confirmed this possibility, although the ability of the aquatic invertebrate populations to recover was also reported.

A number of field trials have been carried out to determine the possible aquatic side effects of the use of pyrethroid insecticides along tracts of African riverine forest in Upper Volta (Molyneux et al., 1978; Takken et al., 1978; Baldry et al., 1978a, 1981), Nigeria (Smies et al., 1980) and Ivory Coast (Everts et al., 1983). Applications were all by air (helicopter) and along approximately 4–15km tracts, spraying the inner edges of the fringing forest of each river bank (single swaths of about 30 m width). Considerable insecticidal contamination of the rivers, whose widths are not given, was inevitable.

In such large-scale field studies of rivers much of the data produced is difficult to translate into a quantitative format. Observations for mortalities were made and effects seen from drift-netting and other collection methods. However, the data which did not always quantify the population effects can mostly be used only in a relative sense over a number of sampling intervals. Furthermore, only one study was sufficiently long to provide an estimate of population recovery. Despite these deficiencies, the efforts of the researchers under difficult conditions do enable a reasonable picture of the effects on this type of aquatic ecosystem to be established. A summary of the results is given in Table 4.29.

Table 4.29. Effect of pyrethroids on aquatic organisms following applications to riverine forest areas[a]

Experimental details (ref. no.)		0·4-2-ha blocks around tributaries of the River Karami in Nigeria; 4-6 km spray-runs; up to 6 post-spray samplings (240, 247).			River Komoe valley in Upper Volta; 15 km spray-runs; studied for 1–10 days after final application (16, 177, 265).			River Marahoue, Ivory Coast; 15 km spray-run; 30-day study period (77)
Pyrethroid application rate (g ai/ha) (applied by helicopter)		Permethrin ULV; 200,300	Cypermethrin ULV; 100	Deltamethrin ULV;20,40	Permethrin ULV;1·9,4·3	Deltamethrin ULV; 0·19, 0·36, 2 × 12·25	Deltamethrin spray 12·5[b]	Deltamethrin 5 × 12·5[c]
Taxonomic group	Effects on organisms[d]							
Annelida	Oligochaeta	–		(–) to ++				
	Hirudinea	–		(++)				
Mollusca	Gastropoda	– to ++	–	–				
Arthropoda (Crustacea)	Decapoda	++		++/R	++	– to ++/R	++	– to ++/R
Arthropoda (Insecta)	Ephemeroptera	(++)	(++)/R	(++)/R	(–)	(–) to (+)	(+)	++/r
	Plecoptera							+/R
	Odonata	++	–	+ to ++	(+)	(–)	(+)	+/R
	Hemiptera	(–) to (+)	(+)	(++)	(+)	(+)	(++)/r	+/R
	Coleoptera	– to (++)	++	(–) to ++	(+)	(+)	(+)	+/R
	Trichoptera	(++)	–	++	(+)	(+)	(+)	+/R
	Diptera	++	–	++	(+)	(+)	(+)	
	Lepidoptera				(+)	(+)	(+)	
	Hymenoptera				(–)	(+)	(+)	
	Orthoptera				(+)	(+)	(+)	
	Homoptera				(+)	(+) to (++)	(+)	
	Dictyoptera				(–)	(+)	(+)	
	Thysanoptera				(–)	(–)	(–)	
	Neuroptera				(+)	(+)	(+)	
Arthropoda (other)	Hydracarina	–		+/R	+	+	(+)	+/R
Teleostii		–			–		(–)	

[a] Riverine forest = narrow bands of forested land along stretches of river bank. All applications to inner edge of both banks of river, except for 2 × 12·5 g ai/ha deltamethrin which was sprayed 15 m further away from the river.

[b] Deltamethrin applied after other non-pyrethroid insecticides.

[c] Single application of permethrin (40 g ai/ha) applied to part of treated riverine forest 2 weeks before first deltamethrin application.

[d] For key to effects see note (a) on Table 4.28.

In none of the studies were adverse effects seen on fish, although in an early trial in which an unspecified rate of permethrin (at 0·5% ai) was ground-sprayed (Smies et al., 1980), some of the fish in a pond were found dead. The authors suggested that this may have been due to a localised high concentration (as can occur with ground-spraying) or to spillage. On one occasion following aerial applications some small fish were seen to be disoriented for a short period after spraying.

Decapod crustaceans were in all trials adversely affected and, where recovery was examined at a later date, did not always re-establish to pre-treatment population levels. Whilst *Caridina africana*, a small shrimp, was again abundant in the year following pyrethroid applications (Smies et al., 1980; Baldry et al., 1981), another species, *Macrobrachium raridus*, did not recover (Baldry et al., 1981). Specimens of the large shrimp *M. vollenhovenii* were found paralysed following spraying with deltamethrin (Everts et al., 1983) but had recovered 2 days later.

Aquatic insect populations were also affected, but in the Upper Volta and Nigerian studies it was difficult to determine to what extent. Smies et al. (1980) reported only presence or absence of insect groups following a form of seine-netting of the bottom of the river. Numbers of organisms were often apparently small but Ephemeroptera and Hemiptera were identified as particularly affected. Ephemeroptera had recovered at the only site observed one year later. In the Ivory Coast River Marahoue study with deltamethrin (Everts et al., 1983), drift netting, artificial substrates and rock samples gave a clear indication of a major effect on the numbers of Ephemeroptera. However, a considerable recovery was evident 4 weeks after the final application. Although there were increases in numbers of Plecoptera, Odonata, Coleoptera, Trichoptera and Gerridae (Hemiptera) caught in drift nets, the overall populations seemed virtually unaffected. The diptera Simulidae and Chironomidae were also found drifting in large numbers for approximately 24 hours after each application. Numbers of simuliid larvae were simultaneously eliminated on rocks (whilst only reduced on artificial substrates), but recolonisation was seen within 1 week.

The aquatic invertebrate fauna of rivers whose banks are sprayed at regular intervals with pyrethroid insecticides for insect pest control are clearly at some risk. If very large lengths of riverine forest are treated each year it is difficult to escape the conclusion that on occasions some species such as the crustacean decapods may be eliminated and may not readily recover. The effects on insects, many of which appear less sensitive than the Crustacea and which can also re-invade by aerial life stages, will probably be much less drastic, recovery of the majority of those affected being likely to occur within a few weeks of the final pyrethroid application. The long-term impact to the ecosystem is not easy to quantify. Trichoptera which are considered by many workers to be important predators of the blackfly larvae (Jenkins, 1964; Burton & McRae, 1972) appeared to be much less sensitive than the Ephemeroptera or

Crustacea, and there was no evidence for significant changes in their numbers, thus any biological control of simuliids by these organisms is unlikely to
be reduced. This confirmed the observations of Muirhead-Thomson (1979)
with *Hydropsyche pellucidula* (found in Africa) and *Simulium* larvae in
laboratory tests. Whilst fish are not directly affected by the treatments,
concern has been expressed that the diminished aquatic invertebrate food
supply may eventually reduce fish populations. However, Kingsbury and
colleagues have shown that fish can adapt in such situations (see following
Section), and in addition the African studies indicate that the overall shortfall
in total aquatic invertebrates may be relatively small and short term.

It is worthy of note that normal stream-drift of organisms from the numerous very small feeder tributaries will (unless these are all also sprayed)
provide an inoculum aiding recovery. Furthermore, many of the rivers and
tributaries can undergo large natural seasonal changes causing widespread
mortality among fish and invertebrates, as a result of drying up, or stagnation.
It could be argued either that most effects of the insecticides are small by
comparison to those caused by natural events or that these compounds
impose an additional stress upon the ecosystem. No doubt further studies
involving larger areas will be carried out, but is is the author's opinion that the
potential benefits of insecticidal control of the insect pests in these riverine
forest areas far outweighs the changes that are likely to occur in the invertebrate populations. The only exception to this is where the numbers of an
economically important organism (such as the shrimp *Macrobrachium vollenhovenii*) may be significantly reduced. Risk–benefit analyses must be
considered in all such situations.

Pine forest applications

Insecticides have been increasingly used in forestry over the past few decades.
Canada has often been at the forefront of forest insecticide developments,
especially those suitable for control of the damage caused by the spruce
budworm, *Choristoneura fumiferana*. The damage from this pest can be
environmentally and economically devastating. Even in 1978 more than 30
million acres of forest became dead or moribund due to budworm attack and
at least another 70 million suffered moderate to severe defoliation. Quite
apart from the economic loss, defoliation on this scale causes massive changes
to the landscape and effects on both terrestrial and aquatic environments are
severe. For example, the animal population can change substantially. Deer
deprived of shelter move to other areas, influxes of rodents and other small
mammals are possible and bird populations change. There is an increased
potential for "wash-off" of soil into water courses. Sedimentation has been
seen to alter water quality and may, for example, adversely affect trout
breeding and fishing, although there appears to have been little study of this,
and the aquatic flora and fauna will inevitably change.

Since the mid-1970s much interest has been focussed on the possible use of pyrethrum and the synthetic pyrethroids for spruce budworm control. However, the high toxicity of these compounds to aquatic organisms under laboratory conditions and the large number of lakes, rivers and streams in forested areas gave rise to considerable economic and environmental concern over their introduction.

A coniferous forest study using natural pyrethrum was carried out in Colorado, USA by Pillmore (1973) who aerially sprayed 12–65ha blocks at rates of 22 and 45 g ai/ha. During the 24 hours after application increases in stream-drift of Ephemeroptera, Diptera and Plecoptera, but not Trichoptera, were observed, with numbers soon returning to pre-spray levels. At the lower rate pyrethrum had no observable effect upon the numbers of organisms in bottom samples. At the higher rate, however, a reduction in the bottom dwellers was seen during the short (30-hour) period of sampling.

A long series of detailed trials with permethrin has been carried out by Kingsbury and his colleagues of the Environment Canada, Forest Pest Management Institute (FPMI). These studies, with permethrin aerially applied as an oil-based formulation, are listed in Table 4.30 and the data are summarised in Table 4.31. The series of investigations over a 5-year period used decreasing application rates and increasing plot areas as the trial programme progressed. Data for effects on the aquatic organisms in different trials and from different sampling sites within trials are inevitably very variable because of the wide range of amounts of pesticide deposited, the result of interception by trees and shrubs. The highest levels were mostly those landing on bodies of water or in clearings where foliage cover is least. However, the actual residue values measured are of little importance, for the studies reflect what in practice would be typical levels of deposition following normal forestry spraying techniques at the applications rates needed to achieve satisfactory spruce budworm control. The most practical dosage of permethrin is considered to be two applications of approximately 17·5 g ai/ha several days apart (DeBoo, 1980). There is a wealth of detail in the FPMI publications, representing an enormous and well conducted programme of work, from which a consistent pattern of environmental effects and recovery does emerge (Table 4.31).

In trials in lakes and streams at least 18 species of fish representing 9 families (catfish, minnows, mud-minnows, perch, salmonids, sculpin, sticklebacks, suckers and sunfish) were exposed to permethrin aerially sprayed at rates between 9 g ai/ha and 140 g ai/ha. Fish mortality was not observed except in the earliest trials at the highest rates applied directly to small lakes. Caged fish died only when at or near the surface, and although a considerable number of the native fish died in a shallow lake directly sprayed at 140 g ai/ha, the author stated that "observations were made of large numbers of apparently unaffected fish of all species present in the lake". Although the deposition rate on the lake was 92% of the emitted spray, *Semotilus corporalis* was seen to be spawning

Table 4.30. Canadian forestry studies of the ecological effects of permethrin[a]

Year	Application area	Application rate (g ai/ha)[b]	Location	Aquatic site	Fish Effects	Fish Diet	Invertebrates Aquatic	Invertebrates Terrestrial	Other	Ref. no.
1976	Lake and surround	140	Quebec	(1)[c] Lake(10 ha × <3 m)	√NC		√N	√	Amphibia	132, 133
		35	Ontario	(2) Lake (30 ha × <14 m)	√NC		√N	√		
1977	5 km length of stream	70	Ontario	(3) Stream (sandy bed)	√NC		√N	√		140
	5–8km lengths of stream valley	70	Quebec	(4) Stream (rocky bed)	√N	√N	√N	√		
		35	Quebec	(5) Stream (rocky bed)	√N'	√N	√N	√		136
1978		17·5	Quebec	(6) Stream (mixed bed)	√N'	√N	√N	√		
	3–5km lengths of stream valley	8·8	Quebec	(7) Stream (gravel bed)	√N'	√N	√N	√		
		2 × 17·5	Quebec	(8) Stream (gravel bed)	√N'	√N	√N	√		135
		2 × 17·5	Quebec	(9) Stream (mixed bed)	√N'	√N	√N	√		
		2 × 17·5	Quebec	(10) Stream (mixed bed)	√N'	√N	√N	√		
	930ha block	2 × 17·5	Ontario	(11) Ponds (0·25–1 m deep)	√C	√N	√NC	√	Bees, birds amphibia, mammals	138
1979	640ha block	17·5	Ontario	(12) Streams (mixed beds) Ponds (1–2 m deep)	√C √C		√C	√	Mammals	137
1980	400ha blocks	17·5	Quebec	(13) Stream (gravel bed)			√N	√		
		17·5	Quebec	(14) Stream (rocky bed)			√N	√		139
		2 × 17·5	Quebec	(15) —						
		2 × 17·5	Quebec	(16) Stream (rocky bed)	√C	√N	√NC	√	Mammals	
	600ha blocks	17·5	New Brunswick	(17) Stream (gravel bed)	√C	√N	√NC	√		134
		2 × 17·5	New Brunswick	(18) Stream (rocky bed)	√C	√N	√NC	√		

[a] Studies carried out by the Forest Pest Management Institute (formerly the Chemical Control Research Institute), Canada, in association with Chipman Chemicals Inc.

[b] Oil concentrate formulations applied in 1–10 l/ha by fixed wing aircraft with Micronair atomiser (except 1980 Quebec, aircraft boom sprayer). Double applications were within a 5–10-day period.

[c] (1)–(18) identify studies in Table 4.31.

[d] N = detailed native population studies; N' = casual observations for mortalities; C = caged studies. √ = study carried out.

Table 4.31. Effect of permethrin on aquatic organisms following application to Canadian pine forests

Study no. (see Table 4.30) Taxonomic group	1	2	3	4	5	6	7	8	9	10	11	12	13	14	16	17	18
						Effects on organisms[a,b]											
Nematoda							(+)/R									(−)	(−)
Platyhelminthes Turbellaria	(−)		(−)													(−)	(−)
Rotifera																	
Annelida Oligochaeta	(−)		(−)			(−)	(−)				−		−			−	−
Hirudinea			(−)								−		−				
Mollusca Gastropoda	(−)		(−)					(−)			−					(−)	(−)
Bivalvia	(−)			(+)/R									−				
Arthropoda (Crustacea) Cladocera	++/r	++/R									++/r						
Ostracoda		++/R									(++)/r						
Copepoda	++/R	++/R									++/R						
Amphipoda	(++)/−			(+)/R							(++)/−						
Isopoda	(++)/−										(++)/−						
Decapoda																	
Arthropoda (Insecta) Ephemeroptera	(++)/−	(+)/r		++/−	++/r	++/R	+/R	++/R	++/R	++/R			++/R	+/R	++/R	+/R	++/R
Plecoptera	(+)	(+)		(+)/R	+/R	+/R	+/R	++/R	++/R	(+)/R		C−	+/R	+/R	C+ / N+/R	(+)/R	(+)/R
Odonata	(++)/R			(+)/R								C(−)	(++)/R	(−)			
Hemiptera																	
Coleoptera	−				(+)	(+)		(+)	(+)	(+)							
Trichoptera	(++)/−	(+)/R		++/r	++/R	(+)/R	(+)/R	+/R	(+)/R	(+)/R	(++)/−		+/R	+/R	(+)/R	(+)/R	(+)/R
Diptera	(++)/−	++/R		++/R	++/R	++/R	++/R	++/R	++/R	++/R	++/R		++/R	+/R	++/R	+/R	++/R
Lepidoptera (emergence traps)	(−)	(−)									(++)/r						
Arthropoda (other) Hydracarina		(+)			++	(+)	(+)	(+)/R	(+)/R	(+)/R			(+)/R	(+)/R	−	+/R	+/R
Teleostii[c]	C+ / N+	C++ / N−	C− / N−								C−	C−				C−	C−
Amphibia T. viridescens	(−)	(−)															
Rana spp. (tadpoles)	(−)										C−						

[a] Invertebrate data combined from sampling techniques and sampling sites; C = caged organisms; N (and all other values) = natural populations.

[b] For key to effects see note (a) on Table 4.28.

[c] Species of fish studied included various unidentified cyprinids and C. cognatus, C. inconstans, Etheostoma spp., I. nebulosus, Lepomis sp., L. gibbosus, L. lota, P. flavescens, S. corporalis, S. fontinalis, S. salar, Umbra spp. (see Table 4.17).

the day after treatment and *Lepomis gibbosus* reproduced successfully from eggs laid in the pre- and post-treatment (8 days) period. Observations for three years following application showed that permethrin had no significant effect on the fish populations. In later studies of forest streams, changes in the aquatic invertebrate fauna were seen to result in shifts in the diet of the fish. Studies of *Salmo salar*, *Salvelinus fontinalis* and *Cottus cognatus* mostly showed an initial gorging on affected aquatic insects followed in some, but not all trials, by a switch to a diet of terrestrial insects and/or the specific aquatic larvae (for example, chironomids in one study) that were least affected by the pyrethroid treatment. Whilst in these studies dietary change was necessitated by the effects of the pyrethroid, *S. fontinalis* have also been noted to increase their consumption of terrestrial arthropods in untreated streams during mid to late summer (Kingsbury & Kreutzweiser, 1979, 1980a; Kingsbury, 1982). However, not all species may be able to adapt so readily. *S. salar* generally inhabits fast flowing areas of streams and rivers and under normal circumstances is less likely to encounter and ingest drifting terrestrial organisms. In one of the most recent studies (Kingsbury, 1982), in which large forest blocks were treated with permethrin at $2 \times 17 \cdot 5$ g ai/ha, *S. fontinalis* disappeared from treated streams about 4 weeks after application. It was suggested that this species emigrated to untreated tributaries following displacement by the more aggressive *S. salar*, as a result of scarcity of food. Similar behaviour has been reported during competition for food in the absence of pesticides (Symons, 1968; Gibson, 1973). After a further 10 weeks, *S. fontinalis* populations were again present in the treated block. There was evidence that *S. salar* parr in treated blocks grew somewhat more slowly than those in control areas in the weeks following treatment (Kingsbury, 1982), this being consistent with the findings of a reduction in the weight of stomach contents. However, soon after, their growth rate increased such that by late summer fish development in treated and control areas could not be distinguished. Pesticide-induced dietary changes, even where aquatic invertebrate recovery was slow (at high rates of permethrin only; see below), did not appear to adversely affect the condition of *S. fontinalis* or ovary production in the year following application.

Amphibia (*Triturus viridescens*, spotted newt; and tadpoles of *Rana* spp.), present in a lake treated at 140 g ai/ha (Kingsbury, 1976a), were unaffected by the pyrethroid. Tadpoles of *Rana clamitans* were also caged in ponds in a later study (Kingsbury & Kreutzweiser, 1979) treated at $2 \times 17 \cdot 5$ g ai/ha. No mortalities occurred with any of these amphibia.

Aerial spraying of permethrin caused substantial disturbance to the aquatic invertebrate populations of ponds and streams. The Ephemeroptera and Trichoptera were the most sensitive groups, effects being apparent at rates as low as 9 g ai/ha, although all the arthropod groups studied (including crustaceans) were affected to some extent. Similar results to those seen in African rivers were reported. The post-treatment stream-drift ranged from one to over three orders of magnitude greater than the pre-spray level and was followed by a depletion, sometimes substantial, of the aquatic invertebrate population; largely due to losses from the most sensitive groups. Reductions in "downstream fauna", up to 2 km outside the treated area, indicated the movement of some residues of the insecticide in the water, although in the water at such sampling stations the levels were usually below the limit of determination. Stream benthos density and diversity returned to normal within one year of a 35 g ai/ha application but remained supressed for up to 16 months following treatment with 70 g ai/ha permethrin. At the lower rates (that is, those more representative of proposed field use) invertebrate populations started to show recovery soon after spraying, and within a few weeks to a few months had returned to normal, being slowest where double applications were used. Observations in the following year also indicated that invertebrate numbers were similar to or exceeded the pre-spray levels of the previous year.

Studies of the permethrin control of spruce budworm carried out to date have used forest blocks of up to 1000 ha. Whilst the aquatic invertebrates in ponds and streams in these trials recovered fairly rapidly from the effect of the pyrethroids it is not easy to predict the overall effects on the aquatic ecosystem, should much larger areas of forest be sprayed. However, following applications of 17·5 g ai/ha it is still likely that recovery of aquatic invertebrates will occur at rates similar to those recorded in the small blocks. Re-invasion of habitats in the small blocks will have occurred at least in part from organisms drifting down from untreated upstream areas and also from the aerial stages of some organisms. Yet the populations were never entirely eliminated and much of the recovery was probably from organisms "indigenous" to the aquatic environments within the sprayed area. The effects on fish populations are less easy to predict. Mortalities from field rates of permethrin will be rare and the fish seem well able to adjust their dietary intake to compensate for the reduction in some groups of aquatic organisms. Furthermore, any reductions in growth rate in the weeks following spraying seem to be substantially made up later in the season. What is more in doubt are the consequences of competition for food by species with differing degrees of aggression. If large areas were treated, the less aggressive fish would be unable to migrate from the sprayed zone.

Different insecticide classes have different environmental effects, thus there is certainly a case for considering the pyrethroids in a spray programme that makes use of their high degree of activity against spruce budworm. This may take the form of only one application of a pyrethroid in a season, further sprayings being with, for example, a carbamate or an organophosphate; or pyrethroids might be used during one season but not the next. Determination of the presence or absence of long lasting environmental or economic effects will only result from a stepwise increase in the areas of forestry that are permitted to be sprayed. Monitoring is expensive, thus specific indicator organisms only should be studied and methodology limited to that which is useful (for example, stream-drift is inevitable thus need not be investigated).

Fish studies should concentrate on measuring community composition, numbers and growth rates. Aquatic invertebrate populations will invariably be affected, therefore investigations should principally monitor recovery, particularly of the most sensitive organisms.

Conclusions

Natural pyrethrum and the synthetic pyrethroids have been extensively studied for their effects on aquatic organisms. In controlled laboratory studies with the insecticides in solution in water free of (or very low in) particulate material, and often using maintained concentrations, these insecticides are highly toxic to fish and many aquatic invertebrates. However, the pyrethroids are lipophilic, have low aqueous solubilities and in the natural environment are rapidly and strongly adsorbed to organic surfaces. In the adsorbed state bio-availability has been shown to be much reduced, such that the "toxic effect" is decreased by up to two to three orders of magnitude. Degradation of the parent molecule results in products of low toxicity to aquatic organisms.

Agricultural and horticultural applications of pyrethroids can result in small amounts of these products entering aquatic environments by spray-drift and/or run-off. Extensive studies of aquatic impact using experimental ponds and field ponds have shown that permethrin, cypermethrin, fenvalerate and deltamethrin are unlikely to have adverse effects on fish populations. Although aquatic invertebrate populations can be affected, and sometimes very visibly, the effects are usually localised, for example, to short lengths of streams or rivers or to the edges of field ponds. Consequently the effects on most organisms are transitory, recovery occurring by re-inoculation from survivors, unaffected regions of a pond, "stream drift" in flowing bodies of water, resistant stages, egg-laying by winged insects, etc. Agricultural and horticultural uses of the pyrethroids studied to date may thus be considered to have at most only transient effects on organisms in aquatic environments.

Forestry and public health riverine forest uses of pyrethroids inevitably result in their direct application to bodies of water. A number of large-scale field trials have been carried out. Fish populations are generally little affected, although the consequences of inter-species aggression during periods of food shortage (depletion of aquatic invertebrates) have not yet been investigated in sufficiently large treatment areas. The effects on aquatic organisms can be considerably greater than from agricultural usage, although again most organisms do recover. Members of the Malacostraca are often at greatest risk, being sensitive to the pyrethroids and less readily able to repopulate than are many other groups of invertebrates.

Bioaccumulation

It is important to establish the relationship between exposure to a chemical and the uptake by organisms. Build-up of chemical residues in organisms can follow several different routes. Intake may be via food, the increase in residue concentrations in animal tissue in successive members of a food chain being referred to as *biomagnification*. Direct uptake can occur from the abiotic environment, for example, from soil, water and bottom and suspended aquatic sediments. The term *bioconcentration* is commonly used to indicate an increase in chemical concentration from water when passing directly into aquatic species, although in its strictest sense it includes direct uptake from any abiotic source. *Bioaccumulation* indicates the combined intake from food and abiotic routes.

Laboratory methods have been used to study the potential for both the bioconcentration and biomagnification of pesticides. In addition, a number of mathematical models have been developed for predicting the bioconcentration of chemicals, using physicochemical properties such as water solubility and partition coefficients between *n*-octanol:water and soil:water.

For aquatic organisms exposed to permethrin and fenvalerate, bioconcentration factors (BCF; concentration in organism/concentration in water) of >1000 have been predicted by Kenaga (1980) and Briggs (1981). However, rate constants for metabolism and excretion were not taken into account.

Pyrethroids are extensively metabolised in all organisms studied (mammals, birds, fish, insects), to products considerably more polar than the parent; for example, *n*-octanol:water partition coefficients ($\log_{10}P$) for permethrin = ~6 and for major metabolites = <3. As concern over environmental bioaccumulation has normally been associated with lipophilic molecules, it would seem unlikely that the pyrethroids metabolites will present a bioaccumulation hazard.

Mammals

Studies with permethrin and cypermethrin have shown that part of the dose which is adsorbed by mammals undergoes prompt and extensive degradation to metabolites that are excreted rapidly in the urine, mainly as polar conjugates. Permethrin, cypermethrin and their metabolites are not accumulated in the fat or other tissues of mammals. This is described in more detail elsewhere (see Chapter 5).

Birds

In an unpublished study (ICI Plant Protection Division), groups of *Anas platyrhynchos* (mallard duck) and *Colinus virginianus* (bobwhite quail) were dosed orally with [*cyclopropane-*] and [*methylene*-^{14}C]permethrin and [*benzyl*-^{14}C]cypermethrin. Dose rates of approximately 0·2 mg/kg bodyweight were administered each day for 4 weeks, equivalent to 1 mg/kg in the diet of ducks and 2·5 mg/kg in the diet of the quails. Birds were killed after 7, 14, 21 and 28 days dosing and after 7 and 14 days depuration (the period during which dosing had ceased).

Within the exposure period, plateau levels of radioactive residues were reached in the fat and all other tissues examined. Plateau residues (as pyrethroid equivalents) in fat were 0·2 mg/kg or less. In liver, kidney and gizzard the concentrations were several-fold lower and in muscle and brain were mostly less than 0·02 mg/kg. During the depuration period the radioactive residues declined so that after 14 days most were below the limits of detection (approximately 0·01 mg/kg).

Hunt et al. (1979) sprayed White Leghorn laying hens with [*methylene-*^{14}C]permethrin on a single occasion. The highest residue levels were found in the skin, kidney, liver and fat, reaching peak concentrations within 7 days. Residues in the tissues declined thereafter with a half-life of approximately 2 weeks or less.

These studies show that permethrin and cypermethrin and their metabolites will not build-up in the tissues of birds.

Aquatic organisms

Pyrethroids may enter aquatic environments directly by spray-drift or indirectly by soil run-off (see p. 189 ff). Spray-drift will be intermittent and the chemical will be rapidly removed from the aqueous phase by adsorption. Entry of pyrethroids by run-off will be in the adsorbed state. Adsorption has been shown to reduce the availability of pyrethroids to aquatic organisms (see p. 211 ff). Furthermore the environmental concentration of the pesticide will decline as a result of microbial degradation (see Chapter 5).

Studies of the bioaccumulation of pyrethroids by aquatic organisms have been carried out using three model systems: (1) in water with maintained concentrations of the chemical, to determine the potential for bioconcentration; (2) in water with chemical incorporated in a biological food source, to study biomagnification; and (3) in water–sediment (soil) systems to study bioconcentration from the adsorbed state.

Bioconcentration from water

Laboratory studies of bioconcentration in flow-through water systems have been carried out with various fish and aquatic invertebrates exposed to permethrin, cypermethrin, fenvalerate and AC 222, 705. A wide range of BCF values have been reported, with concentrations of pyrethroid or total residues (pyrethroid and metabolites) in the organisms two to four orders of magnitude higher than in the water (Table 4.32). However, wherever regular

Table 4.32. Accumulation and depuration of pyrethroids by aquatic organisms in flow-through water systems

Pyrethroid	Organism	Wt (g)	No. of org's	Tank vol. (l)	Volume repl'm /day	Pyrethroid concentration (µg ai/l)	Accumulation period[a] Days	Days to plateau BCF	Plateau BCF Muscle	Plateau BCF Viscera	BCF Total	Days to 50% loss of residues during depuration	Ref. no.
Permethrin	Ictalurus punctatus	3·5	125	30	4	0·7	49	21	76	360	~100 T	<7	*
	Lepomis macrochirus	4·0	125	30	4	0·7	49	21	21	370	~50 T	<3	*
	Lepomis macrochirus	1·5	10	40	8	0·14	30	–	1400	14000	3000 P	–	
	Pimephales promelas	1·0	10	40	8[b]	0·1–0·75	246	<63	10000	7000	20000 P	3–7	246
	Pimephales promelas	fry	15	0·6	30	0·11–0·66	32	e	–	–	~3000 P	–	99
	Cyprinodon variegatus (<24 h)	0·1	20	2·5	b	1·6–10	28	e	–	–	~500 P	–	246
	Helisoma trivolvis	~0·2	10	0·6	30	0·03–0·33	28	e	–	–	~800 P	7	229
	Crassostrea virginica	(~6 cm)	65	80	30	0·16	31	2·5	–	–	1900 P	7	9
	Pteronarcys dorsata	–	10	1·6	27	0·03–0·43	28	e	–	–	~200 P	–	9
Cypermethrin	Salmo gairdneri	<16	100	300	c	~0·1	22	12	–	–	~1000 T	11	†
	Salmo gairdneri	<13	100	400	4	0·19	18	7	–	–	1200 P	8	†
	Salmo gairdneri	~200	10	400	4	0·18	24	e	–	–	~700 P	–	†
Fenvalerate	Cyprinodon variegatus	0·02	20	2·5	b	0·28–3·9	28	e	–	–	~600 P	–	99
	Pimephales promelas	0·1	15	0·6	30	0·14–0·19	28	e	–	–	~3000 P	–	246
	Cyprinus carpio	14	10	–	c	0·8[d]	7	10	–	–	1000 T	4	201
	Helisoma trivolvis	–	10	1·6	27	0·02–0·79	28	e	–	–	~600 P	–	9
	Crassostrea virginica	(~6 cm)	65	80	30	0·19	28	~20	–	–	4700 P	~3	229
AC222,705	Cyprinodon variegatus	0·03	20	2·5	b	~0·03–0·27	28	e	–	–	<1800 P	–	99
	Pimephales promelas	0·13	15	0·6	30	0·02, 0·03	32	e	–	–	~4000 P	–	246
	Crassostrea virginica	(~6 cm)	65	80	30	1·3	29	7	–	–	2300 P	~2	229

[a] Bioconcentration factor (BCF) = concentration in tissues (µg/kg wet wt)/mean measured concentration in water (µg ai/l). P = parent chemical analysed; T = radiolabel study and total ^{14}C-residues measured. BCF values at plateau except as under e below.
[b] Intermittent flow method.
[c] Replacement (24 h) method.
[d] S-acid isomer.
[e] Analysed at end of exposure phase.
* Unpublished data (FMC Corporation).
† Unpublished data (Shell Research).

sampling and analysis was carried out, the concentration in the organism reached a plateau within 3 weeks. Furthermore, on removing the organisms to clean water the residues declined rapidly with 50% loss within a period of 11 days or less.

The considerable variation in plateau BCFs may in some instances reflect the size/stage of organism tested or differences among species in their rates of pyrethroid uptake, redistribution among tissues and metabolism or excretion. However, it is probable that a major factor influencing bioaccumulation in the reported studies was bio-availability of the chemical. In two studies with permethrin and *Lepomis macrochirus*, BCF values were 50 and 3000 (unpublished data, FMC Corporation). In the first study (BCF 50), 500 g total weight of fish were kept in 30 litres water with 4 volume replacements per day. In the other study 25 g of fish were held in 40 litres with 8 volume replacements per day. In the first study the much greater amount of faecal matter in the tank, together with the slower flow rate, would have resulted in a higher proportion of the permethrin being adsorbed to particulate material. Pyrethroids have been shown to be less bio-available in the adsorbed state than in the aqueous phase (see p. 211ff. and below). Consequently bioconcentration in the first study with high "fish loading" was much lower than in the second study where the "load" was much less.

Biomagnification in food chains

It has been proposed (Macek et al., 1979; Ellgehausen et al., 1980) that for most compounds biomagnification of chemical residues within aquatic food chains is quantitatively insignificant when compared with the process of bioconcentration directly from water. It is further suggested that chemicals with a potential for biomagnification through aquatic food chains can be identified by their slow rate of depuration from tissues (Macek et al., 1979). The pyrethroids so far studied (Table 4.32) are rapidly depurated from fish and aquatic invertebrates, therefore biomagnification is unlikely to be of significance.

In a study by FMC Corporation (unpublished data) *Daphnia magna* were exposed for 5 weeks to [*cyclopropane*-^{14}C]permethrin (applied on day 0 only) in a static sediment–water system. On exposure days 2, 4, 8, 16 and 34 daphnids were harvested and fed to *Lepomis macrochirus* as sole food supply (0·4–1·0 g wet wt *D. magna* to 17 g [25 × 0·7 g] fish, on each occasion). The concentration of ^{14}C-labelled residues of permethrin in the *Daphnia* declined from 200 µg/kg (BCF [^{14}C-residues in organisms/^{14}C-residues in water] = ~600) on exposure day 2, to 10 µg/kg (BCF = ~20) by day 34. Residues in the fish were less than 5 µg/kg throughout the feeding study. In a comparable group of fish exposed for the same time period to [^{14}C]permethrin applied to the sediment–water system, ^{14}C-residues in fish were 183 µg/kg on day 2, falling to 4 µg/kg by day 34. It was concluded that uptake of

Table 4.33. Accumulation of [^{14}C]permethrin residues by aquatic organisms in a static water–soil ecosystem; application to water surface at zero time[a]

Organism	Wt (g)	No. of org's	Exposure test method		Exposure period[b]			
			Water: soil ratio	Tank vol. (1)	Days	Days to max. BCF	Max. BCF	BCF at end of exposure
Lepomis macrochirus	0·7	25				≤2	550	7
Daphnia magna	adult	75	6	45	34	≤2	1240	19
Helisoma sp.	—	25				4	380	42
Oedogonium foveolatum	(approx. 1 g)					4	370	43

[a] Unpublished data, FMC Corporation. Nominal concentration of permethrin applied, 1 μg ai/l; mean measured concentration of ^{14}C-residues in water during exposure period, 0·4 μg permethrin equivalents per litre.
[b] Bioconcentration factor (BCF) = concentration in tissue (μg/kg wet wt)/mean measured concentration in water (μg ai/l). Values are given for total residues.

permethrin from food is far less important than uptake from water, as will also be the case for other pyrethroids with similar properties.

Bioconcentration in water–sediment systems

When [*cyclopropane*-^{14}C]permethrin was applied to the water surface of a water–sediment system, the BCF values for total ^{14}C-residues in *Lepomis macrochirus*, *Daphnia magna*, *Helisoma* sp. and *Oedogonium foveolatum* (alga) after 2 days exposure were in the range 400–1200; unpublished data, FMC Corporation (Table 4.33). By exposure-day 34 these had declined to less than 50. Whilst the concentration of permethrin in the water over the 5-week period will have declined due to adsorption, the ^{14}C-residues in the water increased slightly, due no doubt to the degradation of the parent molecule to more water-soluble metabolites.

Studies have been carried out with permethrin, cypermethrin and fenvalerate in which the pyrethroid has been applied to sediment (soil) before flooding and introduction of organisms. Ohkawa et al. (1980) flooded the soil immediately after treatment with [*cyano*-^{14}C]-, [*carboxy*-^{14}C]- or [*methylene*-^{14}C](*S*)-fenvalerate and introduced *Cyprinus carpio*, *Daphnia pulex*, *Cipangopaludina japonica* (snail) and algae 7 days later. Analyses of organisms, water and sediment after 7 and 30 days exposure showed plateau BCF values of 250–500 (Table 4.34). Flooding immediately following [*methylene*-^{14}C](*S*)-fenvalerate application resulted in 3% of the parent chemical being in the water after 7 days (although no attempt was made to determine whether the pyrethroid was in solution or adsorbed to suspended

Table 4.34. Accumulation and depuration of pyrethroid residues by aquatic organisms in static water–soil ecosystems; pyrethroid applied to soil

Pyrethroid/ organism	Wt (g)	No. of org's	Exposure test method		Rate of application of pyrethroid (applied to soil: μg ai/kg dry wt)	Aging period		Exposure period[a]		Plateau or max. BCF[b]			Depuration period: days to 50% loss of residues	Ref. no.
			Water: soil ratio	Tank vol. (l)		Days from application to flooding	Days between flooding and adding organisms	Days	Days to plateau BCF	muscle	viscera	total		
PERMETHRIN														
Ictalurus punctatus	6	125	16	1700	80	30	3	28	[c]	13;4	140;60	30;10	7	*
CYPERMETHRIN														
Ictalurus punctatus	1.5	200	12	1400	500	21	3	23	21	10	55	16	3	†
FENVALERATE														
Cyprinus carpio	65	3	6	100	300	0	7	30	~20	—	—	~275	—	201
Daphnia pulex	(approx. 3 g) 19								[d]	—	—	500;250		
Cipangopaludina japonica		10							[d]	—	—	650;550		
Algae (mixture)	—	—							~20	—	—	~500		

[a] Bioconcentration factor (BCF) = concentration in tissue (μg/kg wet wt)/mean measured concentration in water (μg ai/l). All studies with radiolabelled chemicals.

[b] Values given are for total ^{14}C-residues; plateau BCF given except where stated otherwise.

[c] Maximum BCF values occurred between days 1 and 4. Data given are maximum BCF; BCF at end of exposure phase.

[d] BCF values decreased between exposure days 7 and 30. Data given are 7-day BCF; 30-day BCF.

* Unpublished data (ICI Americas Inc.).

† Unpublished data (ICI Plant Protection Division).

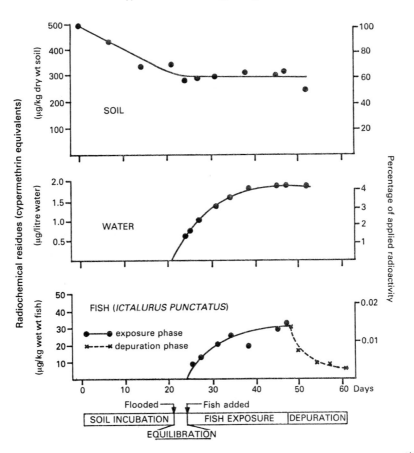

Figure 4.6. Distribution of cypermethrin residues in a soil-water-fish ecosystem. [^{14}C]-Cypermethrin applied to soil 3 weeks before flooding (unpublished data, ICI Plant Protection Division)

particles), declining to 1% by day 30. Bioaccumulation by *C. carpio* in the water–sediment ecosystem was approximately five-fold lower than that observed in water without sediment.

When soil was treated with [*cyclopropane*-^{14}C]permethrin or [*benzyl*-^{14}C]cypermethrin and incubated for 3–4 weeks before flooding and introduction of fish (*Ictalurus punctatus*), very little accumulation occurred (unpublished data, ICI Plant Protection Division and ICI Americas, Inc.). BCF values in whole fish did not exceed 30 during the exposure phase and a rapid decline in tissue residues occurred during depuration, with 50% loss in 7 days or less (Table 4.34 and Figure 4.6). Analysis of the water in the cypermethrin study showed greater than 80% of the radioactivity to be present as pyrethroid metabolites. Parent chemical was below the limit of detection in water and approximately 300 µg/kg in the bottom sediment.

Muir et al. (1983) studied the accumulation and elimination of *trans*-permethrin in *Chironomus tentans* larvae exposed to treated pond and river sediments and sand in aquatic ecosystems. In organisms held in the water phase, steady state concentrations (of total radioactivity) were reached after 12 hours. Plateau BCF values for larvae held in the water or allowed to establish in the "sediment" were less than 70 for pond/river ecosystems and 300 or less for the sand system.

Adsorption of the pyrethroids to soil or sediment therefore reduces their bio-availability to, and hence accumulation by, aquatic organisms.

Conclusions

Pyrethroids being lipophilic molecules are readily taken up by organisms. However, metabolism and excretion reduce tissue concentrations such that even under laboratory conditions of maintained pyrethroid concentrations, plateau residue levels are soon established. Furthermore, on cessation of dosing, residue concentrations in the tissues decline rapidly.

In aquatic environments the presence of bottom sediment and suspended particles, plant surfaces, etc reduce the bio-availability of pyrethroids and hence their potential for accumulation. Some concern has however been expressed that desorption may occur over a period of time and constitute a long term hazard. Amounts of pyrethroids desorbed will be extremely small, as is shown by their adsorption coefficients. In addition, any uptake of desorbed chemical by aquatic organisms will be followed by rapid tissue metabolism and excretion.

Theoretical considerations and practical observations have shown that pyrethroids will not be biomagnified along food chains.

Studies with fish have shown very little accumulation of the major degradation products of permethrin and cypermethrin, as would be predicted for these relatively polar materials resulting principally from ester hydrolysis.

Under the practical conditions of use in agriculture, horticulture and forestry, exposure to pyrethroids will be only occasional, limited and temporary. Chemicals entering the aquatic environment will be rapidly adsorbed to suspended or bottom sediments and to macrophytes, and degraded by microorganisms. Therefore there are no practical hazards to animals from bioaccumulation of the pyrethroids studied.

Acknowledgements

The author acknowledges with gratitude the numerous publications on which he has drawn for source material. He is also indebted to the authors of the many unpublished studies quoted, and to FMC Corporation (USA), Shell

Research (UK), Roussel-Uclaf (France), ICI Plant Protection Division (UK), ICI Americas Inc. (USA), ICI SOPRA (France) and ICI Holland for making data available.

Thanks also go to colleagues in ICI Plant Protection Division who read and commented upon appropriate parts of the manuscript; in particular to J. Cole, H. Gough, M. J. Hamer and W. Wilkinson.

Last, but by no means least, the many hours of typing (from almost illegible handwriting), correcting and proof-reading by Mrs J. Cabell, Miss L. Brown and my wife Lyn are gratefully acknowledged.

References

1. Abram, F. S. H., Evins, C. & Hobson, J. A., 1980, *Permethrin for the Control of Animals in Water Mains*. Water Research Centre, Technical Report TR145.
2. Agrishell, 1982, Sumicidin 10: *Les Insectes Pollinisateurs et l'Entomofaune Utile (Pollinator insects and beneficial fauna)*. Agrishell Recherche et Developpement Report.
3. Ahmed, R. & Muzaffar, N., 1977, Studies on cotton pests and their natural enemies with reference to effects of three insecticides of different persistencies. *Agric. Pakistan*, **28**, 193–203.
4. Akesson, N. B. & Yates, W. E., 1964, Problems relating to application of agricultural chemicals and resulting drift residues. *Ann. Rev. Entomol.*, **9**, 285–318.
5. Ali, A. D., Bakry, N. M., Abdellatif, M. A. & El-Sawaf, S. K., 1973. The control of greater wax moth, *Galleria mellonella* L., by chemicals. I. Susceptibility of the wax moth larvae and honey-bee workers to certain chemicals. *Z. Ang. Ent.*, **74**, 170–77.
6. Aliniazee, M. T. & Cranham, J. E., 1980, Effect of four synthetic pyrethroids on a predatory mite, *Typhlodromus pyri*, and its prey, *Panonychus ulmi*, on apples in southeast England. *Environ. Entomol.*, **9**, 436–9.
7. Anastasi, A. & Bannister, J. V., 1980, Effect of pesticides on Mediterranean fish muscle enzymes. *Acta Adriat.*, **21**, 119–36.
8. Anderson, J. R. & Slinger, J. M., 1975, Europium chelate and fluorescent brightener staining of soil propagules and their photomicrographic counting. I. Methods. *Soil Biol. Biochem.*, **7**, 205–9.
9. Anderson, R. L., 1982, Toxicity of fenvalerate and permethrin to several nontarget aquatic invertebrates. *Environ. Entomol.*, **11**, 1251–7.
10. Anon., 1981, Pyrethroid pesticide with low toxicity to fish proves effective. *Jap. Chem. Weekly*, **22**, 6.
11. Arzone, A. & Vidano, C., 1978, Azione sull'ape di etiofencarb, decamethrine, ciexatin (Action on honeybees of ethiofencarb, decamethrin and cyhexatin). *L'Apicoltore Moderno*, **69**, 157–62.
12. Atkins, E. L., Anderson, L. D. & Tuft, T. O., 1954, Equipment and technique used in laboratory evaluation of pesticide dusts in toxicological studies with honeybees. *J. Econ. Entomol.*, **47**, 965–9.
13. Atkins, E. L., Kellum, D. & Atkins, K. W., 1977a, *Repellent Additives to Reduce Pesticide Hazards to Honeybees*. University of California, Riverside, 1977, Semiannual Report (Project no. 3565-RR[W–139]), pp. 40–62.
14. Atkins, E. L., Kellum, D. & Atkins, K. W., 1978, Integrated pest management strategies for protecting honeybees from pesticides. *Amer. Bee J.*, Aug. 1978, 542–8.

15. Atkins, E. L., Kellum, D. & Neuman, K. J., 1977b, *Effect of Pesticides on Apiculture. Part I: Pesticide Research*. University of California, Riverside, 1976 Annual Report (Project no. 1499), pp. 536–44, 550–62.

16. Baldry, D. A. T., Everts, J., Roman, B., Boon von Ochssee, G. A. & Laveissiere, 1981, The experimental application of insecticides from a helicopter for the control of riverine populations of *Glossina tachinoides* in West Africa. Part VIII: The effects of two spray applications of OMS-570 (endosulfan) and of OMS-1998 (decamethrin) on *G. tachinoides* and non-target organisms in Upper Volta. *Trop. Pest Manage.*, **27**, 83–110.

17. Baldry, D. A. T., Kulzer, H., Bauer, S., Lee, C. W. & Parker, J. D., 1978a, The experimental application of insecticides from a helicopter for the control of riverine populations of *Glossina tachinoides* in West Africa. III. Operational aspects and application techniques. *PANS*, **24**, 423–34.

18. Baldry, D. A. T., Molyneux, D. H. & van Wettere, P. 1978b, The experimental application of insecticides from a helicopter for the control of riverine populations of *Glossina tachinoides* in West Africa. V. Evaluation of decamethrin applied as a spray. *PANS*, **24**, 447–54.

19. Barlow, F. & Hadaway, A. B., 1975, The insecticidal activity of some synthetic pyrethroids against mosquitoes and flies. *PANS*, **21**, 233–8.

20. Barnes, J. M. & Verschoyle, R. D., 1974, Toxicity of new pyrethroid insecticide. *Nature, Lond.*, **248**, 711.

21. Bartlett, B. R. 1964, Integration of chemical and biological control. In *Biological Control of Insect Pests and Weeds*, edited by P. Debach (London: Chapman & Hall), pp. 489–511.

22. Bellec, C., Hebrard, G. & d'Almeida, A., 1983, The effects of helicopter applied adulticides for riverine tsetse control on *Simulium* populations in a West African Savanna habitat. II. Effects as estimated by non-biting stages of *Simulium damnosum* S. I. and other blackfly species caught on aluminium plaque traps. *Trop. Pest Manage.*, **29**, 7–12.

23. Benoit, D. A., Puglisi, F. A. & Olson, D. L., 1982, Fathead minnow (*Pimephales promelas*) early life stage toxicity test method evalution and exposure of four organic chemicals. *Environ. Pollut.*, **28**, 189–97.

24. Bocquet, J. Ch., Pastre, P., Roa, L. & Baumeister, R. 1980, Etude de l'action de la deltamethrine sur *Apis mellifera* en conditions de plein champ. (Study of the action of deltamethrin on *Apis mellifera* under field conditions). *Phytiat.-Phytopharm.*, **29**, 83–92.

25. Bradbury, S. P. & Coats, J. R., 1982, Toxicity of fenvalerate to bobwhite quail *(Colinus virginianus)* including brain and liver residues associated with mortality. *J. Toxicol. Environ. Health*, **10**, 307–19.

26. Bridges, W. R. & Cope, O. B., 1965, The relative toxicities of similar formulations of pyrethrum and rotenone to fish and immature stoneflies. *Pyrethrum Post*, **8**, 3–5.

27. Briggs, G. G., 1981, Theoretical and experimental relationships between soil absorption, octanol–water partition coefficients, water solubilities, bioconcentration factors and the parachor. *J. Agric. Food Chem.*, **29**, 1050–9.

28. Bull, D. L. & House, V. S., 1983, Effects of different insecticides on parasitism of host eggs by *Trichogramma pretiosum* Riley. *Southwest. Entomol.*, **8**, 46–53.

29. Burress, R. M., Gilderhus, P. A. & Cumming, K. B., 1976, *Field Tests of Isobornyl Thiocyanoacetate (Thanite) for Live Collection of Fishes*. US Fish and Wildlife Service, Investigations in Fish Control, No. 71.

30. Burton, G. J. & McRae, T. M., 1972, Observations on trichopteran predators of aquatic stages of *Simulium damnosum* and other *Simulium* species in Ghana. *J. Med. Entomol.* **9**, 289–94.

31. Butcher, J., Boyer, N. & Fowler, C. D., 1975, Impact of Dursban and Abate on microbial numbers and some chemical properties of standing ponds. *Proceedings of the 10th Canadian Symposium on Water Pollution Research, Canada*, pp. 33–41.

31. Butcher, J., Boyer, N. & Fowler, C. D., 1975, Impact of Dursban and Abate on microbial numbers and some chemical properties of standing ponds. *Proceedings of the 10th Canadian Symposium on Water Pollution Research, Canada*, pp. 33–41.
32. Butcher, J., Boyer, N. & Fowler, C. D., 1977, Some changes in pond chemistry and photosynthetic activity following treatment with increasing concentrations of chlorpyrifos. *Bull. Environ. Contam. Toxicol.*, **17**, 752–8.
33. Cameron, P. J., Thomas, W. P. & Hill, R. L., 1980, Introduction of lucerne aphid parasites and a preliminary evaluation of the natural enemies of *Acyrthosiphon* spp. (Hemiptera: Aphididae) in New Zealand. *Proceedings of the Second Australian Conference on Grassland Invertebrate Ecology, 1979*, edited by T. K. Crosby & R. P. Potlinger, pp. 219–23.
34. Caro, J. H., Freeman, H. P. & Turner, B. C., 1974, Persistence in soil and run-off of soil-incorporated carbaryl in a small watershed. *J. Agric. Food Chem.*, **22**, 860–3.
35. Caro, J. H. & Taylor, A. W., 1971, Pathways of loss of dieldrin from soils under field conditions. *J. Agric. Food Chem*, **19**, 379–84.
36. Cazenave, A., Debray, P. & Esteulle, M., 1980, Etude du comportement d'abeilles *Apis mellifera* sur des plantes en fleurs traitees avec le fenvalerate (Study of the behaviour of the bee *Apis mellifera* on plants and flowers treated with fenvalerate). *Phytiat. -Phytopharm.*, **29**, 93–106.
37. Cheng, H. H., 1980, Toxicity and persistence of pyrethroid insecticides as foliar sprays against darksided cutworm (Lepidoptera: Noctuidae) on tobacco in Ontario. *Canad. Entomol.*, **112**, 451–6.
38. Coats, J. R. & O'Donnell-Jeffery, N. L., 1979, Toxicity of four synthetic pyrethroid insecticides to rainbow trout. *Bull. Environ. Contam. Toxicol.*, **23**, 250–5.
39. Coats, S. A., Coats, J. R. & Ellis, C. R., 1979, Selective toxicity of three synthetic pyrethroids to eight coccinellids, a eulophid parasitoid, and two pest chrysomelids. *Environ. Entomol.*, **8**, 720–2.
40. Cole, J. F. H. & Wilkinson, W., 1984, Permethrin, dimethoate and pirimicarb; effects of spring applications on arthropods in the cereal ecosystem. *Faune et Flore Auxiliaresen Agriculture, ACTA Symposium.*, Paris 1983. pp. 229–237.
41. Croft, B. A., 1976, Establishing insecticide-resistant phytoseiid mite predators in deciduous tree fruit orchards. *Entomophaga*, **21**, 383–99.
42. Croft, B. A. & Meyer, R. H., 1973, Carbamate and organophosphorous resistance patterns in populations of *Amblyseius fallacis*. *Environ. Entomol.*, **2**, 486–88.
43. Croft, B. A. & Wagner, S. W., 1981, Selectivity of acaricidal pyrethroids to permethrin-resistant strains of *Amblyseius fallacis*. *J. Econ. Entomol.*, **74**, 703–6.
44. Croft, B. A., Wagner, S. W. & Scott, J. G., 1982, Multiple- and cross-resistances to insecticides in pyrethroid-resistant strains of the predatory mite, *Amblyseius fallacis*. *Environ. Entomol.*, **11**, 161–4.
45. Croft, B. A. & Whalon, M. E., 1982, Selective toxicity of pyrethroid insecticides to arthropod natural enemies and pests of agricultural crops. *Entomophaga*, **27**, 3–21.
46. Croft, B. A. & Whalon, M. E., 1983, Inheritance and persistence of permethrin resistance in the predatory mite, *Amblyseius fallacis* (Acarina: Phytoseiidae). *Environ. Entomol.*, **12**, 215–8.
47. Crossland, N. O., Aquatic toxicology of cypermethrin. II. Fate and biological effects in pond experiments. *Aquat. Toxicol.*, **2**, 205–22.
48. Crossland, N. O., Shires, S. W. & Bennett, D., 1982, Aquatic toxicology of cypermethrin. III. Fate and biological effects of spray drift deposits in fresh water adjacent to agricultural land. *Aquat. Toxicol.*, **2**, 253–70.
49. Crowder, L. A., Tollefson, M. S. & Watson, T. F., 1979, Dosage-mortality studies of synthetic pyrethroids and methyl parathion on the tobacco budworm in central Arizona. *J. Econ. Entomol.*, **72**, 1–3.

50. Crowell, H. H., 1977, Chemical control of terrestrial slugs and snails. *Oreg. Agric. Exp. Stn. Bull.*, **628**, 1–70.
51. Currier, W. W., Maccollom, G. B. & Baumann, G. L., 1982, Drift residues of air-applied carbaryl in an orchard environment. *J. Econ. Entomol.*, **75**, 1062–8.
52. Damgaard-Pedersen, F., 1980, Permethrin – Giftighed overfor bier og andre nyttige artropoder (Permethrin – toxicity to bees and other useful arthropods). *Vaxtskydds Rapport*, **12**, 34–43.
53. Darwazeh, H. A. & Mulla, M. S., 1974, Biological activity of organophosphorus compounds and synthetic pyrethroids against immature mosquitoes. *Mosquito News*, **34**, 151–5.
54. Darwazeh, H. A., Mulla, M. S. & Sjogren, R. D., 1975, Evaluation of mosquito adulticides in irrigated pastures using nonthermal aerosols. *Proc. Pap. Calif. Mosq. Cont. Assoc. Ann. Conf.*, **43**, 169–71.
55. Darwazeh, H. A., Mulla, M. S. & Whitworth, B. T., 1978, Synthetic pyrethroids for the control of resistant mosquitoes in irrigated pastures. *Proc. Pap. Calif. Mosq. Cont. Assoc. Ann. Conf.*, **46**, 121–2.
56. David, D., 1981. Laboratory evaluation of repellent properties against birds of the synthetic pyrethroid decamethrin. *Poult. Sci.*, **60**, 1149–51.
57. Davies, J. B.,Gboho, C., Baldry, D. A. T., Bellec, C., Sawadogo, R. & Tiao, P. C., 1982, The effects of helicopter applied adulticides for riverine tsetse control on *Simulium* populations in a West African Savanna habitat. I. Introduction, methods and the effect on biting adults and aquatic stages of *Simulium damnosum* S. I. *Trop. Pest Manage.*, **28**, 284–90.
58. Davies, J. B., Walsh, J. F., Baldry, D. A. T. & Bellec, C., 1983, The effects of helicopter applied adulticides for riverine tsetse control on *Simulium* populations in a West African Savanna habitat. III. Conclusions: the possible role of adulticiding in onchocerciasis control in West Africa. *Trop. Pest Manage.*, **29**, 13–5.
59. DeBoo, R. F., 1980, *Experimental Aerial Applications of Permethrin for Control of Choristoneura fumiferana in Quebec, 1976–77*. Forest Pest Management Institute (Environment Canada) Report FPM-X-41.
60. Debray, P., 1981, Effet de la Sumicidin 10 (fenvalerate) sur l'environnement: le point actuel des etudes (Effect of Sumicidin 10 [fenvalerate] on the environment: summary of present studies). *La Defense des Vegetaux*, **209**, 229–36.
61. Dejoux, C., 1979, Emploi des pesticides et pollution des eaux continentales tropicales (Use of pesticides and pollution of tropical continental waters). *C.R. Congres Lutte Contre les Insectes en Milieu Tropical*, Marseille, pp. 861–72.
62. Doherty, J. D., Salem, N. Jr., Lauter, C. J. & Eberhard, G. T., 1981, Mn^{2+} and Ca^{2+} ATPases in lobster axon plasma membranes and their inhibition by pesticides. *Comp. Biochem. Physiol.*, **69C**, 185–190.
63. Dowson, R. J., & Garvie, W. P., 1979, Ripcord: a new wide spectrum synthetic pyrethroid. *Pyrethroid Symposium*, Warsaw 1978. *Pestycydy 1979, Special Issue*, pp. 46–52.
64. Draughon, F. A. & Ayres, J. C., 1978, Effect of selected pesticides on growth and citrinin production by *Penicillium citrinum. J. Food Sci.*, **43**, 576–8.
65. Draughon, F. A. & Ayres, J. C., 1980, Insecticide inhibition of growth and patulin production in *Penicillium expansum, Penicillium urticae, Aspergillus clavatus, Aspergillus terreus*, and *Byssochlamys nivea. J. Agric. Food Chem.*, **28**, 1115–7.
66. Dunning, R. A., Cooper, J. M., Wardman, J. M. & Winder, G. H., 1982, Susceptibility of the carabid *Pterostichus melanarius* (Illiger) to aphicide sprays applied to the sugar-beet crop. *Annals Appl. Biol.*, **100** (Tests of Agrochemicals and Cultivars, Supplement), 32.
67. Eaton, J. G., McKim, J. M. & Holcombe, G. W., 1978, Metal toxicity to embryos and larvae of seven fresh-water fish species – 1. Cadmium. *Bull. Environ. Contam. Toxicol.*, **19**, 95–103.

68. Edwards, P. J. & Brown, S. M., 1982, Use of grassland plots for testing the effects of pesticides on earthworms. *Pedobiologia*, **24**, 145–50.
69. Edwards, P. J. & Wilkinson, W., 1983, A laboratory toxicity test for carabid beetles. *Tenth International Congress on Plant Protection, Brighton, UK*, p. 719.
70. Edwards, R. W. & Brown, V. M., 1967, Pollution and fisheries: a progress report. *Wat. Pollut. Control*, **66**, 63–78.
71. Eller, L. L., 1975, Gill lesions in freshwater teleosts. In *The Pathology of Fishes*, edited by W. E. Ribelin & G. Migaki (Wisconsin: University of Wisconsin Press), pp. 305–30.
72. Ellgehausen, H., Guth, J. A. & Esser, H. O., 1980, Factors determining the bioaccumulation potential of pesticides in the individual compartments of aquatic food chains. *Ecotox. Environ. Safety*, **4**, 134–57.
73. Elliott, M., Janes, N. F. & Potter, C., 1978, The future of pyrethroids in insect control. *Ann. Rev. Entomol.*, **23**, 443–69.
74. Elliott, M., Janes, N. F., Stevenson, J. H. & Walters, J. H. H., 1980, Selectivity of insecticides between beneficial and pest insects. *Rothamsted Experimental Station Annual Report*, 127–8.
75. Elliott, M., Janes, N. F., Stevenson, J. H. & Walters, J. H. H., 1983, Insecticidal activity of the pyrethrins and related compounds. Part XIV: Selectivity of pyrethroid insecticides between *Ephestia kuhniella* and its parasite *Venturia canescens*. *Pestic. Sci.*, **14**, 423–6.
76. Ernst, B., Julien, G., Doe, K. & Parker, R., 1980, *Environmental Investigations of the 1980 Spruce Budworm Spray Program in New Brunswick*. Environ. Canada Environ. Prot. Serv. Surveill. Rep. EPS-5-AR-81-3.
77. Everts, J. W., van Frankenhuyzen, Ki., Roman, B. & Koeman, J. H., 1983, Side-effects of experimental pyrethroid applications for the control of tsetse flies in a riverine forest habitat (Africa). *Arch. Environ. Contam. Toxicol.*, **12**, 91–7.
78. Ferguson, D. E. & Bingham, C. R., 1966, The effects of combinations of insecticides on susceptible and resistant mosquito fish. *Bull. Environ. Contam. Toxicol.*, **1**, 97–103.
79. Francois, Y., Lhoste, J. & Rupaud, Y., 1982, Facteurs influencant la toxicite de la deltamethrine, insecticide pyrethrinoide, sur la truise *(Salmo fario L.)* (Factors affecting the toxicity of deltamethrin, a pyrethroid insecticide, on trout [*Salmo fario L.*]). *C. R. Seances Acad. Agric. Fr.*, **68**, 652–7.
80. Friesen, M. K., 1981, Effects of the insecticide permethrin on the aquatic life stages of the burrowing mayfly *Hexagenia rigida* (Ephemeroptera: Ephemeridae). M.Sc. Thesis, University of Guelph, Canada.
81. Friesen, M. K., Galloway, T. D. & Flannagan, J. F., 1983, Toxicity of the insecticide permethrin in water and sediment to nymphs of the burrowing mayfly *Hexagenia rigida* (Ephemeroptera: Ephemeridae). *Canad. Entomol.*, **115**, 1007–14.
82. Frost, K. R. & Ware, G. W., 1970, Pesticide drift from aerial and ground applications. *Agric. Eng.*, **51**, 460–4.
83. Garnas, R. L. & Schimmel, S. C., 1981, Toxicity, bioconcentration and persistence of pyrethroid insecticides in the marine environment. *181st Amer. Chem. Soc. Natl. Meet., Divn. Pestic. Chem. Abstr.* 30.
84. Gerig, V. L., 1979, Bienerigiftigkeit der synthetisches pyrethrine (The toxicity of synthetic pyrethroids to foraging bees). *Schweiz. Bienen.-Z.* **101 NF**, 228–36.
85. Gerig, V. L., 1981, Bienengiftigkeit der synthetisches pyrethrine (2 Teil). (The toxicity of synthetic pyrethroids to foraging bees [2nd part]). *Schweiz. Bienen.-Z.* **104 NF**, 155–74.
86. Ghassemi, M., Painter, P. & Powers, M., 1982, Estimating drift and exposure due to aerial application of insecticides in forests. *Environ. Sci. Technol.*, **16**, 510–14.
87. Gibson, R. J., 1973, Interactions of juvenile Atlantic Salmon (*Salmo salar L.*) and

brook trout (*Salvelinus fontinalis* Mitchell). In *International Atlantic Salmon Symposium*, 1973, Vol 4. (The International Atlantic Salmon Foundation), pp. 181–202.

88. Glickman, A. H., Hamid, A. A. R., Rickert, D. E. & Lech, J. J., 1981, Elimination and metabolism of permethrin isomers in rainbow trout. *Toxicol. Appl. Pharmacol.*, **57**, 88–98.

89. Glickman, A. H. & Lech, J. J., 1981, Hydrolysis of permethrin, a pyrethroid insecticide, by rainbow trout and mouse tissues *in vitro*: a comparative study. *Toxicol. Appl. Pharmacol.*, **60**, 186–92.

90. Glickman, A. H. & Lech, J. J., 1982, Differential toxicity of *trans*-permethrin in rainbow trout and mice. II. Role of target organ sensitivity. *Toxicol. Appl. Pharmacol.*, **66**, 162–71.

91. Glickman, A. H., Shono, T., Casida, J. E. & Lech, J. J., 1979, *In vitro* metabolism of permethrin isomers by carp and rainbow trout liver microsomes. *J. Agric. Food Chem.*, **27**, 1038–41.

92. Glickman, A. H., Weitman, S. D. & Lech, J. J., 1982, Differential toxicity of *trans*-permethrin in rainbow trout and mice. I. Role of biotransformation. *Toxicol. Appl. Pharmacol.*, **66**, 153–61.

93. Gray, A. J., Conners, T. A., Hoellinger, H. & Hoang-Nam, N., 1980, The relationship between the pharmacokinetics of intravenous cismethrin and bioresmethrin and their mammalian toxicity. *Pestic. Biochem. Physiol.*, **13**, 281–93.

94. Gromisz, Z., 1979, Szkodliwosc pyrethroidcur dla psczcs (Toxicity of pyrethroids to bees). *Pyrethroid Symposium*, Warsaw 1978. *Pestycydy 1979, Special Issue*, pp. 119–22.

95. Hagley, E. A. C., Pree, D. J. & Holliday, N. J., 1980, Toxicity of insecticides to some orchard carabids (Coleoptera: Carabidae). *Canada. Entomol.*, **112**, 457–62.

96. Hagley, E. A. C., Pree, D. J. & Simpson, C. M., 1981, Toxicity of insecticides to parasites of the spotted tentiform leafminer (Lepidoptera: Gracillariidae). *Canad. Entomol.*, **113**, 899–906.

97. Hall, F. R., 1979, Effects of synthetic pyrethroids on major insect and mite pests of apple. *J. Econ. Entomol.*, **72**, 441–6.

98. Hamelink, J., 1979, Bioavailability of chemicals in aquatic environments. In *Biotransformation and Fate of Chemicals in the Aquatic Environment*, edited by A. W. Maki, K. L. Dickson & J. Cairns, Jr. (Washington, DC: American Society of Microbiology), pp. 56–62.

99. Hansen, D. J., Goodman, L. R., Moore, J. C. & Higdon, P. K., 1983, Effects of the synthetic pyrethroids AC 222,705, permethrin and fenvalerate on sheepshead minnows in early life stage toxicity tests. *Environ. Toxicol. Chem.*, **2**, 251–8.

100. Hargreaves, J. R. & Cooper, L. P., 1979, Phytotoxicity tests with pyrethroid insecticides on glasshouse grown tomato seedling. *Queensl. J. Agric. Anim. Sci.*, **36**, 151–4.

101. Harvey, J. & Pease, H. L., 1973, Decomposition of methomyl in soil. *J. Agric. Food Chem.*, **21**, 784–6.

102. Hashimoto, Y. & Nishiuchi, Y., 1981, Establishment of bioassay methods for the evaluation of acute toxicity of pesticides to aquatic organisms. *J. Pestic. Sci.*, **6**, 257–64.

103. Haskins, J. R., Grothaus, R. H., Batchelor, R., Sullivan, W. N. & Schechter, M. S., 1974, Effectiveness of three synthetic pyrethroids against mosquitoes. *Mosquito News*, **34**, 385–8.

104. Hattori, J., 1977, Sumicidin (fenvalerate). *Japan Pesticide Info.*, **33**, 13–9.

105. Heinrichs, E. A., Reissig, W. H., Valencia, S. & Chelliah, S., 1982, Rates and

effect of resurgence-inducing insecticides on populations of *Nilaparvata lugens* (Homoptera: Delphacidae) and its predators. *Environ. Entomol.*, **11**, 1269–73.

106. Herbert, D. W. M., 1965, Pollution and fisheries. In *Ecology and the Industrial Society*, edited by G. T. Goodman, R. W. Edwards & J. M. Lambert (Oxford: Blackwell), pp. 173–95.

107. Hill, E. F., Spann, J. W. & Williams, J. D., 1975, *Lethal Dietary Toxicities of Environmental Pollutants to Birds*. US Fish and Wildlife Service Report, Wildlife No. 191.

108. Hill, I. R. & Wright, S. J. L., 1978, The behaviour and fate of pesticides in microbial environments. In *Pesticide Microbiology; Microbiological Aspects of Pesticide Behaviour in the Environment*, edited by I. R. Hill & S. J. L. Wright (London: Academic Press), pp. 79–136.

109. Holcombe, G. W., Phipps, G. L. & Fiandt, J. T., 1982*a*, Effects of phenol, 2,4-dimethyphenol, 2,4-dichlorophenol, and pentachlorophenol on embryo, larval, and early juvenile fathead minnows (*Pimephales promelas*). *Arch. Environ. Contam. Toxicol.*, **11**, 73–8.

110. Holcombe, G. W., Phipps, G. L. & Tanner, D. K., 1982*b*, The acute toxicity of kelthane, dursban, disulfoton, pydrin and permethrin to fathead minnows *Pimephales promelas* and rainbow trout *Salmo gairdneri*. *Environ. Pollut. Ser. A*, **29**, 167–78.

111. Hoy, M. A., Castro, D. & Cahn, D., 1982, Two methods for large scale production of pesticide-resistant strains of the spider mite predator *Metaseiulus occidentalis* (Nesbitt) (Acarina, Phytoseiidae). *Z. Ang. Ent.*, **94**, 1–9.

112. Hoy, M. A., Flaherty, D., Peacock, W. & Culver, D., 1979, Vineyard and laboratory evaluations of methomyl, dimethoate and permethrin for a grape pest management program in the San Joaquin Valley of California. *J. Econ. Entomol.*, **72**, 250–5.

113. Hoyt, S. C., Westigard, P. H. & Burts, E. C., 1978, Effects of two synthetic pyrethroids on the codling moth, pear psylla, and various mite species in northwest apple and pear orchards. *J. Econ. Entomol.*, **71**, 431–4.

114. Hull, L. A. & Starner, V. R., 1983, Impact of four synthetic pyrethroids on major natural enemies and pests of apple in Pennsylvania. *J. Econ. Entomol.*, **76**, 122–30.

115. Hunt, L. M., Gilbert, B. N. & Lemeilleur, C. A., 1979, Distribution and depletion of radioactivity in hens treated dermally with ^{14}C-labelled permethrin. *Poultry Sci.*, **58**, 1197–201.

116. Hurlbert, S. H., Mulla, M. S., Keith, J. O., Westlake, W. E. & Dush, M. E., 1970, Biological effects and persistence of dursban in a freshwater pond. *J. Econ. Entomol.*, **63**, 43–52.

117. Hurlbert, S. H., Mulla, M. S. & Willson, H. R., 1972, Effects of an organophosphorus compound on the phytoplankton, zooplankton and insect populations of freshwater ponds. *Ecolog. Monogr.*, **42**, 269–99.

118. Ide, J., Nakada, Y., Endo, R., Muramatsu, S., Konishi, K., Mizuno, T., Ohno, S., Yamazaki, Y., Endo, H., Fujita, K. & Tsuji, H., 1983, Chloromethylvinyl pyrethroids. *Agric. Biol. Chem.*, **47**, 927–8.

119. Iftner, D. C., 1982, The effects of synthetic pyrethroids on the feeding behaviour and dispersal of the two spotted spider mite. *Ohio J. Sci.*, **82**, 3.

120. International Commission for Bee Botany, 1980, *Symposium on the Harmonisation of Methods for Testing the Toxicity of Pesticides to Bees*, 1980, Wageningen, Holland.

121. International Commission for Bee Botany, 1982, *Second Symposium on the Harmonisation of Methods for Testing the Toxicity of Pesticides to Bees*, 1982, Hohenheim, W. Germany.

122. Jarvinnen, A. W. & Tanner, D. K., 1982, Toxicity of selected controlled release and corresponding unformulated technical grade pesticides to the fathead minnow (*Pimephales promelas*). *Environ. Pollut. Ser. A*, **27**, 179–95.
123. Jenkins, D. W., 1964, Pathogens, parasites and predators of medically important arthropods. *Bull. WHO*, **30** (Suppl.), 1–150.
124. Johansen, C. A. & Mayer, D., 1976, Bee poisoning hazard, Pullman 1975. *Insecticide and Acaricide Tests*, **1**, 78–9.
125. Jolly, A. L. & Avault, J. W., 1978, Acute toxicity of permethrin to several aquatic animals. *Trans. Amer. Fish. Soc.*, **107**, 825–7.
126. Jolly, A. L., Graves, J. B., Avault, J. W. & Koonce, K. L., 1978, Effects of a new insecticide on aquatic animals. *Louisiana Agriculture*, **21**, 3–4.
127. Kalushkov, P., 1982, The effect of five insecticides on coccinellid predators (Coleoptera) of aphids *Phorodon humuli* and *Aphis fabae* (Homoptera). *Acta Ent. Bohemoslov.*, **79**, 167–80.
128. Karickhoff, S. W. & Brown, D. S., 1979, *Determination of Octanol/Water Distribution Coefficients, Water Solubilities, and Sediment/Water Partition Coefficients for Hydrophobic Organic Pollutants.* USA Environmental Protection Agency, Report No. EPA-600/4-79-032, Washington DC.
129. Kenaga, E. E., 1979, Acute and chronic toxicity of 75 pesticides to various animal species. *Down to Earth*, **35**, 25–31.
130. Kenaga, E. E., 1980, Predicted bioconcentration factors and soil sorption coefficients of pesticides and other chemicals. *Ecotox. Environ. Safety*, **4**, 26–38.
131. Kindt, T. & Stark, J., 1981, Studier ar bin och biforgiftringar i ett mindre vaxthus forsok (Studies of bees and toxicity to bees in a small-scale greenhouse experiment). *Vaxtskydds Rapport*, **14**, 133–7.
132. Kingsbury, P. D., 1976a, *Studies of the Impact of Aerial Applications of the Synthetic Pyrethroid NRDC-143 on Aquatic Ecosystems.* Chemical Control Research Institute (Environment Canada), Report CC-X-127.
133. Kingsbury, P. D., 1976b, Effects of an aerial application of the synthetic pyrethroid permethrin on a forest stream. *Manitoba Entomol.*, **10**, 9–17.
134. Kingsbury, P. D., 1982, *Permethrin in New Brunswick Salmon Nursery Streams.* Forest Pest Management Institute (Environment Canada), Report FPM-X-52.
135. Kingsbury, P. D. & Kreutzweiser, D. P., 1979, *Impact of Double Applications of Permethrin on Forest Streams and Ponds.* Forest Pest Management Institute (Environment Canada), Report FPM-X-27.
136. Kingsbury, P. D. & Kreutzweiser, D. P., 1980a, *Dosage-effect Studies on the Impact of Permethrin on Trout Streams.* Forest Pest Management Institute (Environment Canada), Report FPM-X-31.
137. Kingsbury, P. D. & Kreutzweiser, D. P., 1980b, *Environmental Impact Assessment of a Semi-Operational Permethrin Application.* Forest Pest Management Institute (Environment Canada), Report FPM-X-30.
138. Kingsbury, P. D. & McLeod, B. B., 1979, *Terrestrial Impact Studies in Forest Ecosystems Treated with Double Applications of Permethrin.* Forest Pest Management Institute (Environment Canada), Report FPM-X-28.
139. Kreutzweiser, D. P., 1982, *The Effects of Permethrin on the Invertebrate Fauna of a Quebec Forest.* Forest Pest Management Institute (Environment Canada), Report FPM-X-50.
140. Kreutzweiser, D. P. & Kingsbury, P. D., 1982, *Recovery of Stream Benthos and its Utilization by Native Fish following High Dosage Permethrin Applications.* Forest Pest Management Institute (Environment Canada), Report FPM-X-59.
141. Kumaraguru, A. K. & Beamish, F. W. H., 1981, Lethal toxicity of permethrin (NRDC-143) to rainbow trout, *Salmo gairdneri*, in relation to body weight and water temperature. *Water Research*, **15**, 503–5.

142. Kumaraguru, A. K., Beamish, F. W. H. & Ferguson, H. W., 1982, Direct and circulatory paths of permethrin (NRDC-143) causing histopathological changes in the gills of rainbow trout, *Salmo gairdneri, Richardson. J. Fish Biol.*, **20**, 87–91.
143. Lagier, R. F., Johansen, C. A., Kleinschmidt, M. G., Butler, L. I., McDonough, L. M. & Jackson, D. S., 1974, *Adjuvants Decrease Insecticide Hazard to Honey Bees.* College of Agric. Res. Center, Washington State Univ., Pullman, Bulletin 801.
144. Lawson, T. J., 1979, Some factors affecting the dispersal of aerial sprays. *Agricultural Aviation Group All-day Symposium*, Feb. 1979.
145. Leake, L. D., 1977, The action of (*S*)-3-allyl-2-methyl-4-oxocyclopent-2-enyl (1*R*)-*trans* chrysanthemate, (*S*)-bioallethrin, on single neurones in the central nervous system of the leech, *Hirudo medicinalis. Pestic. Sci.*, **8**, 713–21.
146. Leake, L. D., Lauckner, S. M. & Ford, M. G., 1980, Relationship between neurophysiological effects of selected pyrethroids and toxicity to the leech *Haemopsis sanguisuga* and the locust *Schistocerca gregaria*. In *Insect Neurobiol. Pestic. Action (Neurotox. 79), Proc. Symp. Soc. Chem. Ind. 1979*, 423–30.
147. Legner, E. F. & Medved, R. A., 1981, Pink bollworm, *Pectinophora gossypiella* (Diptera: Gelechiidae), Suppression with gossyplure, a pyrethroid, and parasite releases. *Canad. Entomol.*, **113**, 355–7.
148. Lhoste, J., Francois, Y. & Rupaud, Y., 1979, Ichtyotoxicite de la decamethrine vis-à-vis de *Salmo trutta* L., en fonction de l'age et des conditions experimentales (Toxicity of decamethrin to *Salmo trutta* in relation to age and experimental conditions). *C.R. Congres Lutte Contre les Insectes en Milieu Tropical*, Marseille, pp. 885–901.
149. Lhoste, J. & L'Hotellier, M., 1982, Effects of deltamethrin on the environment. In *Deltamethrin Monograph*, Chapter 9. Roussel-Uclaf.
150. Linden, E., Bengtsson, B. E., Svanberg, O. & Sundstrom, G., 1979, The acute toxicity of 78 chemicals and pesticide formulations against two brackish water organisms, the bleak (*Alburnus alburnus*) and the harpacticoid *Nitocra spinites. Chemosphere*, **11/12**, 843–51.
151. Lloyd, R. & Jordan, D. H. M., 1963, Predicted and observed toxicities of several sewage effluents to rainbow trout. *J. Inst. Sewage. Purif.*, Part 2, 167–173.
152. Lofs-Holmin, A., 1982, Influence of routine pesticide spraying on earthworms (Lumbricidae) in field experiments with winter wheat. *Swedish J. Agric. Res.*, **12**, 121–3.
153. Loubaresse, J. P., Labonne, V., Jolie, H. & van Offeren, A., 1977, Resultats dessais en France et au Benelux avec le fenvalerate, nouvel insecticide du groupe des pyrethroids de synthese (Results of French and Belgian studies with fenvalerate, a new insecticide from the synthetic pyrethroid group). *Meded. Fac. Landb. Rijk. Gent*, **42**, 1825–38.
154. Lund, A. E. & Narahashi, T., 1981a, Kinetics of sodium channel modification by the insecticide tetramethrin in squid axon membranes. *J. Pharmacol. Exp. Ther.*, **Z19**, 464–73.
155. Lund, A. E. & Narahashi, T., 1981b, Modification of sodium channel kinetics by the insecticide tetramethrin in crayfish giant axons. *Neurotoxicol.* **2**, 213–9.
156. Lynch, J. & Poole, N. J., 1979, *Microbial Ecology – a Conceptual Approach* (Oxford: Blackwell).
157. Lynch, T. R. & Johnson, H. E., 1982, Availability of a hexachlorobiphenyl isomer to benthic amphipods from experimentally contaminated natural sediments. In *Aquatic Toxicology and Hazard Assessment*, (ASTM 766), 5th Conference, 1980, edited by J. G. Pearson, R. B. Foster & W. E. Bishop (Philadelphia: American Society for Testing and Materials), pp. 251–68.

158. Macek, K. J., 1975, Acute toxicity of pesticide mixtures to bluegills. *Bull. Environ. Contam. Toxicol.*, **14**, 648–52.
159. Macek, K. J. Petrocelli, S. R. & Sleight, B. H., 1979, Considerations in assessing the potential for significance of biomagnification of chemical residues in aquatic food chains. In *Aquatic Toxicology* (ASTM 667), 2nd conference, 1977, edited by L. L. Marking & R. A. Kimerle (Philadelphia: American Society for Testing and Materials), pp. 251–68.
160. Macek, K. J., Walsh, D. F., Hogan, J. W. & Holz, D. D., 1972, Toxicity of the insecticide dursban to fish and aquatic invertebrates in ponds. *Trans. Amer. Fish Soc.*, **101**, 420–7.
161. Marking, L. L. & Bills, T. D., 1976, *Toxicity of rotenone to fish in standardized laboratory tests*. U.S. Fish and Wildlife Service, Investigations in Fish Control, No. 72.
162. Marking, L. L. & Mauck, W. L., 1975, Toxicity of paired mixtures of candidate forest insecticides to rainbow trout. *Bull. Environ. Contam. Toxicol.*, **13**, 518–23.
163. Matsumura, F., 1977, Absorption, accumulation and elimination of pesticides by aquatic organisms. In *Pesticides in Aquatic Environments*, edited by M. A. Q. Khan (New York: Plenum Press), pp. 77–105.
164. Mauck, W. L., Olsen, L. E. & Marking, L. L., 1976, Toxicity of natural pyrethrin and five pyrethroids to fish. *Arch. Environ. Contam. Toxicol.*, **4**, 18–29.
165. Maurizio, A., 1961, Lebensdauer und altern bei der honigbiene (Lifespan and aging of honeybees). *Gerontologia*, **5**, 110–28.
166. Mayer, F. L., Street, J. C. & Neuhold, J. M., 1972, DDT intoxication in rainbow trout as affected by dieldrin. *Toxicol. Appl. Pharmacol.*, **22**, 347–54.
167. McKim, J. M., 1977, Evaluation of tests with early life stages of fish for predicting long-term toxicity. *J. Fish. Res. Board, Canada*, **34**, 1148–54.
168. McKim, J. M., Arthur, J. W. & Thorslund, T. W., 1975, Toxicity of a linear alkylate sulphonate detergent to larvae of four species of freshwater fish. *Bull. Environ. Contam. Toxicol.*, **14**, 1–7.
169. McKim, J. M., Eaton, J. G. & Holcombe, G. W., 1978, Metal toxicity to embyros and larvae – early juveniles of eight species of freshwater fish. II. Copper. *Bull. Environ. Contam. Toxicol.*, **19**, 608–16.
170. McLeese, D. W., Metcalfe, C. D. & Pezzack, D. S., 1980a, Uptake of PCBs from sediment by *Nereis virens* and *Crangon septemspinosa*. *Arch. Environ. Contam. Toxicol.*, **9**, 507–18.
171. McLeese, D. W., Metcalfe, C. D. & Zitko, V., 1980b, Lethality of permethrin, cypermethrin and fenvalerate to salmon, lobster and shrimp. *Bull. Environ. Contam. Toxicol.*, **25**, 950–5.
172. Mikami, N., Takahash, N., Hayash, K. & Miyamoto, J., 1980, Photodegradation of fenvalerate (Sumicidin) in water and soil surface. *J. Pestic. Sci.*, **5**, 225–36.
173. Miura, T. & Takahashi, R. M., 1976, Effects of a synthetic pyrethroid, SD43775, on non-target organisms when utilized as a mosquito larvicide. *Mosquito News*, **36**, 322–6.
174. Miyamoto, J., 1976, Degradation, metabolism and toxicity of synthetic pyrethroids. *Environ. Health Perspect.*, **14**, 15–28.
175. Moffet, J. O., Stoner, A. & Ahring, R. M., 1982, Effect of fenvalerate applications on honeybees in flowering rape. *Southwest. Entomol.*, **7**, 111–5.
176. Mohsen, Z. H. & Mulla, M. S., 1981, Toxicity of blackfly larvicidal formulation to some aquatic insects in the laboratory. *Bull. Environ. Contam. Toxicol.*, **26**, 696–703.
177. Molyneux, D. H., Baldry, D. A. T., van Wettere, P., Takken, W. & de Raadt, P., 1978, The experimental application of insecticides from a helicopter for the control of riverine populations of *Glossina tachinoides* in West Africa. I. Objectives, experimental area and insecticides evaluated. *PANS*, **24**, 391–403.

178. Moore, R. F., 1980, Behavioural and biological effects of NRDC-161 as a factor in control of the boll weevil. *J. Econ. Entomol.*, **73**, 256–67.
179. Motoyama, N., Rock, G. C. & Dauterman, W. C., 1970, Organophosphorus resistance in an apple orchard population of *Typhlodromus (Amblyseius) fallacis*. *J. Econ. Entomol.*, **63**, 1439–42.
180. Mount, G. A. & Pierce, N. W., 1975, Toxicity of pyrethroids and organosphosphorus adulticides to five species of mosquitoes. *Mosquito News*, **35**, 63–6.
181. Mowlam, M. D., Highwood, D. P., Dowson, R. J. & Hattori, J., 1977, Field performance of fenvalerate, a new synthetic pyrethroid insecticide. *Proceedings of the British Crop Protection Conference – Pests and Diseases*, pp. 649–56.
182. Muir, D. C. G., Townsend, B. E. & Lockhart, W. L., 1983, Bioavailability of six organic chemicals to *Chironomus tentans* larvae in sediment and water. *Environ. Toxicol. Chem.*, **2**, 269–81.
183. Muirhead-Thomson, R. C., 1970, The potentiating effect of pyrethrins and pyrethroids on the action of organophosphorus larvicides in *Simulium* control. *Trans. Roy. Soc. Trop. Med. Hyg.*, **64**, 895–906.
184. Muirhead-Thomson, R. C., 1977, Comparative tolerance levels of black fly *(Simulium)* larvae to permethrin (NRDC 143) and temephos. *Mosquito News*, **37**, 172–9.
185. Muirhead-Thomson, R. C., 1978, Lethal and behavioral impact of permethrin (NRDC 143) on selected stream macroinvertebrates. *Mosquito News*, **38**, 185–90.
186. Muirhead-Thomson, R. C., 1979, Experimental studies on macroinvertebrate predator–prey impact of pesticides. The reactions of *Rhyacophila* and *Hydropsyche* (*Trichoptera*) larvae to *Simulium* larvicides. *Can. J. Zool.*, **57**, 2264–70.
187. Muirhead-Thomson, R. C., 1981a, Relative toxicity of decamethrin, chlorphoxim and temephos (Abate) to *Simulium* larvae. *Tropenmed. Parasit.*, **32**, 189–93.
188. Muirhead-Thomson, R. C., 1981b, Tolerance levels of selected stream macroinvertebrates to the simulium larvicides, chlorphoxim and decamethrin. *Tropenmed. Parasit.*, **32**, 265–8.
189. Mulla, M. S., Arias, J. R., Sjogren, R. D. & Akesson, N. B., 1973, Aerial application of mosquito adulticides in irrigated pastures. *Proc. and Papers Calif. Mosq. Cont. Assoc.*, **41**, 51–6.
190. Mulla, M. S. & Darwazeh, H. A., 1976, Field evaluation of new mosquito larvicides and their impact on some nontarget insects. *Mosquito News*, **36**, 251–6.
191. Mulla, M. S., Darwazeh, H. A. & Dhillin, M. S., 1980, New pyrethroids as mosquito larvicides and their effects on nontarget organisms. *Mosquito News*, **40**, 6–12.
192. Mulla, M. S., Darwazeh, H. A. & Dhillon, M. S., 1981, Impact and joint action of decamethrin and permethrin and freshwater fishes on mosquitoes. *Bull. Environ. Contam. Toxicol.*, **26**, 689–95.
193. Mulla, M. S., Darwazeh, H. A. & Majori, G., 1975, Field efficacy of some promising mosquito larvicides and their effects on nontarget organisms. *Mosquito News*, **35**, 179–85.
194. Mulla, M. S., Navvab-Gojrati, H. A. & Darwazeh, H. A., 1978a, Biological activity and longevity of new synthetic pyrethroids against mosquitoes and some nontarget insects. *Mosquito News*, **38**, 90–6.
195. Mulla, M. S., Navvab-Gojrati, H. A. & Darwazeh, H. A., 1978b, Toxicity of mosquito larvicidal pyrethroids to four species of freshwater fishes. *Environ. Entomol.*, **7**, 428–30.
196. Narahashi, T., 1982, Modification of nerve membrane sodium channels by the insecticide pyrethroids. *Comp. Biochem. Physiol.*, **72C**, 411–4.
197. Narahashi, T. & Lund, A. E., 1980, Giant axons as models for the study of the mechanism of action of insecticides. In *Insect Neurobiol. Pestic. Action (Neurotox. 79), Proc. Symp. Soc. Chem. Ind. 1979*, 497–505.

198. Neuenschwander, P., Hagen, K. S. & Smith, R. F., 1975, Predation on aphids in California's alfalfa fields. *Hilgardia*, **43**, 53–78.
199. Nishiuchi, Y., 1978, Toxicity of formulated pesticides to fresh water organisms – LIII. *Suisan. Zoshoku. (The Aquiculture)*, **26**, 109–13.
200. Nishiuchi, Y., 1979, Toxicity of formulated pesticides to fresh water organisms –LVIII. *Suisan. Zoshoku. (The Aquiculture)*, **27**, 42–7.
201. Ohkawa, H., Kikuchi, R. & Miyamoto, J., 1980, Bioaccumulation and biodegradation of the (*S*)-acid isomer of fenvalerate (Sumicidin) in an aquatic model ecosystem. *J. Pestic. Sci.*, **5**, 11–22.
202. Osborne, M. P., 1980, The insect synapse: structural and functional aspects in relation to insecticidal action. In *Insect Neurobiol. Pestic. Action (Neurotox. 79), Proc. Symp. Soc. Chem. Ind. 1979*, 29–40.
203. Penman, D. R., Chapman, R. B. & Jesson, K. E., 1981, Effects of fenvalerate and azinphosmethyl on two-spotted spider mite and phytoseiid mites. *Ent. Exp. Appl.*, **30**, 91–7.
204. Pike, K. S., Mayer, D. F., Glazer, M. & Kious, C., 1982, Effects of permethrin on mortality and foraging behaviour of honey bees in sweet corn. *Environ. Entomol.*, **11**, 951–3.
205. Pillmore, R. E., 1973, Toxicity of pyrethrum to fish and wildlife. In *Pyrethrum*, edited by J. Casida (New York: Academic Press), pp. 143–65.
206. Pionke, H. B. & Chesters, G., 1973, Pesticide–sediment–water interactions. *J. Environ. Qual.*, **2**, 29–45.
207. Plapp, F. W. Jr. & Bull, D. L., 1978, Toxicity and selectivity of some insecticides to *Chrysopa carnea*, a predator of the tobacco budworm. *Environ. Entomol.*, **7**, 431–4.
208. Plapp, F. W. Jr. & Vinson, S. B., 1977, Comparative toxicities of some insecticides to the tobacco budworm and its ichneumonid parasite, *Campoletis sonorensis*. *Environ. Entomol.*, **6**, 381–4.
209. Pree, D. J., 1979, Toxicity of phosmet, azinphosmethyl, and permethrin to the oriental fruit moth and its parasite, *Macrocentrus ancylivorus*. *Environ. Entomol.*, **8**, 969–72.
210. Pree, D. J. & Hagley, E. A. C., 1977, Toxicity of some insecticides to eggs of the oriental fruit moth and codling moth. *Proc. Entomol. Soc. Ont.*, **108**, 69–74.
211. Press, J. W., Flaherty, B. R. & McDonald, L. L., 1978, Toxicity of five insecticides to the predaceous bug *Xylocoris flavipes* (Hemiptera: Anthocoridae). *J. Georgia Entomol. Soc.*, **13**, 181–4.
212. Rajakulendran, S. V. & Plapp, F. W., 1982a, Comparative toxicities of five synthetic pyrethroids to the tobacco budworm (Lepidoptera: Noctuidae), an ichneumonid parasite, *Campoletis sonorensis*, and a predator, *Chrysopa carnea*. *J. Econ. Entomol.*, **75**, 769–72.
213. Rajakulendran, S. V. & Plapp, F. W., 1982b, Synergism of five synthetic pyrethroids by chlordimeform against the tobacco budworm (Lepidoptera: Noctuidae) and a predator, *Chrysopa carnea* (Neuroptera: Chrysopidae). *J. Econ. Entomol.*, **75**, 1089–92.
214. Raw, F., 1959, Estimating earthworm populations by using formalin. *Nature, Lond.*, **184**, 1661–2.
215. Rawn, G. P., Muir, D. C. G. & Webster, G. R. B., 1981, Uptake and persistence of permethrin by fish, vegetation and hydrosoil. *8th Annual Aquatic Toxicology Workshop*, p. 9, Guelph, Canada.
216. Rawn, G. P., Webster, G. R. B. & Muir, D. C. G., 1982, Fate of permethrin in model outdoor ponds. *J. Environ. Health*, **B17**, 463–86.
217. Reissig, W. H., Heinrichs, E. A. & Valencia, S. L., 1982, Insecticide-induced resurgence of the brown planthopper, *Nilaparvata lugens*, on rice varieties with different levels of resistance. *Environ. Entomol.*, **11**, 165–8.
218. Rettich, F., 1979, The toxicity of four synthetic pyrethroids to mosquito larvae and

pupae (Diptera, Culicidae) in Czechoslovakia. *Acta Entomol. Bohemoslov.*, **76**, 395–401.

219. Rettich, F., 1980, Field evaluation of permethrin and decamethrin against mosquito larvae and pupae (Diptera, Culicidae). *Acta Entomol. Bohemoslov.*, **77**, 89–96.
220. Riedl, H. & Hoying, S. A., 1980, Impact of fenvalerate and diflubenzuron on target and non-target arthropod species on Bartlett pears in Northern California. *J. Econ. Entomol.*, **73**, 117–22.
221. Roach, S. H. & Hopkins, A. R., 1981, Reduction in arthropod predator population in cotton fields treated with insecticides for *Heliothis* spp. control. *J. Econ. Entomol.*, **74**, 454–7.
222. Rock, G. C., 1979, Relative toxicity of two synthetic pyrethroids to a predator *Amblyseius fallacis* and its prey *Tetranychus urticae*. *J. Econ. Entomol.*, **72**, 293–4.
223. Roush, R. T. & Hoy, M. A., 1978, Relative toxicity of permethrin to a predator, *Metaseiulus occidentalis*, and its prey, *Tetranychus urticae*. *Environ. Entomol.*, **7**, 287–8.
224. Ruscoe, C. N. E., 1979, The Impact of the photostable pyrethroids as agricultural insecticides. *Proceedings of the British Crop Protection Conference: Pests and Diseases*, pp. 803–14.
225. Rzehak, von H. & Basedow, T., 1982, Die Auswirkungen verschiedener Insektizide auf die epigaischen Raubarthropoden in Winterrapsfeldern (The effects of various insecticides on the predator arthropods in winter rape crops). *Anz. Schadlingskde. Pflanzenschutz, Umweltschutz*, **55**, 71–5.
226. Salibian, A. & Fichera, L. E., 1980, Ecotoxicologia del insecticida piretroide decametrina: Effectos del "Decis" sobre juveniles de *Astyanax (A.) fasciatus fasciatus* (Tetragonopteridae, Pisces) en cautiverio. *Resum. VIII Reunion Argent. Ecologia*, p. 15.
227. Salibian, A. & Fichera, L. E., 1981, Ecotoxicology of pyrethroid insecticides: Short term effects of Decis 2–5 on juvenile *Astyanax (Astyanax) fasciatus fasciatus* (Tetragonopteridae, Pisces) in captivity. *Comp. Biochem. Physiol.*, **70C**, 265–8.
228. Salkeld, E. H. & Potter, C., 1953, The effect of age and stage of development of insect eggs on their resistance to insecticides. *Bull. Entomol. Res.*, **44**, 527–80.
229. Schimmel, S. C., Garnas, R. L., Patrick, J. M., Jr. & Moore, J. C., 1983, Acute toxicity, bioconcentration, and persistence of AC222,705, benthiocarb, chlorpyrifos, fenvalerate, methyl parathion, and permethrin in the estuarine environment. *J. Agric. Food Chem.*, **31**, 104–13.
230. Schulten, G. G. M., van de Klashorst, G., 1977, Genetics of resistance to parathion and demethon-S-methyl in *Phytoseiulus persimilis* A.H. (Acari: Phytoseiidae). *Proc. IV Int. Congress Acarol.* Saalfelden, Austria.
231. Schulten, G. G. M., van de Klashorst, G. & Russell, V. M., 1976, Resistance of *Phytoseiulus persimilis* A.H. (Acari: Phytoseiidae) to some insecticides. *Z. Ang. Ent.*, **80**, 337–41.
232. Shacklock, P. F. & Croft, G. B., 1981, Effect of grazers on *Chondrus crispus* in culture. *Aquaculture*, **22**, 331–42.
233. Shannon, C. E. & Weaver, W., 1949, *The Mathematical Theory of Communication* (Urbana: University of Illinois Press).
234. Shinkaji, N., 1976, Toxicity of some pesticides to *Phytoseiulus persimilis* Athias-Henriot (Acarina: Phytoseiidae). *Bull. Fruit Tree Res. Stn., Japan*, **E1**, 103–16.
235. Shinkaji, N. & Adachi, T., 1978, The effect of certain pesticides on the predacious mite *Amblyseius longispinosus* (Evans) (Acarina: Phytoseiidae). *Bull. Fruit Tree Res. Stn., Japan*, **E2**, 99–108.
236. Shires, S. & Debray, P., 1982, Pyrethroids and the bee problem. *Shell Agric.*, May 1982, 1–3.
237. Shour, M. H. & Crowder, L. A., 1980, Effects of pyrethroid insecticides on the common green lacewing. *J. Econ. Entomol.*, **73**, 306–9.

238. Singh, O. P. & Rawat, R. R., 1981, Note on the safety of some insecticides for *Diaeretus rapae* M'Intosh, a parasite of the mustard aphid. *Ind. J. Agric. Sci.*, **51**, 204–5.

239. Smart, L. E. & Stevenson, J. H., 1982, Laboratory estimation of toxicity of pyrethroid insecticides to honeybees: relevance to hazard in the field. *Bee World*, **63**, 150–2.

240. Smies, M., Evers, R. H. J., Peijnenburg, F. H. M. & Koeman, J. H., 1980, Environmental aspects of field trials with pyrethroids to eradicate tsetse fly in Nigeria. *Ecotoxicol. Environ. Safety*, **4**, 114–28.

241. Smith, F. F., Henneberry, T. J. & Boswell, A. L., 1963, The pesticide tolerance of *Typhlodromus fallacis* (Garman) and *Phytoseiulus persimilis* A.H. with some observations on the predator efficiency of *P. persimilis*. *J. Econ. Entomol.*, **56**, 274–8.

242. Smith, K., Kaushik, N. K. & Solomon, K. R., 1981, A comparison of the effects of three pesticides in a lake ecosystem using large volume (125 m^3) *in situ* enclosures. *8th Annual Aquatic Toxicology Workshop*, p. 13. Guelph, Canada.

243. Soderlund, D. M. & Casida, J. E., 1977, Effects of pyrethroid structure on rate of hydrolysis and oxidation by mouse liver microsomal enzymes. *Pestic. Biochem. Physiol.*, **7**, 391–401.

244. Solomon, K. R., Smith, K., Guest, G., Yoo, J. Y. & Kaushik, N. K., 1980, Use of limnocorrals in studying the effects of pesticides in the aquatic ecosystem. *Can. Tech. Rep. Fish. Aquat. Sci.*, **975**, 1–9.

245. Spehar, R. L., Tanner, D. K. & Gibson, J. H., 1982, Effects of kelthane and pydrin on early life stages of fathead minnows (*Pimephales promelas*) and amphipods (*Hyalella azteca*). In *Aquatic Toxicology and Hazard Assessment* (ASTM 766), 5th Conference, 1980, edited by J. G. Pearson, R. B. Foster & W. E. Bishop (Philadelphia: American Society for Testing and Materials), pp. 234–44.

246. Spehar, R. L., Tanner, D. K. & Nordling, B. R., 1983, Toxicity of the synthetic pyrethroids, permethrin and AC222705 and their accumulation in early life stages of fathead minnows and snails. *Aquat. Toxicol.*, **3**, 171–82.

247. Speilberger, U., Na'Isa, B. K., Koch, K., Manno, A., Skidmore, P. R. & Coutts, H. H., 1979, Field trials with the synthetic pyrethoid insecticides, permethrin, cypermethrin and decamethrin against *Glossina (Diptera, Glossinidae)* in Nigeria. *Bull. Entomol. Res.*, **69**, 667–89.

248. Sprague, J. B., 1969, Measurement of pollutant toxicity to fish, 1. Bioassay methods for acute toxicity. *Water Res.*, **3**, 793–821.

249. Stenseth, C., 1979, Effect of fungicides and insecticides on an OP-resistant strain of *Phytoseiulus persimilis* Athias-Henriot. *Forsk. Fors. Landbruket*, **30**, 77–83.

250. Stephenson, R. R., 1982, Aquatic toxicology of cypermethrin. 1. Acute toxicity to some freshwater fish and invertebrates in laboratory tests. *Aquat. Toxicol.*, **2**, 175–85.

251. Stephenson, R. R. Choi, S. Y. & Olmos-Jerez, A, 1984, Determining the toxicity and hazard to fish of a rice insecticide. *Crop Protection*, **3**, 151–165.

252. Stevenson, J. H., 1978, The acute toxicity of unformulated pesticides to worker honeybees (*Apis mellifera* L.). *Plant Pathol.*, **27**, 38–40.

253. Stevenson, J. H., Needham, P. H. & Walker, J., 1978, Poisoning of honeybees by pesticides : investigations of the changing pattern in Britain over 20 years. *Rothamsted Experimental Station Annual Report*, 1977 Part 2, 55–72.

254. Stratton, G. W. & Corke, C. T., 1981, Interaction of permethrin with *Daphnia magna* in the presence and absence of particulate material. *Environ. Pollut., Ser. A*, **24**, 135–44.

255. Stratton, G. W. & Corke, C. T., 1982a, Comparative fungitoxicity of the insecticide permethrin and ten degradation products. *Pestic. Sci.*, **13**, 679–85.

256. Stratton, G. W. & Corke, C. T., 1982b, Toxicity of the insecticide permethrin and some degradation products towards algae and cyanobacteria. *Environ. Pollut. Ser. A*, **29**, 71–80.
257. Strickler, K. & Croft, B. A., 1982, Selection for permethrin resistance in the predatory mite *Amblyseius fallacis*. *Ent. Exp. Appl.*, **31**, 339–45.
258. Sukhoruchenko, G. I., Smirnova, A. A., Vikar, Ye. V. & Kapitan A. I., 1982, The effect of pyrethroids on the arthropods of a cotton agrobiocenosis (In *Entomol. Obozr.* 1981). *Entomol. Rev.*, **60**, 1–10.
259. Sun, Y-P., 1968, Dynamics of insect toxicology – mathematical and graphical evaluation of the relationship between insect toxicity and rates of penetration and detoxification of insecticides. *J. Econ. Entomol.*, **61**, 949–55.
260. Surulivelu, T. & Menon, M. V., 1982, Contact toxicity of synthetic pyrethroids organophosphorus and carbamate insecticides to adults of the parasite *Chelonus blackburni* Cameron. *J. Agric. Sci., Camb.*, **98**, 331–4.
261. Symons, P. E. K., 1968, Increase in aggression and in strength of the social hierarchy among juvenile Atlantic salmon deprived of food. *J. Fish. Res. Board Can.*, **25**, 2387–401.
262. Syrett, P. & Penman, D. R., 1980, Studies of insecticide toxicity to lucerne aphids and their predators. *NZ J. Agric. Res.*, **23**, 575–80.
263. Tag el-Din, A., Abbas, M. M., Aly, H. A., Tantawy, G. & Askar, A., 1981, Acute toxicities to *Mugil cephalus* fry caused by some herbicides and new pyrethroids. *Meded. Fac. Landb. Rijk. Gent*, **46**, 387–91.
264. Tagatz, M. E. & Ivey, J. M., 1981, Effects of fenvalerate on field and laboratory developed estuarine benthic communities. *Bull. Environ. Contam. Toxicol.*, **27**, 256–67.
265. Takken, W., Balk, F., Jansen, R. C. & Koeman, J. H., 1978, The experimental application of insecticides from a helicopter for the control of riverine populations of *Glossina tachinoides* in West Africa. VI. Observations on side-effects. *PANS*, **24**, 455–66.
266. Thompson, G. D. & Meisch, M. V., 1977, Efficacy of permethrin as a larvicide and adulticide against ricefield mosquitoes. *J. Econ. Entomol.*, **70**, 771–4.
267. Tu, C. M., 1980, Influence of five pyrethroid insecticides on microbial populations and activities in soil. *Microb. Ecol.*, **5**, 321–27.
268. Tu, C. M., 1982, Effect of pyrethroid insecticides on soybean and its pathogen *Rhizoctonia solani* Kuehn. *J. Environ. Sci. Health*, **B17**, 43–50.
269. Waddill, van H., 1978, Contact toxicity of four synthetic pyrethroids and methomyl to some adult insect parasites. *Fla. Entomol.*, **61**, 27–30.
270. Walsh, G. E. & Alexander, S. V., 1980, A marine algal bioassay method : results with pesticides and industrial wastes. *Water, Air and Soil Pollution*, **13**, 45–55.
271. Wang, C. M., Narahashi, T. & Scuka, M., 1972, Mechanism of negative temperature coefficient of nerve blocking of allethrin. *J. Pharmacol. Exp. Ther.*, **182**, 443–53.
272. Ware, G. W., Estesen, B. J., Cahill, W. P., Gerhardt, P. D. & Frost, K. R., 1969a, Pesticide drift. I. High-clearance *vs* aerial applications of sprays. *J. Econ. Entomol.* **62**, 840–3.
273. Ware, G. W., Apple, E. J., Cahill, W. P., Gerhardt, P. D, & Frost, K. R., 1969b, Pesticide drift. II. Mist-blower *vs* aerial applications of sprays. *J. Econ. Entomol.* **62**, 844–6.
274. Warner, L. A. & Croft, B. A., 1982, Toxicities of azinphosmethyl and selected orchard pesticides to an aphid predator, *Aphidoletes aphidimyza*. *J. Econ. Entomol.*, **75**, 410–15.
275. Wauchope, R. D., 1978, The pesticide content of surface water draining from agricultural fields – a review. *J. Environ. Qual.*, **7**, 459–72.

276. Wauchope, R. D., 1980, Runoff studies and pesticide registration. In *Test Protocols for Environmental Fate and Movement of Toxicants*, Proceedings of the Symposium of the Association of Official Analytical Chemists, 94th Annual Meeting. Washington D.C. pp. 200–5.

277. Weaver, J. B., Jr., All, J. N., Weaver, D. B. & Hornyak, E. P., 1979, Influence of various insecticides on yield parameters of two cotton genotypes. *J. Econ. Entomol.*, **72**, 119–23.

278. van Wettere, P., Baldry, D. A. T., Molyneux, D. H., Clarke, J. H., Lee, C. W. & Parker, J. D., 1978, The experimental application of insecticides from a helicopter for the control of riverine populations of *Glossina tachinoides* in West Africa. IV. Evaluation of insecticides applied as aerosols. *PANS*, **24**, 435–46.

279. Whalon, M. E., Croft, B. A. & Mowry, T. M., 1982, Introduction and survival of susceptible and pyrethroid-resistant strains of *Amblyseius fallacis* (Acari: Phytoseiidae) in a Michigan apple orchard. *Environ. Entomol.*, **11**, 1096–9.

280. Wilkinson, J. D., Biever, K. D. & Ignoffo, C. M., 1979, Synthetic pyrethroid and organophosphate insecticides against the parasitoid *Apanteles marginiventris* and the predators *Geocoris punctipes*, *Hippodamia convergens*, and *Podisus maculiventris*. *J. Econ Entomol.*, **72**, 473–5.

281. Wilkinson, W. & Bull, J. M., 1984, The toxicity to honeybees of permethrin spray on top fruit and oilseed rape. *Faune et Flore Auxiliaresen Agriculture, ACTA Symposium*, Paris, pp. 335–65.

282. Wilkinson, W. & Cole, J. F. H., 1984, A field study to show the effects of permethrin and azinphos-methyl on the arthropod fauna of apple trees. In *Influence of Pesticides on the Beneficial Fauna in Fruit Trees*. Colmar 1981, IOBC WPRS Bulletin, 56–58.

283. Willis, G. H. & Hamilton, R. A., 1973, Agricultural chemicals in surface runoff, ground water and soil. 1. Endrin. *J. Environ. Qual.*, **2**, 463–6.

284. Willis, G. H., Rogers, R. L. & Southwick. E. M., 1975, Losses of diuron, linuron, fenac and trifluralin in surface drainage water. *J. Environ. Qual.*, **4**, 399–401.

285. Wong, S. W. & Chapman, R. B., 1979, Toxicity of synthetic pyrethroid insecticides to predaceous phytoseiid mites and their prey. *Aust. J. Agric. Res.*, **30**, 497–501.

286. Wuhrmann, K., 1952, Sur quelques principes de la toxicologie du poisson (On some principles of fish toxicology). *Bull. Centre Belge et Document. Eaux*, **15**, 49–60.

287. Yates, W. E. & Akesson, N. B., 1973, Reducing pesticide chemical drift. In *Pesticide Formulations*, edited by W. van Valkenberg (New York: Marcel Dekker), pp 275–341.

288. Yates, W. E., Akesson, N. B. & Bayer, D. E., 1978, Drift of glyphosate sprays applied with aerial and ground equipment. *Weed Sci.*, **26**, 597–604.

289. Yates, W. E., Akesson, N. B. & Brazelton, R. W., 1981, Systems for the safe use of pesticides. *Outlook Agric.*, **10**, 321–6.

290. Yunus, A. & Soon, L. G., 1971, A problem in the use of insecticides in paddy fields in West Malaysia – a case study. *Malaysian Agric. J.*, **48**, 167–78.

291. Zherebkina, M. V. & Schagun, Y. L., 1971, Some physiological changes in honeybees during their preparation for winter. *Vest. Nauchno-issle. Noisled. Inst. Pchel*, No. 20.

292. Zitko, V., Carson, W. G. and Metcalfe, C. D., 1977, Toxicity of pyrethroids to juvenile Atlantic salmon. *Bull. Environ. Contam. Toxicol.*, **18**, 35–41.

293. Zitko, V., McLeese, D. W., Metcalfe, C. D. & Carson, W. G., 1979, Toxicity of permethrin, decamethrin, and related pyrethroids to salmon and lobster. *Bull. Environ. Contam. Toxicol.*, **21**, 338–43.

5. Metabolism and environmental degradation

J. P. Leahey

The use of natural pyrethrins and the pyrethroids as household, public health and agricultural insecticides has stimulated an enormous amount of work on the fate of these compounds in target species, mammals and, where relevant, in crops and in the general environment. Much of this work, which has been carried out by industry, government research institutions and universities, has been published, although there is undoubtedly also additional unpublished data in the files of industry and government registration authorities.

In any attempt to summarise the very extensive literature on pyrethroid metabolism and environmental degradation, the task is made more difficult by the complex stereochemistry of these molecules. This stereochemistry has already been fully discussed in Chapter 1 and will therefore normally not be discussed in the text of this chapter. However, where relevant, the stereochemistry of any particular compound will be indicated in figures showing metabolic and degradation pathways. In cases where experiments were carried out with mixtures of isomers, or complete isomeric characterisation has not been achieved, a generalised structure will be used.

Virtually all of the work discussed in this chapter will have utilised radiolabelled compounds. However, the positions of radiolabelling utilised will not be indicated in either the text or in figures, unless it contributes to an understanding of the information being presented, for example when degradation results in evolution of $[^{14}C]$-carbon dioxide.

Conjugation of the metabolites formed from the various pyrethroids is, as with most xenobiotics, a major metabolic process in mammals, insects and plants. Such conjugation processes will be indicated on the figures showing metabolic pathways, but only novel or unusual conjugates will be referred to in the text.

Photodegradation

Perhaps the most relevant place to start a summary of the degradation of the pyrethroid insecticides is with the photodegradation of these compounds, since extreme photosensitivity was the reason why the natural pyrethrins and their early synthetic analogues were limited to non-agricultural pest control. However, an understanding of the photosusceptibility of these compounds was an important contribution to the eventual synthesis of photostable analogues. These analogues are now a very important part of the modern insecticide armoury, and have changed pyrethroids from compounds limited to public health and domestic usage into compounds of major agricultural importance.

The natural pyrethrins

The recognition that light was the key factor in the rapid deactivation of the natural pyrethrins was recognised as early as 1932 (Tattersfield, 1932), and there were then many attempts made to stabilise the pyrethrins by the addition of ultra-violet screening agents and antioxidants (Abe et al., 1972; Miskus & Andrews, 1972). However, very little work has been reported on the nature of the photoproducts produced. Most attempts to elucidate the photochemical pathways for pyrethrins have been carried out on model compounds, mainly chrysanthemic acid and simple derivatives of this acid (Sasaki et al., 1970; Ueda & Matsui, 1971; Bullivant & Pattenden, 1976). The results obtained by these workers indicated that the major photochemical reaction for *trans*-chrysanthemic acid is homolytic cleavage of the 1,3-bond of the cyclopropane ring. The further reactions of the diradical thus formed are summarised in Figure 5.1.

However, the results from these model compound studies did not appear to predict correctly the phototransformation reactions which were detected in the one reported study of the photodegradation of a natural pyrethrin (Chen & Casida, 1969). In this study irradiation of pyrethrin I with a 275 watt sun lamp produced at least 11 photoproducts within 8 hours. Most, if not all, of these photoproducts were considered still to have retained their ester linkage, although no identification of any of these compounds was made. However, saponification of these esters released a large number of acidic compounds, and identification of some of these acids allowed the postulation of a partial photodegradation pathway for pyrethrin I (see Figure 5.2). Positive attempts were also made in this study to see if any *trans* to *cis* interconversion had occurred, since such isomerisation would be expected from the results obtained with chrysanthemic acid. But no acids with a *cis* configuration were detected, although it was emphasised by these workers that *cis*-isomers could be present in the unidentified material.

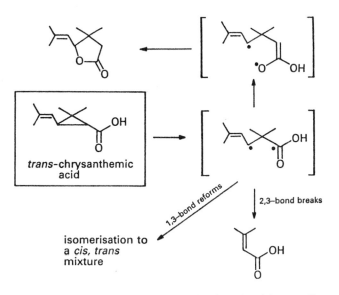

Figure 5.1. Phototransformation of chrysanthemic acid (postulated intermediates shown in square brackets)

Photolabile synthetic pyrethroids

The first synthetic analogues of the natural pyrethrins were nearly all based on modifications of the alcohol moiety, and hence they still contained the photolabile chrysanthemic acid moiety. Early studies (Chen & Casida, 1969) with the *trans*-isomers of allethrin, tetramethrin (phthalthrin) and dimethrin (see Figure 5.3) showed that the phototransformations established for the acid moiety of pyrethrin I also occurred with these analogues. Photomodification of the alcohol halves of these molecules was also detected, but the nature of these modifications was not elucidated. As with pyrethrin I, no *trans* to *cis* isomerisation or ester cleavage was detected in these studies.

However, rather more detailed investigations have since been carried out by Casida and his co-workers on allethrin (Ruzo et al., 1980; Kimmel et al., 1982; Ando et al., 1983). The phototransformations detected in these later studies (see Figure 5.4) were rather different to the earlier results obtained by Chen & Casida, especially in that photoisomerisation and ester cleavage were detected. In addition, epoxidation reactions, and the isomerisation of the alcohol moiety propenyl side chain to a cyclopropane ring, were shown to occur. This latter phototransformation was initially characterised by Bullivant & Pattenden (1973) in studies with bio-allethrin (the most active isomer of allethrin), but in this work isomerisation to yield the cyclopropane ring was the only reaction detected (95% yield), with no trace of any of the other phototransformations reported by Casida and his co-workers. This variation in the results obtained

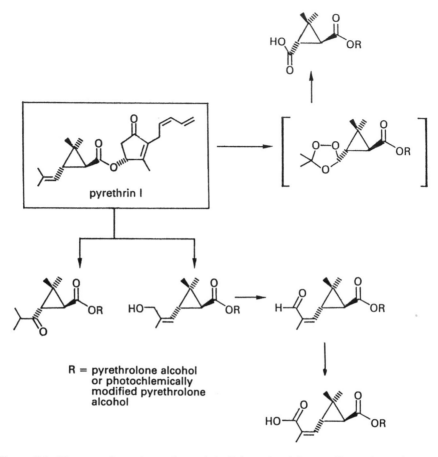

Figure 5.2. Phototransformations of pyrethrin I (postulated intermediates shown in square brackets)

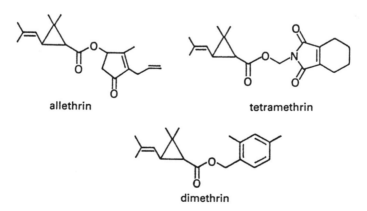

allethrin

tetramethrin

dimethrin

Figure 5.3. Structures of allethrin, tetramethrin and dimethrin

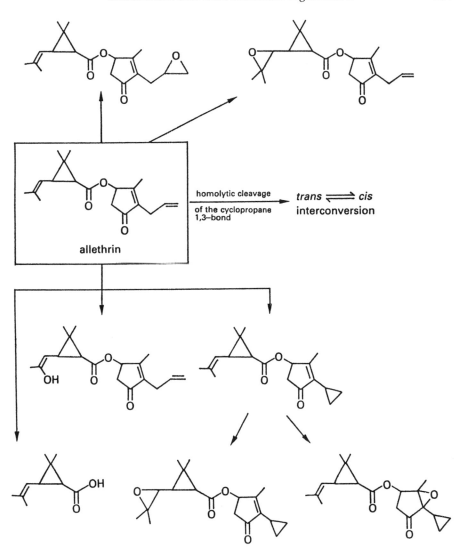

Figure 5.4. Phototransformations of allethrin

with allethrin appears at first sight to be rather confusing, but these variable results may well be explained by the different irradiation systems used in these studies, that is in solution in various solvents, as thin films on glass and with radiation sources ranging from sunlight to a variety of artificial light sources.

Recent work by Ruzo et al. (1982) with *trans*-tetramethrin and *trans*-phenothrin has further characterised the photoreactions of the chrysanthemic acid moiety. This study confirms the epoxidation, isobutenyl methyl group oxidation, ester cleavage and *trans* to *cis* isomerisation reactions established

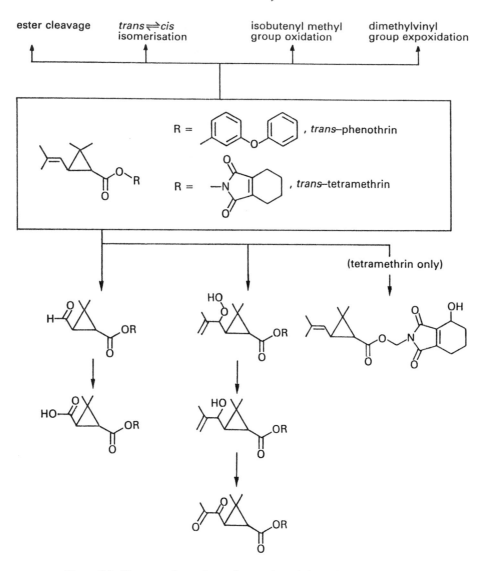

Figure 5.5. Phototransformations of *trans*-phenothrin and *trans*-tetramethrin

with either pyrethrin I or allethrin. However, additional reactions, illustrated in Figure 5.5, for the chrysanthemic acid group were also elucidated. In addition, hydroxylation of the alcohol moiety of tetramethrin was detected in this study.

Two other photolabile pyrethroids, resmethrin and kadethrin, have also been subjected to detailed photochemical investigation (Ueda et al., 1974; Ohsawa & Casida, 1979). The results obtained with these two compounds are summarised in Figures 5.6 and 5.7. Both of these compounds contain the same alcohol

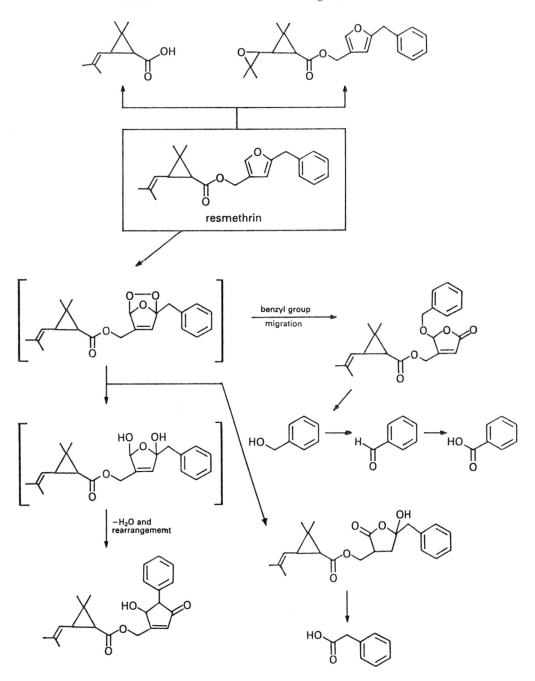

Figure 5.6. Phototransformations of resmethrin (postulated intermediates in square brackets)

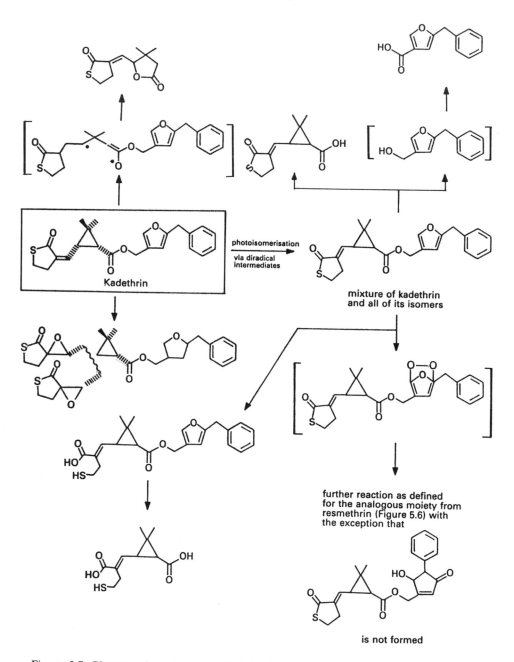

Figure 5.7. Phototransformation of kadethrin (postulated intermediates shown in square brackets)

moiety, and similar reactions, with the unexplained exception noted on Figure 5.7, were detected for this fragment in both molecules, with oxidation of the furan ring initiating most of the phototransformations. Ester cleavage was also detected for both molecules, as was epoxidation of the carbon–carbon double bond within the acid moiety. In the case of kadethrin, reactions involving hydrolysis of the thiolactone ring and isomerisation via diradical intermediates were also observed.

Photostabilised synthetic pyrethroids

The extreme photolability of the natural and early synthetic pyrethroids was eventually overcome with the synthesis of permethrin (Elliott et al., 1973a). In this compound the photosensitive dimethylvinyl group of chrysanthemic acid has been replaced by the more photostable dichlorovinyl group; and 3-phenoxybenzyl alcohol provides a less photosensitive alcohol moiety. The resulting "photostabilised" pyrethroid was found to have a half-life, in sunlight, measurable in days (Elliott et al., 1973b) rather than in hours, as had been the case with the natural pyrethrins and their early synthetic analogues. Other photostabilised pyrethroids soon followed on from permethrin, and a new class of agricultural insecticides was thus created.

However, the achievement of increased photostability did not end the interest in the photochemistry of these compounds. In fact, with the conversion of pyrethroids from domestic to agricultural insecticides came the requirement for even more detailed photochemical investigations, as part of the need to understand the environmental fate of these compounds.

Permethrin

The photochemistry of permethrin has been studied extensively, both in solution and on a soil surface (Holmstead & Casida, 1977; Holmstead et al., 1978a). In solution in methanol, hexane, water and water/acetone mixtures, photoisomerisation about the 1,3-bond of the cyclopropane ring and ester cleavage are the major reactions. Minor pathways are the replacement of one chlorine atom by hydrogen, fission of the cyclopropane ring to yield 3-phenoxydimethylacrylate, oxidation of 3-phenoxybenzyl alcohol to the corresponding aldehyde and carboxylic acid, and cleavage of the ether linkage. Unidentified polar material was also formed and this was especially important in aqueous solution. Similar pathways of phototransformation were detected on a soil surface, but the rate of degradation was much slower than in solution, and photoisomerisation was not so important. These photodegradation reactions for permethrin are summarised in Figure 5.8.

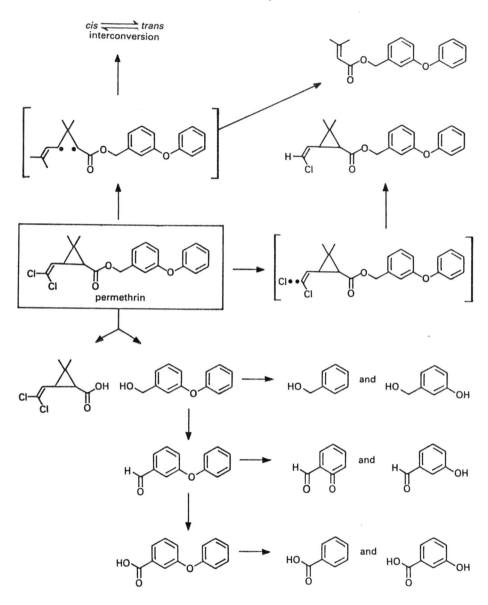

Figure 5.8. Phototransformations of permethrin (postulated intermediates shown in square brackets)

Deltamethrin

The photochemistry of deltamethrin (originally decamethrin) has been studied in a variety of organic solvents, in aqueous acetonitrile, and on glass plates and silica gel (Ruzo et al., 1977). These studies have elucidated a complex series of reactions and photoisomerisations (see Figure 5.9). The relative importance of the various processes summarised in Figure 5.9 depends on the irradiation conditions used, but, in general, isomerisation, ester cleavage and reductive debromination are the dominant reactions. As might be expected, many of the phototransformations of deltamethrin are similar to those elucidated for permethrin (see Figure 5.8). However, with deltamethrin a fairly minor reaction occurs which is not paralleled by permethrin. This process is initiated by homolytic cleavage of the oxygen, α-carbon bond of the alcohol moiety, followed by elimination of carbon dioxide and recombination of the remaining fragments to yield 1-(α-cyano-3-phenoxybenzyl)-2,2-dimethyl-3-(2,2-dibromovinyl)cyclopropane.

Fenvalerate

The phototransformation of fenvalerate has been studied by two groups of workers (Holmstead et al., 1978b; Mikami et al., 1980). In the earlier study irradiations were carried out in organic solvents, in aqueous acetonitrile, as a thin film on glass and on cotton leaves. In the later study, sterile distilled water, aqueous acetone (98 : 2), river water, sea water and soil surfaces were used. Within this range of experimental systems there were a number of both quantitative and qualitative differences in the photoproducts produced. However, with both studies, decarboxylation to yield 3-(4-chlorophenyl)-4-methyl-2-(3-phenoxybenzyl)-valeronitrile was a major process. This is an interesting contrast to deltamethrin where the equivalent reaction is very minor. A complex range of additional reactions were also detected in either one or both of these studies. These phototransformations, which involve ester and ether cleavage, hydrolysis of the cyanide group and other radical initiated reactions, are summarised in Figure 5.10. The release of hydrogen cyanide was established as a fairly major process by Mikami et al.; but these workers also showed that this compound is then very rapidly photodegraded to carbon dioxide.

Tralomethrin and tralocythrin

Tralomethrin and tralocythrin can be considered as derivatives of deltamethrin and the *cis*, 1R, (S)-α-cyano isomer of cypermethrin, in which the dihalovinyl group has been saturated by the addition of a molecule of bromine. The major initial photodegradation reaction which occurs with these compounds is debromination to yield deltamethrin and cypermethrin (Ruzo & Casida, 1981). Loss of hydrogen bromide to yield mono-brominated analogues of deltamethrin and cypermethrin also occurs, but this is a much less important process. The other significant reactions detected are *cis* to *trans* isomerisation

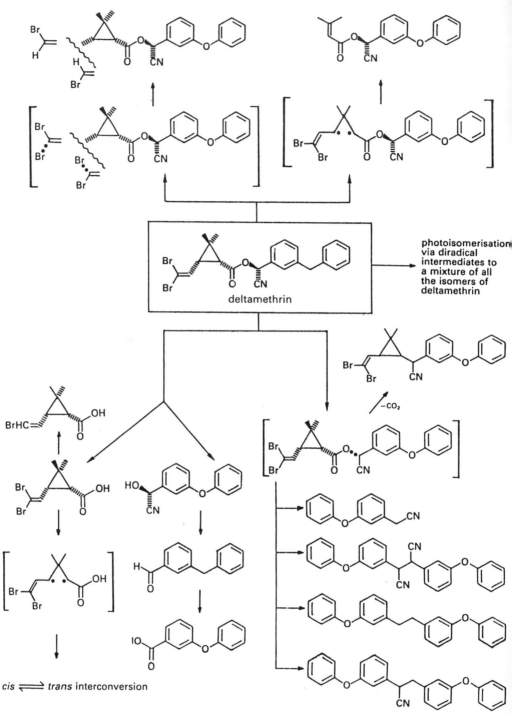

Figure 5.9. Phototransformations of deltamethrin (postulated intermediates shown in square brackets)

and ester cleavage. These photodegradation pathways are illustrated in Figure 5.11.

General discussion

Photochemical studies have been carried out on the pyrethroids under a variety of conditions, that is in solution in numerous organic solvents, in water and on plant, soil and inert surfaces. A range of radiation sources have also been used. It is beyond the scope of this chapter to give a detailed comparison of the results obtained with these different irradiation conditions. However, an investigation of the literature referenced in this section does show that significant qualitative and quantitative differences are detected for any one compound, depending on the irradiation conditions used. Thus the photodegradation studies which have been carried out should be considered as a source of information on the photosensitivity of a given compound and on its potential photodegradation pathways. However, it must be realised that the results from laboratory studies, using artificial light sources, may not necessarily give results relevant to degradation by sunlight in the natural environment.

Metabolism in animals

Perhaps the most comprehensively studied area of pyrethroid metabolism is that in animals. Much of this work has been done on rats and mice, and with *in vitro* preparations (e.g. isolated liver fractions) from these animals. Rats and mice have predominated in metabolism studies because of their importance as models for toxicity assessment; a knowledge of the metabolism and elimination of the pyrethroids by these animals is an essential part of this assessment. Other laboratory animals used in toxicity studies have also been investigated, but to a lesser extent. In addition, the fate of many of the agriculturally important pyrethroids will have been studied in livestock (e.g. cows, goats and chickens). Such studies are important since pyrethroid-treated crops may well form part of these animals' diets, and may thence enter the human diet in meat, milk and eggs. However, actual publication of data for farm animals has not been as comprehensive as for laboratory animals, although some information is available.

The natural pyrethrins

The metabolism of the two most important constituents of the six insecticidal esters extracted from the pyrethrin flower, pyrethrin I and II, has been studied after oral administration to rats, and in rat and mouse isolated liver fractions

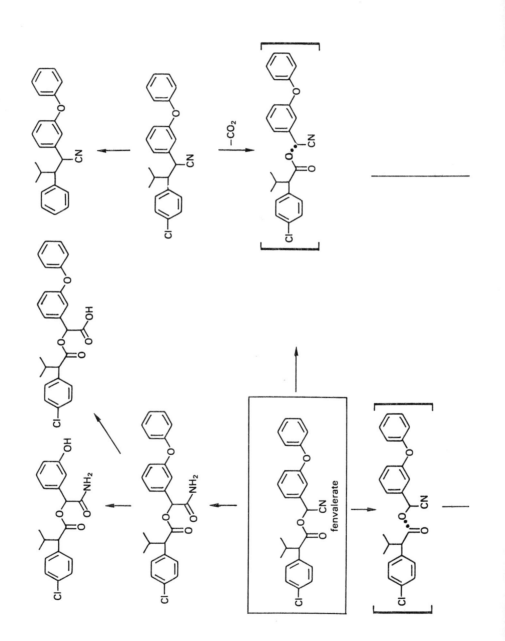

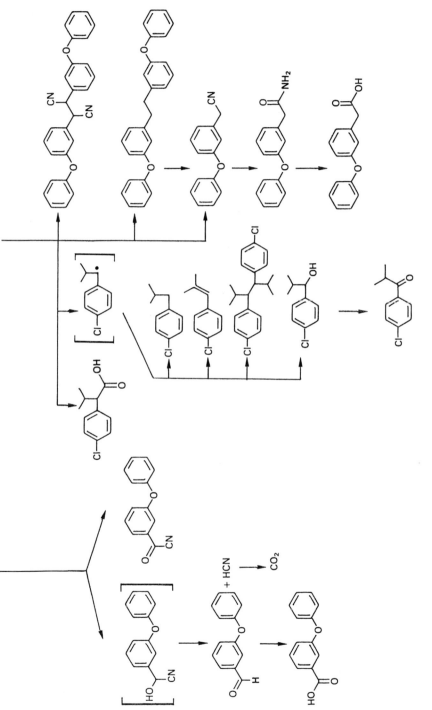

Figure 5.10. Phototransformations of fenvalerate (postulated intermediates shown in square brackets)

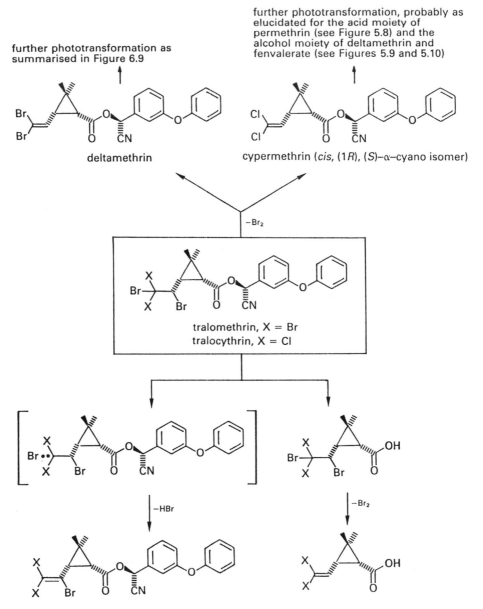

further phototransformation as
summarised in Figure 6.9

further phototransformation, probably as
elucidated for the acid moiety of
permethrin (see Figure 5.8) and the
alcohol moiety of deltamethrin and
fenvalerate (see Figures 5.9 and 5.10)

deltamethrin

cypermethrin (*cis*, (1*R*), (*S*)–α–cyano isomer)

−Br₂

tralomethrin, X = Br
tralocythrin, X = Cl

−HBr

−Br₂

Figure 5.11. Phototransformations of tralomethrin and tralocythrin (postulated intermediates
shown in square brackets)

(Yamamoto et al., 1971; Casida et al., 1971; Elliott et al., 1972). These studies,
the results from which are summarised in Figure 5.12, showed that the
trans-methyl group of the isobutenyl side chain of pyrethrin I is readily oxidised

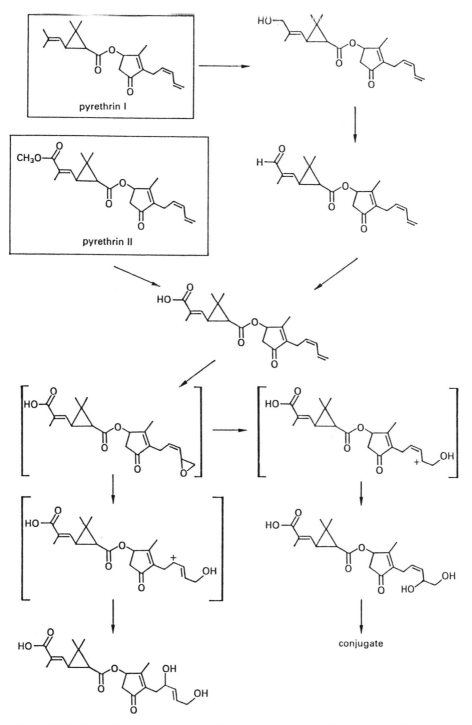

Figure 5.12. Metabolism of pyrethrin I and II in rats, and in isolated liver fractions from rats and mice (postulated intermediates shown in square brackets)

to a carboxylic acid, via the corresponding alcohol and aldehyde. Hydrolysis of the methyl ester of pyrethrin II yields the same acid. This acid is then further modified by oxidation of the pentadienyl group of the alcohol moiety, probably initiated by epoxidation of the terminal double bond. Isomeric diols result from hydrolysis of this epoxide, one of which reacts further to form a conjugate with an unidentified aromatic acid.

Hydrolysis of the ester linkage with pyrethrolone alcohol did not appear to be at all significant. The evolution of trace amounts of $[^{14}C]$-carbon dioxide ($\sim 1\%$ of the dose) from pyrethrin I, radiolabelled in the carboxyl group, being the only process detected where cleavage of this ester would have occurred.

Early synthetic pyrethroids (unimportant in agriculture)

Allethrin

The studies, discussed above, on the metabolism of pyrethrin I and II in rats, also included an investigation of the fate of allethrin (Yamamoto et al., 1971; Casida et al., 1971; Elliott et al., 1972). As with the natural pyrethrins, oxidation of the isobutenyl side chain is a major process for allethrin, as is oxidation of the unsaturated side chain of the alcohol moiety. However, unlike the natural pyrethrins, metabolites resulting from cleavage of the central ester linkage are also formed, although at a fairly low level. In addition, hydroxylation of one of the geminal dimethyl groups on the cyclopropane ring was detected. These metabolic pathways for allethrin in rats are summarised in Figure 5.13.

Terallethrin

The metabolism of terallethrin after both oral and subcutaneous administration to rats has been studied by Mihara et al. (1981). The doses given (dose rate 5 mg/kg) were eliminated readily from the rat, regardless of the route of administration, so that within 2 days 95% of the dose had been excreted. The route of elimination was approximately equally divided between urine and faeces.

With either oral or subcutaneous administration metabolism of terallethrin was rapid, with no more than 0·3% of the dose being excreted unchanged. Oxidation of a cyclopropane ring methyl group, initially to an alcohol and then to a carboxylic acid, and ester cleavage were the major metabolic reactions elucidated. Reactions involving the formation of a diol on the unsaturated side chain of the alcohol moiety and hydroxylation of the methyl group attached to the cyclopentenyl ring, were also detected. A summary of the metabolic pathways resulting from these reactions and combinations thereof is given in Figure 5.14.

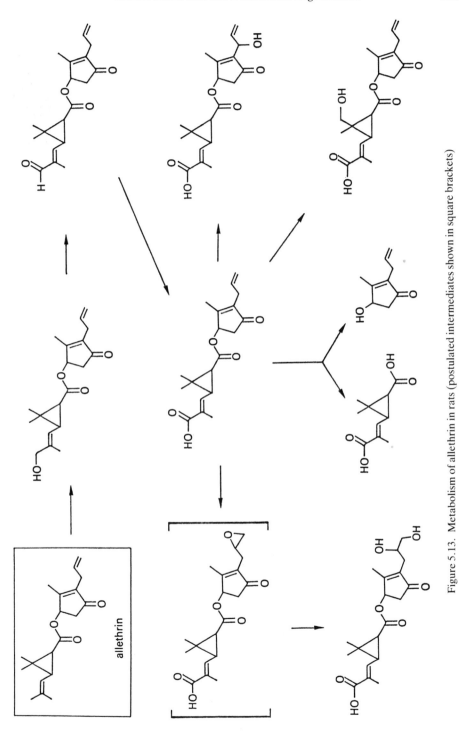

Figure 5.13. Metabolism of allethrin in rats (postulated intermediates shown in square brackets)

J. P. Leahey

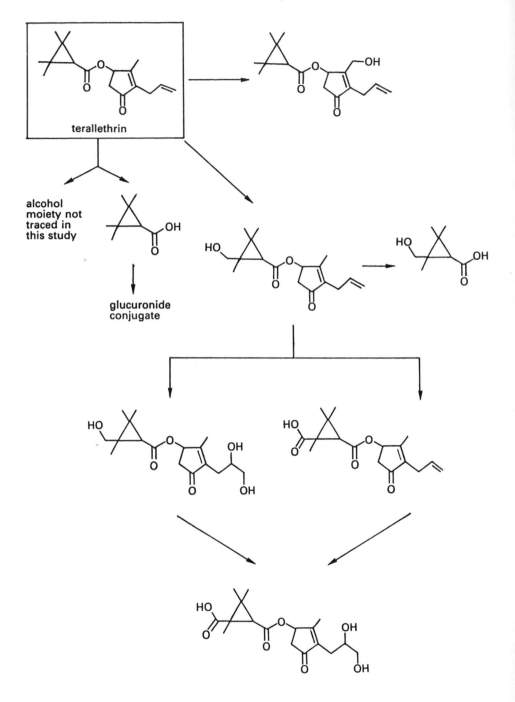

Figure 5.14. Metabolism of terallethrin in rats

Tetramethrin (also called phthalthrin)

The metabolism of tetramethrin after oral dosing to rats at 500 mg/kg has been studied by Miyamoto et al. (1968). This dose was rapidly eliminated by the rat (89% within 2 days) approximately equally in both the urine and faeces. Some unchanged tetramethrin was detected in the faeces (16% of the dose), but only metabolites of tetramethrin were present in the urine.

Ester cleavage was the major initial metabolic reaction detected for tetramethrin. The 3,4,5,6-tetrahydrophthalimidomethanol released is then readily converted to 3,4,5,6-tetrahydrophthalimide, probably by oxidation of the alcohol to an unstable carboxylic acid, which readily decarboxylates. This compound is then reduced to cyclohexane-1,2-dicarboximide, followed by hydroxylation to 3-hydroxycyclohexane-1,2-dicarboximide (see Figure 5.15). The fate of the acid moiety generated by ester cleavage was not traced in this study. However, data from other pyrethroids containing chrysanthemic acid (e.g. allethrin, see Figure 5.13) would suggest that the isobutenyl *trans*-methyl group will be oxidised. Studies with rat liver microsomes (Suzuki & Miyamoto, 1974) confirm that such oxidation does occur. More recent studies with tetramethrin (Smith & Casida, 1981) have also established that decarboxylation of chrysanthemic acid is a minor pathway in rats. This reaction is initiated by epoxidation of the isobutenyl group followed by a non-enzymatic decarboxylation as illustrated in Figure 5.15.

Barthrin and dimethrin

The metabolism of barthrin and dimethrin was studied by Masri et al. (1964) in rabbits. Radiolabelled material was not used in this work, so only minimal metabolic pathways (see Figure 5.16) for these two compounds were worked out, by isolating metabolites from the urine. Ester cleavage and oxidation of the alcohol fragments released to carboxylic acids were the processes elucidated.

Proparthrin (also called kikuthrin)

Nakanishi et al. (1971) dosed rats at a rate of 100 mg/kg with tritium-labelled proparthrin. They showed that much of the administered radioactivity was readily excreted (40% in the urine, 35% in the faeces) within 4 days. They also established that unmetabolised proparthrin was not present in either the urine or the bile, indicating that the compound, once absorbed, is readily metabolised. The glucuronide of 3-hydroxymethyl-5-(2-propynyl)-furan was identified as a major metabolite in the urine and bile (see Figure 5.17).

Resmethrin

Studies by two groups (Miyamato et al., 1971; Ueda et al., 1975a) have yielded

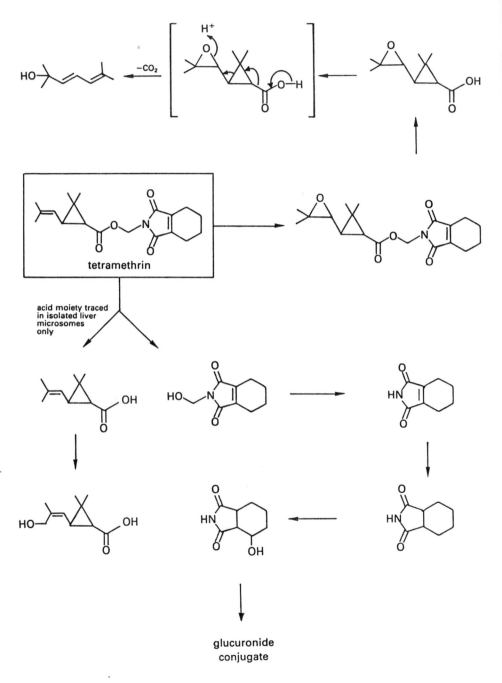

Figure 5.15. Metabolism of tetramethrin in rats and rat liver microsomes (postulated intermediates shown in square brackets)

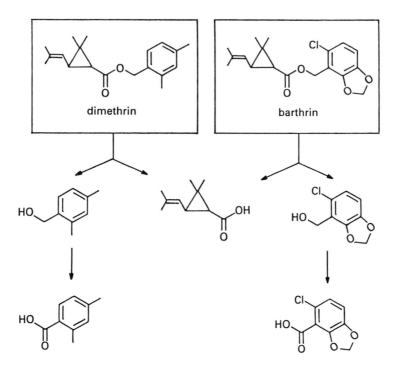

Figure 5.16. Metabolism of dimethrin and barthrin in rabbits.

a fairly thorough understanding of the metabolism of resmethrin after oral administration to rats. *In vitro* studies with isolated mouse and rat liver microsomes have also been carried out (Ueda et al., 1975b). The combined results from these three studies are summarised in Figure 5.18.

In both *in vivo* studies it was found that metabolism of orally administered resmethrin was very rapid, so that no unmetabolised resmethrin was detected in either the urine or faeces. Ester cleavage was a major metabolic process, with the fragment from the acid moiety being excreted more readily than those from the alcohol moiety. Enterohepatic circulation was implicated as a possible reason for this difference.

In the *in vivo* work of Ueda et al. the resolved (+)-*trans* and (+)-*cis* isomers of resmethrin were administered separately to the rats. This allowed the detection of an interesting metabolic process in that isomerisation at the cyclopropane C-3 carbon was found to occur. Abstraction of a proton from the cyclopropane ring C-3 carbon of the aldehyde intermediate formed in the oxidation of the isobutenyl group was postulated as the process initiating this isomer interconversion.

In the *in vitro* studies the separated (+)-*trans*, (−)-*trans*, (+)-*cis* and (−)-*cis*

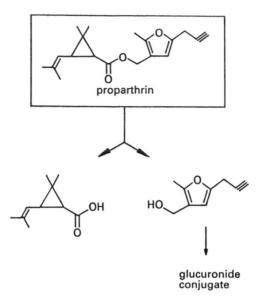

Figure 5.17. Metabolism of proparthrin in rats

isomers of resmethrin were incubated with rat or mouse liver microsomes. In this way additional insight into the effect of the stereochemistry of resmethrin on its metabolic fate was obtained. Thus it was established that, although in general the isobutenyl *trans*-methyl group is preferentially oxidised, the reverse is true for the (−)-*cis*-resmethrin isomer. It was also shown that two mechanisms are important for cleavage of the ester linkage; one mechanism is catalysed by liver esterases and the other by oxidases. The former mechanism is important for the *trans*-isomers, but much less so for the *cis*-isomers; whereas an oxidase catalysed process is the major ester cleavage mechanism for the *cis*-isomers. This difference in mechanism of ester cleavage has also been found to occur with other cyclopropane-based pyrethroids.

Kadethrin

The metabolism of kadethrin has been studied after oral administration to rats and by incubation with mouse liver microsomes (Ohsawa & Casida, 1980). As with the closely related compound, resmethrin, metabolism in the rat is very rapid, so that only 1.5% of the dose is excreted unchanged. Excretion of fragments derived from the acid moiety is rapid, with almost complete elimination within 3 days. Fragments from the alcohol moiety are, in agreement with the results obtained with resmethrin, less readily excreted.

The metabolic conversions elucidated for kadethrin are summarised in Figure 5.19. The major initial process is cleavage of the central ester bond. The

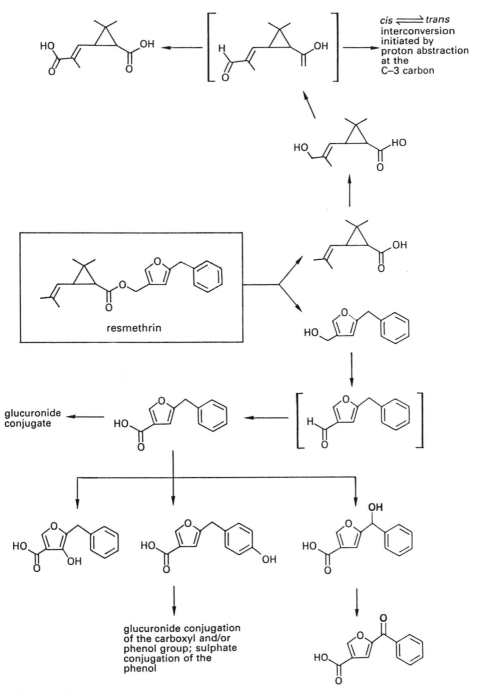

Figure 5.18. Metabolism of resmethrin in rats, and in rat and mouse liver microsomes (postulated intermediates shown in square brackets)

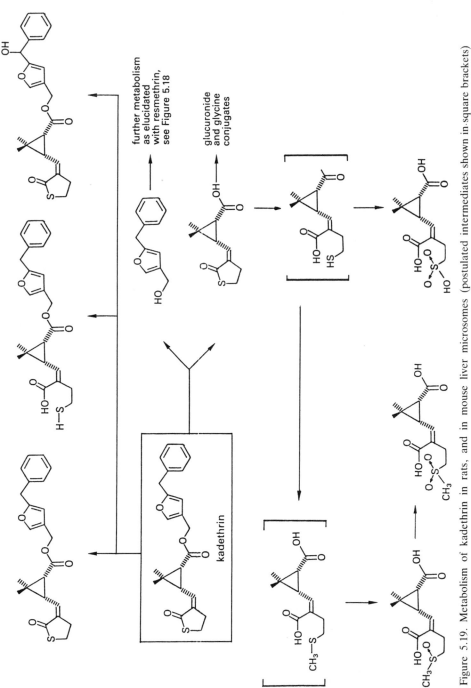

Figure 5.19. Metabolism of kadethrin in rats, and in mouse liver microsomes (postulated intermediates shown in square brackets)

alcohol fragment generated is then further metabolised as elucidated in the studies with resmethrin (see Figure 5.18). Further metabolism of the acid fragment initially involves hydrolysis of the thiolactone ring. The mercaptan thus released is then either oxidised directly to a sulphonic acid or is first methylated prior to oxidation to a methyl sulphoxide and then a methyl sulphone. Hydroxylation reactions and thiolactone hydrolysis also occur on the intact molecule.

Phenothrin

Miyamoto et al. (1974) have studied the metabolism of the (+)-*trans*-isomer of phenothrin by rats and in isolated liver fractions from a variety of mammals. After oral administration to rats phenothrin was readily excreted, 95% of the dose within 1 day and 100% within 2 days (57% in urine and 43% in faeces). Phenothrin radiolabelled in the alcohol moiety was used in this study so that only the fate of compounds retaining this fragment could be traced. Ester cleavage, followed by rapid oxidation of the 3-phenoxybenzyl alcohol released to a benzoic acid, was shown to be a major metabolic process, as was hydroxylation of the terminal aromatic ring at the 4'-position. Ester cleavage will also release chrysanthemic acid and this can be expected to be further metabolised as elucidated for this fragment with resmethrin (see Figure 5.18).

Qualitatively similar pathways of metabolism were also detected in isolated liver fractions from rats, mice, guinea pigs, rabbits and dogs. However, in these *in-vitro* experiments 3-phenoxybenzyl alcohol was detected as a major product, whereas, *in-vivo*, rapid oxidation to 3-phenoxybenxoic acid precludes the isolation of this compound.

Additional studies, in rats, have also been carried out with the (+)-*cis*-isomer of phenothrin (Suzuki et al., 1976). In this work three additional metabolites resulting from hydroxylation and oxidation of the intact molecule were detected. These metabolic pathways elucidated for phenothrin in the above two studies are summarised in Figure 5.20.

Photostabilised synthetic pyrethroids (compounds developed for use in agriculture)

Permethrin

The metabolism of permethrin has been studied in detail in a wide variety of animals, so that a very comprehensive understanding of the metabolism of this compound by animals has been obtained. A preliminary study in rats was carried out by Elliott et al. (1976) and this was followed by very detailed investigations in rats (Gaughan et al., 1977) and in isolated rat and mouse

J. P. Leahey

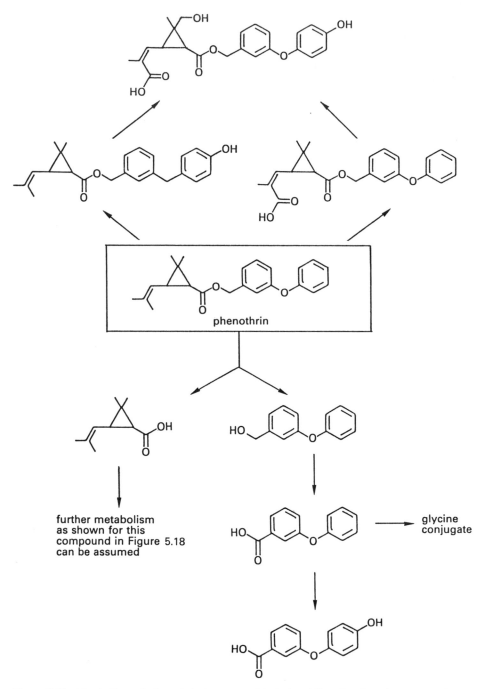

Figure 5.20. Metabolism of phenothrin in rats, and in isolated liver fractions from rats, mice, guinea pigs, rabbits and dogs

liver microsomes (Shono et al., 1979). The metabolic pathways detected in these three studies are summarised in Figure 5.21.

The *in vivo* investigations established that permethrin is extensively metabolised by the rat so that very little unchanged permethrin is excreted (~3% of the dose for the *trans*-isomers and ~6% for the *cis*-isomers). Excretion was fairly rapid, the *trans*-isomers being eliminated more readily than the *cis*-isomers, and with a higher urinary route of excretion for the *trans*-isomers (81−90% of the dose) compared with the *cis*-isomers (45−54% of the dose). However, with both isomers virtually total elimination of the dose was achieved within 12 days. These differences between the *cis*- and *trans*-isomers were attributed to a much greater rate of metabolism for the *trans*-isomers. Such a differential rate of metabolism is confirmed by *in vitro* studies (Soderlund & Casida, 1977) where mouse liver microsomes were shown to metabolise *trans*-permethrin approximately four times faster than *cis*-permethrin. Furthermore, this difference was shown to be due to a great susceptibility to esterase attack for the *trans*-isomers, with virtually no susceptibility for the *cis*-isomers; the rate of metabolism by oxidase catalysed processes was virtually identical for both isomers.

Despite this different susceptibility to esterase attack, the major *in vivo* route of metabolism for both the *cis*- and *trans*-isomers is via ester cleavage. However, the cleavage of the *cis*-isomers is catalysed by an oxidase, not an esterase. This is consistent with earlier results obtained with *cis*- and *trans*-resmethrin. The other important metabolic process is hydroxylation of the 4′-position of the terminal aromatic ring which, coupled with ester cleavage and oxidation, yields 4′-hydroxy-3-phenoxybenzoic acid as a major metabolite. Less important reactions are hydroxylation at the 2′-position of the 3-phenoxybenzyl moiety and hydroxylation of the geminal dimethyl group on the cyclopropane ring. This latter process, coupled with ester cleavage, yields isomeric hydroxycarboxylic acids, and, where a *cis*-configuration exists between the hydroxyl and carboxyl groups, these compounds are isolated as lactones. It is not known if this lactonisation is an *in vivo* process or an artefact of isolation. Further oxidation of these hydroxy-acids was detected in the *in vitro* studies, although not *in vivo*, to yield, via the aldehydes, the corresponding diacids.

In vivo, only one metabolite with the central ester linkage intact was detected with *trans*-permethrin; but this metabolite, formed by hydroxylation at the 4′-position, was present in only trace amounts (0.1% of the dose). With *cis*-permethrin, however, up to 10% of the permethrin ingested was eliminated as metabolites with the ester bond unbroken. These metabolites are formed by monohydroxylation at the 2′- or 4′-positions or of the geminal dimethyl group. A dihydroxylated compound, formed by oxidation at the 4′-position and of the geminal dimethyl group, is also formed. This approximately 100-fold difference in the formation of "ester-metabolites" between *cis*- and *trans*-permethrin is consistent with their large difference in susceptibility to esterase catalysed hydrolysis.

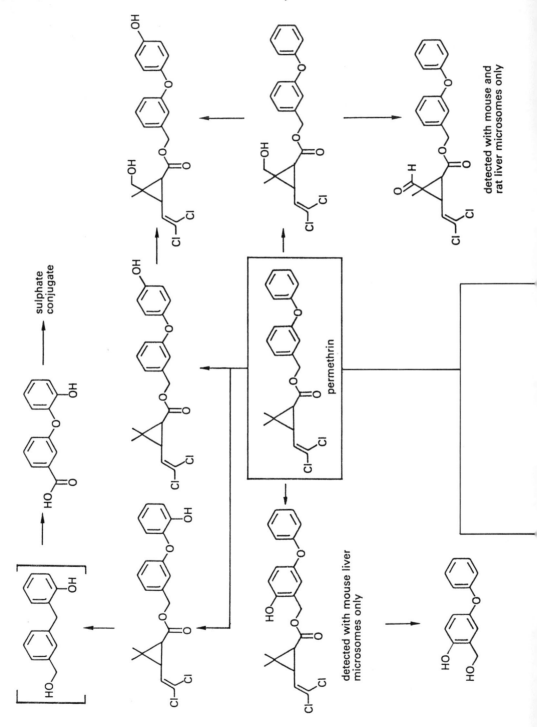

detected with mouse and rat liver microsomes only

sulphate conjugate

permethrin

detected with mouse liver microsomes only

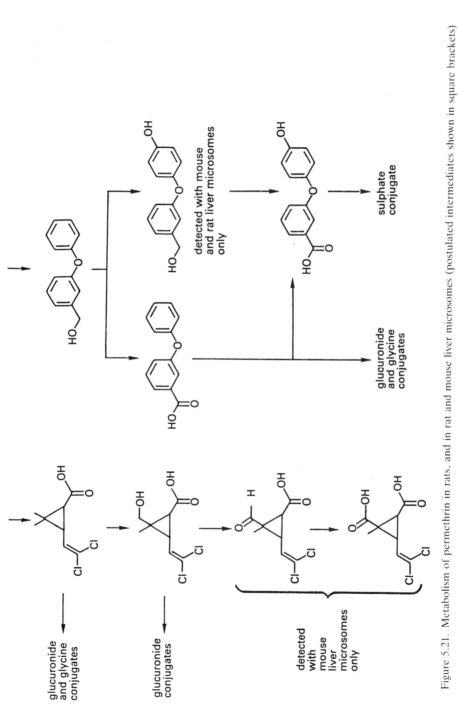

Figure 5.21. Metabolism of permethrin in rats, and in rat and mouse liver microsomes (postulated intermediates shown in square brackets)

In addition to these very thorough studies in rats and isolated liver fractions, information has also been published on the fate of permethrin in ruminants (cows and goats), chickens and fish. With cows an oral dose of 1 mg/kg, administered daily for three consecutive days, was eliminated fairly readily so that only low residues were present in the tissues of treated animals 13 days after initiation of dosing (Gaughan et al., 1978a). The highest tissue residues were detected in the fat (0·33 mg/kg) and most of this residue was characterised as unchanged permethrin. Residues ranging from 0·072 to 0·21 mg/kg were detected in the liver, and part of this residue was characterised as permethrin and the acid and alcohol fragments generated by ester cleavage. Unextractable and unidentified polar compounds were also present. The residues in the other tissues were very low and were not characterised. Residues of up to 0·25 mg/kg were present in the milk, and this was characterised as mainly unchanged permethrin. Low levels of an "ester metabolite" resulting from hydroxylation of the geminal dimethyl group of *cis*-permethrin were also detected in the milk.

The metabolic pathways elucidated from analysis of cows' excreta showed that similar metabolic processes to those occurring *in vivo* in the rat (see Figure 6.21) also occur in the cow. However, in the cow, for both *cis*- and *trans*-permethrin, a larger proportion of the excreted metabolites still retain the central ester bond (30% in the cow, 10% in the rat). In addition, hydroxylation of the 4'-position of the alcohol moiety is much less important in the cow compared with the rat; whereas hydroxylation of the geminal dimethyl group is much more important. Hydroxylation at the 2'-position of the 3-phenoxy-benzyl group was not detected in the cow. Conjugation reactions in the cow also varied slightly from those in the rat in that some glutamic acid conjugates were detected in addition to glucuronides, as characterised in rats.

In general, a similar pattern of metabolism and excretion to that in the cow has been established for goats (Hunt & Gilbert, 1977; Ivie & Hunt, 1980). However, in the studies with goats, the milk was shown to contain glycine conjugates of 3-phenoxybenzoic acid and 4'-hydroxy-3-phenoxybenzoic acid, in addition to permethrin and the "ester metabolite" of permethrin which were detected in cows' milk.

Studies with laying hens (Gaughan et al., 1978b), in which oral doses of 10 mg/kg were administered on three successive days, gave data similar to those obtained with the cow in terms of ready elimination of the dose and similarity of metabolic pathways. Nine days after initiation of dosing the highest tissue residue (0.2−1.36 mg/kg) appeared in the fat and was shown to be mainly unchanged permethrin. Residues of 0.08−0.27 mg/kg were detected in the liver, but no successful characterisation was achieved with this residue. Slightly higher residues (0.24−0.34 mg/kg) were present in the kidney and part of this residue was characterised, after an acid hydrolysis step, as a variety of oxidised derivatives of the acid and alcohol moieties (i.e. geminal dimethyl hydroxylation, benzyl alcohol to benzoic acid oxidation and hydroxylation at the 4'-position of the alcoholic moiety). Residues of 0.61– 2.75 mg/kg were present in the yolks of eggs 6 days after the first dose; residues in the albumen were much lower (0.12–0.18 mg/kg). Approximately half of the residue in the yolk was characterised as permethrin. The remainder of the yolk residue was a complex mixture of ester metabolites and metabolites generated by ester cleavage. Geminal dimethyl group hydroxylation, hydroxylation at the 4'-position, and benzyl alcohol to benzoic acid oxidation are the reactions involved in the formation of these metabolites. A variety of sulphate, glucuronide and uncharacterised conjugates of these metabolites were also detected.

Analysis of the excreta showed that the metabolism of permethrin in chickens is virtually identical to the metabolism in the cow. Minor differences were the ability of the chicken to form conjugates with taurine, but not with glutamic acid, and decarboxylation by the chicken (up to 5% of the dose), but not by the cow, of the carboxylic acids generated after ester cleavage.

In the permethrin studies discussed above *cis*-permethrin generated higher residue levels than *trans*-permethrin in the fat of all the animals investigated. Generally, except for the cow, residues in all other tissues analysed were also higher with *cis*-permethrin. This pattern is consistent with the greater overall rate of metabolism, and hence ready elimination, for *trans*-permethrin which results from its much greater susceptibility, relative to *cis*-permethrin, to esterase catalysed hydrolysis.

Although permethrin is very non-toxic to mammals it is highly toxic to fish, that is at parts per billion levels in water. This extreme difference has naturally generated an interest in the metabolism of permethrin by fish, especially in comparison to mammalian metabolism. *In vitro* studies with trout and mouse tissue preparations indicate that the trout has a much lower ability to hydrolyse the permethrin ester bond than the mouse (Glickman & Lech, 1981). This is in agreement with *in vivo* studies with trout (Glickman et al., 1979; Glickman et al., 1981) where an "ester metabolite" (permethrin hydroxylated at the 4'-position) was characterised as the major metabolite in trout, this metabolite being excreted as a glucuronide in the bile. In addition, ester cleavage products were not detected as significant metabolites in the trout excreta. However, additional *in vitro* studies by Glickman et al. (1982) could not totally account for the differential toxicity of *trans*-permethrin to trout and mice solely in terms of a differential rate of metabolism. Therefore other factors, such as greater physiological sensitivity by fish, must also be important.

Cypermethrin and deltamethrin

The very close structural similarity of these two compounds allows their metabolism by animals to be discussed jointly. Both compounds have been studied in rats and mice and also in isolated liver microsomes (Hutson & Casida, 1978; Ruzo et al., 1978, 1979; Shono et al., 1979; Crawford et al., 1981;

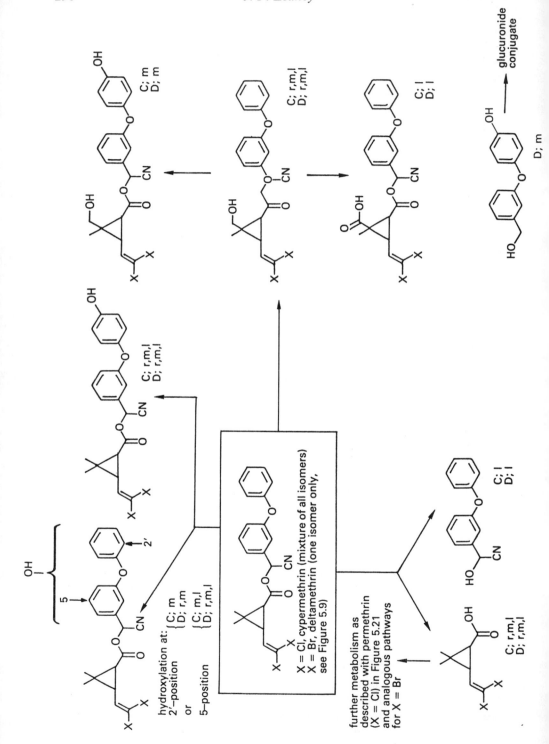

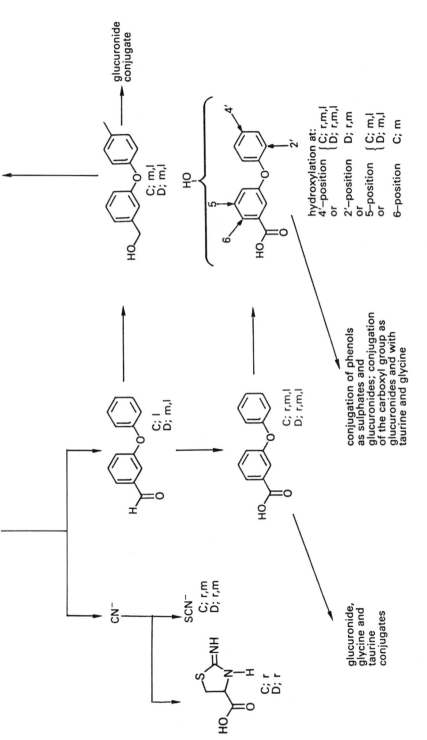

Figure 5.22. Metabolism of cypermethrin and deltamethrin in rats, mice and isolated liver microsomes (the letters below each metabolite indicate where that particular metabolite has been detected: C= cypermethrin, D = deltamethrin, r = rat, m = mouse, l = liver microsomes)

Hutson et al., 1981). The results from these studies are summarised in Figure 5.22.

Both cypermethrin and deltamethrin are readily excreted by rats and mice so that only low residues remain in the animals eight days after dosing (unless experiments are carried out with $[^{14}C]$-radiolabelling in the cyanide group, see below). Metabolism of both compounds is fairly extensive, with mice tending to give a wider range of metabolic pathways than the rat. The major initial pathways of metabolism both *in vivo* and in isolated liver microsomes are ester cleavage and hydroxylation at the 4'-position of the alcohol moiety. The α-cyano-3-phenoxybenzyl alcohol released by ester cleavage is very unstable, and is rapidly converted to the corresponding aldehyde with the release of cyanide ion. The aldehyde group is, in turn, readily oxidised to a carboxylic acid, so that 3-phenoxybenzoic acid and 4'-hydroxy-3-phenoxybenzoic acid are, as with permethrin, major metabolites for both cypermethrin and deltamethrin. The acid moiety released by ester cleavage is either identical to that released from permethrin (i.e. for cypermethrin) or is its dibromo analogue (i.e. for deltamethrin). These acids, not unexpectedly, are further metabolised as described for the identical/analogous acid from permethrin (see Figure 5.21). However, such further metabolism, other than by conjugation, is generally only a minor pathway, except for the metabolism of decamethrin by mice, where hydroxylation of the geminal dimethyl group is a significant process.

Additional minor pathways of metabolism for these two compounds involve hydroxylation at the 2'-, 5- and 6-positions of the alcohol moiety and also the formation of 3-phenoxybenzyl alcohol (presumably via reduction of 3-phenoxybenzaldehyde). Not all of these reactions occurred in all of the biological systems studied, that is rats, mice and isolated enzymes. However, an indication is given in Figure 5.21 as to the biological system where each reaction was detected.

With both cypermethrin and deltamethrin cyanide ion is a major product from metabolism. This is converted mainly to thiocyanate ion, but minor amounts of 2-iminothiazolidine-4-carboxylic acid are also formed. This latter compound is rapidly excreted in the urine, but the thiocyanate ion is eliminated much more slowly, with significant residues remaining in the body (equivalent to 20% of the dose) eight days after dosing. These residues are localised mainly in the stomach and skin. Such metabolic detoxification has previously been established for the cyanide ion in mammals (Wood & Cooley, 1956; Williams, 1959).

Since 3-phenoxybenzoic acid is a major metabolite from cypermethrin and deltamethrin, it is relevant to refer also to the work of Crayford & Hutson (1980a,b) who have orally administered this compound to rats. Pathways of metabolism established with cypermethrin and deltamethrin studies were confirmed for this moiety in this work. However, an interesting, although very minor, metabolic process was also detected, in which 3-phenoxybenzoic acid is conjugated with glycerol dipalmitate to form a lipophilic, triglyceride conjugate.

Fenpropathrin

Crawford & Hutson (1977) have studied the metabolism of fenpropathrin by rats and rat liver microsomes. The metabolic pathways elucidated by these workers are summarised in Figure 5.23. Fenpropathrin is rapidly metabolised and excreted by rats, with ester cleavage, hydroxylation at the 4'-position of the alcohol moiety and hydroxylation of a methyl group *trans* to the carboxyl group on the cyclopropane ring being the major metabolic processes. After ester cleavage the α-cyano-3-phenoxybenzyl alcohol released is further metabolised, as previously discussed with cypermethrin and deltamethrin, to 3-phenoxybenzoic acid, with the release of cyanide ion. Low levels of 3-phenoxybenzyl alcohol are also formed.

The *in vitro* studies with this compound indicated that, like *cis*-permethrin, ester cleavage was an oxidase-catalysed reaction, with esterase catalysed hydrolysis being unimportant. This is consistent with the presence of a substituent group (i.e. a methyl group) *cis* to the carboxyl group at both the 2- and 3-positions on the cyclopropane ring.

Fenvalerate

The metabolism of fenvalerate has been studied in both rats and mice (Ohkawa et al., 1979; Kaneko et al., 1981) and the results from these studies are summarised in Figure 5.24. Despite its very different acid moiety this compound is "handled" by rats and mice in a way similar to that of the cyclopropane-based pyrethroids; that is, excretion is fairly rapid, metabolism is extensive and ester cleavage and hydroxylation of the 4'-position of the alcohol moiety are the major initial metabolic reactions. The alcohol moiety generated by ester cleavage is identical to that from cypermethrin and deltamethrin, and is further metabolised as previously described (see Figure 5.22), with a few exceptions as noted on Figure 5.24. The acid fragment released by ester cleavage is mainly excreted unchanged, or as a glucuronide conjugate. However, mono- and dihydroxylation of this fragment is also a fairly significant process, and the hydroxy-acids thus formed are then further modified by lactonisation, oxidation and dehydration reactions (see Figure 5.24 for details).

The fate of the cyanide ion released after ester cleavage was also investigated in these studies. As previously discussed (see cypermethrin and deltamethrin, p. 298) conversion to thiocyanate ion was the major detoxification process, but significant conversion of the cyanide ion to carbon dioxide (up to 7.2% of the dose) was also detected. Such a conversion was not reported with cypermethrin and deltamethrin.

J. P. Leahey

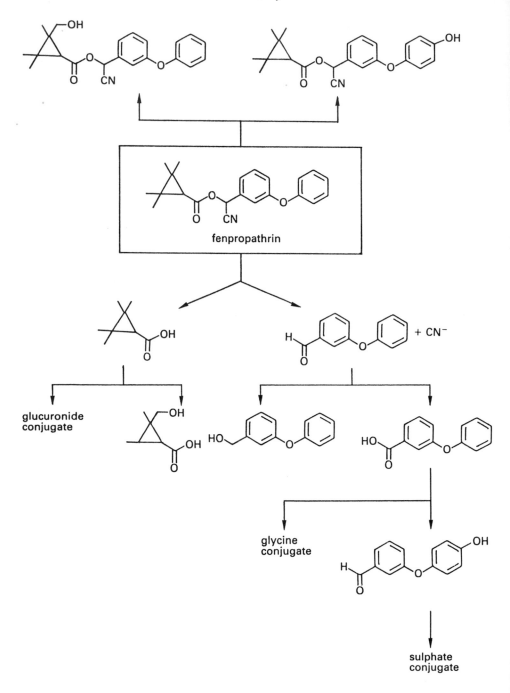

fenpropathrin

glucuronide
conjugate

glycine
conjugate

sulphate
conjugate

+ CN⁻

Figure 5.23. Metabolism of fenpropathrin in rats and rat liver microsomes

Some information on the metabolism of fenvalerate by fish has been obtained in a model ecosystem study (Ohkawa et al., 1980a). This study is discussed later in this chapter (degradation in soil and natural water, p. 333) and the metabolites detected in fish are shown in Figure 5.41.

Fluvalinate

The metabolism of fluvalinate has been extensively studied in rats (Quistad et al., 1983), rhesus monkeys (Quistad & Selim, 1983), cows (Quistad et al., 1982a) and chickens (Staiger et al., 1982). All of these studies were carried out with fluvalinate radiolabelled in the acid moiety only, therefore the fate of the α-cyano-3-phenoxybenzyl alcohol fragment was not traced. However, the fate of this compound is already well understood from studies with other pyrethroids which contain this alcohol (e.g. cypermethrin and deltamethrin, see p. 295).

With all the species investigated similar pathways of metabolism were generally found, and these pathways are summarised in Figure 5.25. Like fenvalerate, the acid moiety of fluvalinate is not cyclopropane-based. However, unlike fenvalerate, the fate of this compound in rats is, in some respects, rather different to that of the cyclopropane-based pyrethroids. Much (75−88%) of an oral dose is excreted in the faeces, mainly (~50%) as unchanged fluvalinate, whereas much lower levels of unchanged pyrethroid are excreted by rats dosed with the other pyrethroids reviewed in this chapter. Thus fluvalinate appears to be less readily absorbed than most other pyrethroids, and it is fairly resistant to degradation in the gastro-intestinal tract. However, once absorbed the metabolism of fluvalinate is rapid, so that no unchanged compound is present in the urine. Another unexpected difference between fluvalinate and other pyrethroids is the minimal amount of hydroxylation that occurs at the 4'-position of the intact ester. Thus 4'-hydroxy-fluvalinate is a very minor metabolite (<0.1% of the dose), whereas the equivalent compound is fairly significant for many other pyrethroids, for example, 4'-hydroxy-deltamethrin, ~7% of the dose, 4'-hydroxy-cypermethrin, ~9% of the dose.

Ester cleavage is, as with all synthetic pyrethroids, the major metabolic process for fluvalinate in the rat. The acid moiety released is then either excreted unchanged or is further modified as follows:

(a) by methyl group hydroxylation;

(b) by carbon−nitrogen bond cleavage to yield 2-chloro-4-(trifluoro-methyl)-aniline, and then hydroxylation to give 2-chloro-4-trifluoro-methyl-6-hydroxyaniline;

(c) by conversion of the carboxylic acid to an amide;

(d) by conjugation with a wide variety of compounds, many of which have either never, or only rarely, been reported as conjugates with xenobiotics, for example, bile acid conjugates (cholic acid, taurocholic

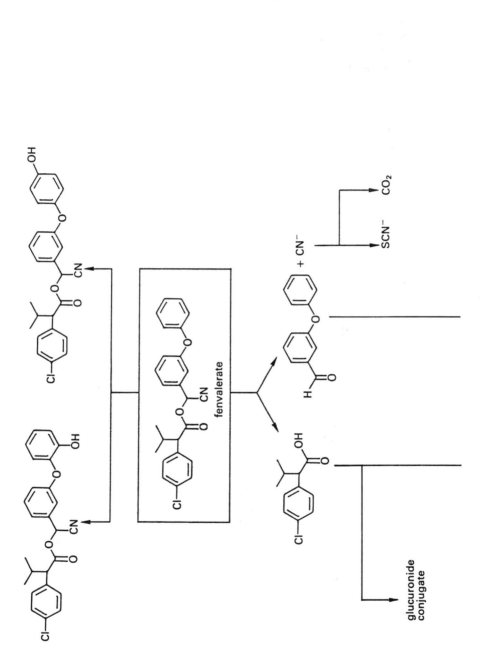

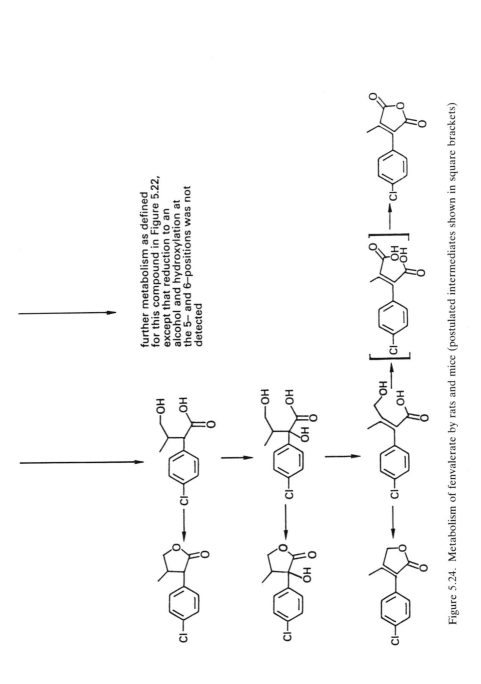

further metabolism as defined for this compound in Figure 5.22, except that reduction to an alcohol and hydroxylation at the 5– and 6–positions was not detected

Figure 5.24. Metabolism of fenvalerate by rats and mice (postulated intermediates shown in square brackets)

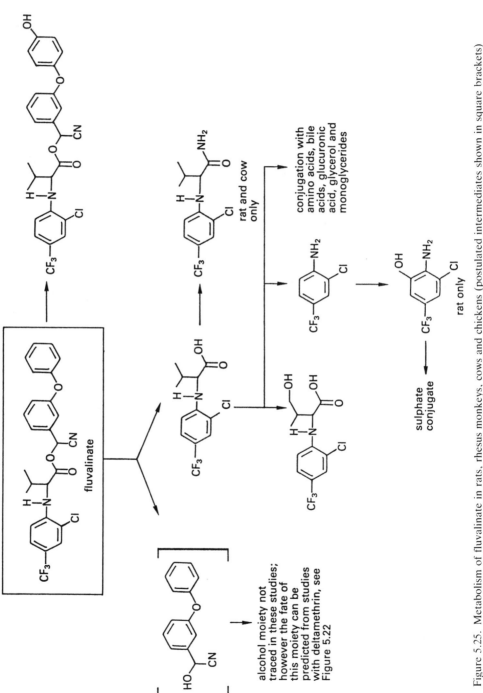

Figure 5.25. Metabolism of fluvalinate in rats, rhesus monkeys, cows and chickens (postulated intermediates shown in square brackets)

acid and taurochenodeoxycholic acid), and conjugates with glycerol and monoglycerides.

The rhesus monkey, like the rat, excretes much ($\sim$55%) of an oral dose of fluvalinate in the faeces, mainly ($\sim$45% of the dose) as unchanged pyrethroid. However, once absorbed, again in agreement with the rat, fluvalinate is extensively metabolised. With cows and chickens, metabolism of this compound is more extensive than for rats and rhesus monkeys, so that only 20% (cows) and 9% (chickens) of a dose of fluvalinate is excreted unchanged. The primary metabolic pathways in rhesus monkeys, cows and chickens are, with the few exceptions noted on Figure 5.25, similar to those elucidated for the rat. However, conjugation reactions of the acid moiety generated by ester cleavage show the following major differences for the four species studied:

(a) glucuronide conjugation is a major reaction in the rhesus monkey and the cow, whereas little or no glucuronides are detected in rats and chickens;

(b) conjugation with bile acids is a significant reaction in the rat, cow and chicken, but it is only a very minor process in the rhesus monkey;

(c) conjugation with glycerol and monoglycerides occurs in the rat, but is not detected in the other three species.

With the cow a single oral dose, at 1 mg/kg, gave rise to a maximum residue of 1.2 mg/kg in the milk; most of this residue ($\sim$70%) was due to unchanged fluvalinate. At slaughter, eight days after dosing, low residues were detected in a whole range of tissues, the highest residue being in the renal fat (0·16 mg/kg). Fluvalinate accounted for at least 80% of this fat residue.

With chickens a single oral dose of 1 mg/kg resulted in low residues in the eggs (albumen <0·01 mg/kg; yolk $\sim$0·13 mg/kg). Approximately one-third of the yolk residue was identified as fluvalinate, and the acid generated by ester cleavage accounted for a further 40−60% of the residue. This latter compound was present mainly as its taurochenodeoxycholate conjugate. Residues in the tissues of chickens were found to be very low, even when killed just 1 day after dosing. The highest residue was present in the liver (0·1 mg/kg), but this residue was not characterised. Residues up to 0·05 mg/kg were detected in the fat, and 72% of this residue was identified as fluvalinate.

Tralomethrin and tralocythrin

A comparative study has been carried out on the fate of tralomethrin, tralocythrin, deltamethrin and cypermethrin in rats (Cole et al., 1982a). Rapid, and virtually complete, debromination of tralomethrin and tralocythrin to yield deltamethrin and cypermethrin (*cis*, (1*R*), (*S*)-α-cyano isomer) was shown to occur. The latter two compounds are then further metabolised as described previously (see Figure 5.22).

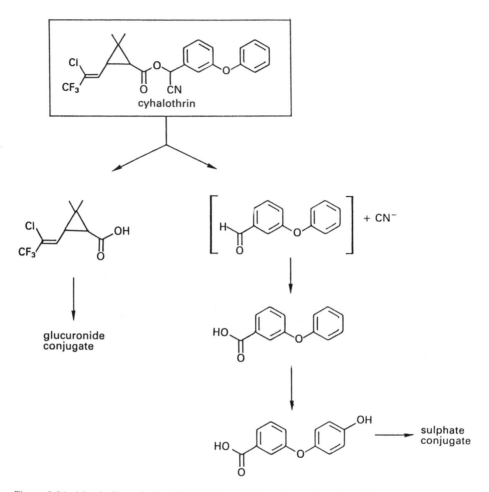

Figure 5.26. Metabolism of cyhalothrin in the rat, dog and cow (postulated intermediates shown in square brackets)

Cyhalothrin

The major pathways of metabolism for cyhalothrin in the rat, dog and cow (Harrison et al., 1983) are, as would be expected, analogous to those established for deltamethrin and cypermethrin (see Figure 5.26).

General discussion

The metabolism of pyrethroids by mammals, especially rats and mice, is certainly the most thoroughly investigated aspect of the metabolism of these compounds. With so much data available it is possible to make a number of generalisations concerning the mammalian metabolism of pyrethroids.

1. Although the pyrethroids are highly lipophilic compounds, they are nevetheless not stored to any significant extent in the fatty tissues, or any other tissues, of mammals. This is a reflection of the ready metabolism of these compounds with the generation of less lipophilic metabolites which are easily excreted, or which can be conjugated and excreted.

2. The most important metabolic process in mammals, for most pyrethroids, is cleavage of the central ester linkage. The only exceptions to this rule are the natural pyrethrins and their very close analogue, allethrin. With these compounds oxidative attack on the intact esters predominates.

3. Hydroxylation of alkyl carbons and aromatic rings is the second most important metabolic process in mammals. Where primary alcohols result from this hydroxylation they may then be further oxidised, via aldehydes, to carboxylic acids.

4. For pyrethroids containing a cyclopropane-based acid moiety, the stereochemistry of the 1,3-bond of this ring greatly influences the metabolism of these compounds. Where there is a substituent *trans* to the carboxyl group at C-1, ester cleavage is most rapid and esterases are the main enzymes which catalyse this cleavage. However, the presence of a *cis* configuration about the 1,3-bond greatly reduces the rate of ester cleavage. Furthermore, oxidases rather than esterases are found to be the enzymes catalysing this process (Soderlund & Casida, 1977).

5. The addition of a cyanide group to the α-carbon of the pyrethroids derived from 3-phenoxybenzyl alcohol reduces the susceptibility of the molecule to both hydrolytic and oxidative metabolism (Soderlund & Casida, 1977).

It can be seen from the above generalisations that deltamethrin and *cis*-cypermethrin are potentially the most metabolically stable of the cyclopropane-based pyrethroids. However, these compounds are nevertheless sufficiently rapidly metabolised to allow efficient excretion by mammals. The non-cyclopropane derived pyrethroid, fenvalerate, exhibits a similar level of metabolic stability (Soderlund & Casida, 1977), but it is also metabolised and excreted efficiently.

Comparative metabolic data for a wide range of pyrethroids are not available in non-mammalian animals. So it is not possible to extend the above generalisations to all animals. However, studies with permethrin and fluvalinate in chickens do indicate that metabolism of pyrethroids in birds is similar to that in mammals. Furthermore, studies in which non-domestic birds, the quail and mallard duck, were continually dosed for 28 days with either permethrin or cypermethrin, have shown that these two pyrethroids are rapidly and totally excreted, and hence are almost certainly rapidly metabolised (Leahey et al., 1980).

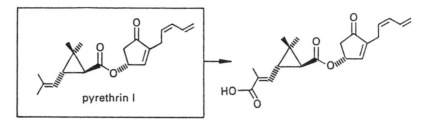

Figure 5.27. Metabolism of pyrethrin I by houseflies

Metabolism in insects

Although insects are the target for the pyrethroids, the metabolism of these compounds by insects has in no way received the same detailed level of investigation as have mammals. This is not surprising since most metabolism studies with pyrethroids have been stimulated by the need to assess the hazard these compounds might pose to mankind and his environment. Since pyrethroids are designed to pose a hazard to insects, such a reason does not exist for insect metabolism studies. Furthermore, since insects are extremely sensitive to intoxication by pyrethroids, *in vivo* studies with insects are extremely challenging, although *in vitro* studies (e.g. using insect abdomen homogenates) have been used to circumvent this problem. However, the metabolism of pyrethroids by insects is of interest in terms of attempting to understand the large toxicity differential between insects and mammals, the action of synergists with these compounds and the mechanism of resistance that has been generated in insects under pressure from pyrethroids.

The natural pyrethrins

Only a limited amount of information is available on the metabolism of the natural pyrethrins by insects. Initial studies, carried out by a number of workers (summarised by Yamamoto, 1973), indicated fairly ready metabolism by insects but did not unequivocally establish any metabolic pathways. Later *in vivo* and *in vitro* studies of pyrethrin I with houseflies (Yamamoto & Casida, 1966; Yamamoto et al., 1969) showed a complex pattern of metabolism for this compound; but only one metabolic process, oxidation of the isobutenyl *trans*-methyl group, was elucidated (see Figure 5.27).

Early synthetic pyrethroids (unimportant in agriculture)

Allethrin and dimethrin

The *in vivo* and *in vitro* metabolism of allethrin and dimethrin by houseflies has been investigated by Yamamoto and co-workers (Yamamoto & Casida,

1966; Yamamoto et al., 1969). In this work the most detailed characterisation of metabolites was achieved with allethrin. Hydroxylation of both the *cis-* and *trans*-isobutenyl methyl groups was the major metabolic reaction detected, with hydroxylation of the *trans*-methyl predominating. The alcohols thus formed are then either conjugated, probably as glucosides, or are further oxidised to aldehydes and carboxylic acids. The latter two reactions are fairly minor *in vivo* processes. Under *in vitro* conditions (housefly abdomen homogenates) conjugation was not detected, but further oxidation to carboxylic acids became a major process. These modifications to the acid moiety of allethrin accounted for most of the metabolic changes occurring in the housefly, with no significant attack on the alcohol fragment or significant ester cleavage being detected.

Dimethrin, which contains the same acid moiety as allethrin, is probably metabolised in a fashion analogous to allethrin. However, characterisation of the metabolites formed with this compound was not as extensive as with allethrin, so that the carboxylic acid resulting from oxidation of the isobutenyl *trans*-methyl group was the only metabolite actually identified.

Figure 5.28 summarises the metabolic pathways elucidated for these two compounds in houseflies.

Tetramethrin (phthalthrin)

Yamamoto and his co-workers (Yamamoto & Casida, 1966; Yamamoto et al., 1969) carried out preliminary investigations of the metabolism of tetramethrin in houseflies concurrent with their studies of pyrethrin I, allethrin and dimethrin, as discussed above. However, these early studies have since been superceded by some rather more detailed work (Miyamoto & Suzuki, 1973; Suzuki & Miyamoto, 1974). This work confirmed the earlier workers' result that houseflies oxidise the isobutenyl *trans*-methyl group of tetramethrin, but they also established that ester cleavage is the major metabolic process in houseflies. After ester cleavage the alcohol moiety is further modified by the loss of its hydroxymethyl group. This latter reaction is especially significant in *in vitro* experiments with housefly abdomen homogenates, but was also found to occur in control incubations without enzymes present. Loss of this hydroxymethyl group may therefore be an artefact formed during analysis, rather than a metabolic process. The metabolic pathways elucidated for tetramethrin in houseflies are summarised in Figure 5.29.

Photostabilised synthetic pyrethroids (compounds developed for use in agriculture)

Permethrin

The metabolism of permethrin has been studied in a number of insect species, but the most detailed evaluation of metabolic pathways has been achieved in a

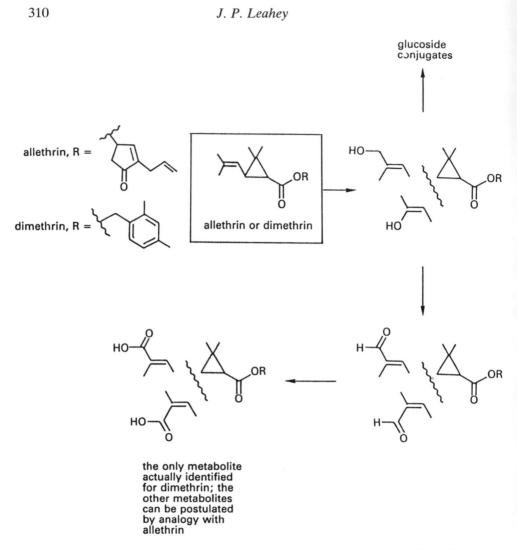

allethrin, R =

dimethrin, R =

glucoside conjugates

allethrin or dimethrin

the only metabolite
actually identified
for dimethrin; the
other metabolites
can be postulated
by analogy with
allethrin

Figure 5.28. Metabolism of allethrin and dimethrin by houseflies and housefly abdomen homogenates

study with cockroach adults, housefly adults and cabbage looper larvae (Shono et al., 1978). The results from this study are illustrated in Figure 5.30. In all three insect species *cis*-permethrin is metabolised less readily than *trans*-permethrin, and this difference is especially significant with the housefly and cabbage looper. Ester cleavage and hydroxylation at the 4'-position of the alcohol moiety are major metabolic processes in all three insects, as is oxidation of 3-phenoxybenzyl alcohol (generated by ester cleavage) to 3-phenoxybenzoic acid. Hydroxylation of the geminal dimethyl group of the acid moiety is also a major reaction in the cockroach, although it is of minor

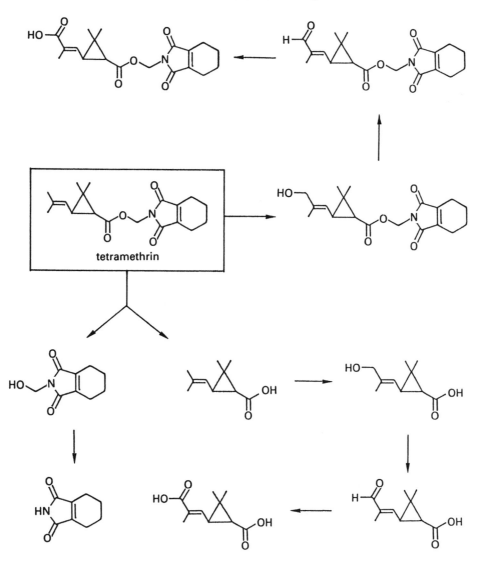

Figure 5.29. Metabolism of tetramethrin by houseflies and housefly abdomen homogenates

importance in the housefly and cabbage looper. Hydroxylation of the 6'-position of the alcohol moiety was also detected, but only as a minor process in houseflies. Similar results have also been obtained in *in vitro* studies utilising isolated enzymes from houseflies and cabbage loopers (Shono & Casida, 1978; Shono et al., 1979). Other insect species studied are: tobacco budworm and the bollworm (Bigley & Plapp, 1978), the porina moth (Chang & Jordan, 1982), the cotton leaf-worm (Holden, 1979) and the cattle tick (Schnitzerling et al.,

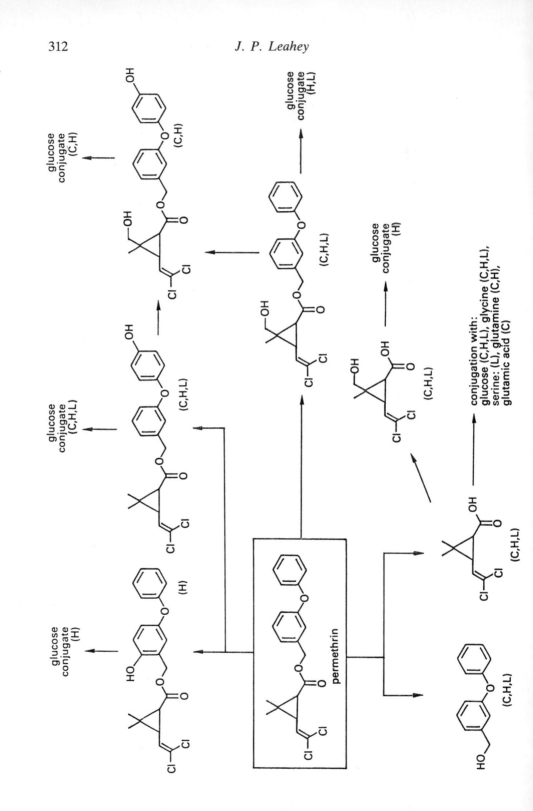

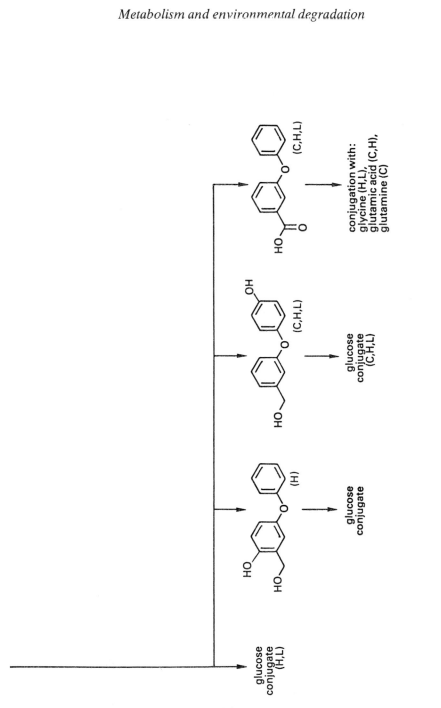

Figure 5.30. Metabolism of permethrin by the cockroach (C), housefly (H) and cabbage looper (L); the insects in which a metabolite has been found are indicated below each structure

1983). In general, a less detailed evaluation of metabolic pathways was reported with these insects, compared with that achieved by Shono et al. However, the major pathways of metabolism did appear to be the same for all the species studied, with an additional pathway, that is hydroxylation at the 2'-position of the alcohol moiety, being observed with the tobacco budworm and the bollworm.

Cypermethrin

Really detailed investigations of the metabolism of cypermethrin by insects have not been reported. However, the data that is available does indicate that ester cleavage and hydroxylation at various positions of the molecule are the major metabolic processes (Holden, 1979; Chang & Jordan, 1982; Schnitzerling et al., 1983).

Tralomethrin and tralocythrin

The metabolism of tralomethrin and tralocythrin has been investigated following topical application to houseflies, ingestion by cabbage looper larvae and *in vitro* incubation with housefly homogenates and cockroach nerve cords (Ruzo et al., 1981). The major initial process detected with all of these insects was debromination to yield either deltamethrin or cypermethrin (such debromination is illustrated in Figure 5.11). This reaction appears to be mediated by naturally occurring thiols, rather than by enzymes, since thiol-containing tissue homogenates gave ready debromination even after denaturing by heat. As would be expected the deltamethrin and cypermethrin formed are in turn metabolised by ester cleavage and presumably also by hydroxylation and oxidation, although detailed characterisation of such further metabolism was not reported in this study.

General discussion

The data available on the metabolism of pyrethroids by insects, though limited, does indicate that the major primary metabolic pathways are generally similar to those in mammals. Esterases and oxidases are, as with mammals, the enzymes which mediate insect metabolism, and in the few cases where comparisons have been made these insect enzymes have been shown to generate slower reaction rates than mammalian enzymes (Jao & Casida, 1974; Shono et al., 1979). Such differential rates of metabolism probably explain, at least partially, why pyrethroids are much more toxic to insects than mammals. Again the "rule" established with mammals that *trans*-isomers are generally much more readily hydrolysed by esterases than *cis*-isomers is usually also true in insects. However, with insects there are some exceptions to this generalisation (see a review by Soderlund et al., 1983), notably with the green lacewing larva, which has unusually high esterase levels, and these esterases hydrolyse the *cis*-isomers of permethrin and cypermethrin more rapidly than the *trans*-isomers (Ishaaya & Casida, 1981). Furthermore these esterases hydrolyse deltamethrin even more rapidly than *cis*-cypermethrin. This level of esterase activity is undoubtedly a major factor in the insensitivity of the lacewing larva to pyrethroids.

The importance of metabolism in the detoxification of pyrethroids by insects is confirmed by the increased potency achieved by mixing specific esterase and oxidase inhibitors with pyrethroids. A number of workers have investigated the mechanism and effects of such "synergists" (summarised by Soderlund et al., 1983) and a range of effective pyrethroid synergists have been discovered.

The wide usage of pyrethroids in insect control has caused the emergence of resistant strains in insect populations, and early in the life of pyrethroids cross-resistance to DDT-resistant strains was also noted. Since metabolism is such an important factor in pyrethroid detoxification by insects, increased rates of metabolism by resistant insects could be the mechanism involved. In some cases increased rates of metabolism do seem to contribute to the mechanism of resistance, and synergists may then be employed, at least partially, to overcome this resistance (Farnham, 1971, 1973; Liu et al., 1981). However, a number of workers have demonstrated that a differential metabolism rate is certainly not the major resistance mechanism in resistant strains of a number of insects (summarised by Soderlund et al., 1983). In such cases a reduction in the sensitivity of the nervous system itself and/or an ability to prevent penetration of the pyrethroid to its target site have been invoked as the resistance mechanism.

Metabolism in plants

A knowledge of the metabolism of pyrethroid insecticides by plants is obviously only of interest for those compounds which are used, or are being developed for use, in agriculture. For such pyrethroids a detailed knowledge of their fate on plants is a very necessary part of the evaluation of their safety. Undoubtedly, for all the pyrethroids currently in use in agriculture, very thorough studies have been completed on a large range of crops. Although all of this work may not have been published, there is nevertheless still a fairly comprehensive "data bank" in the literature on the metabolism of pyrethroids on plants.

Phenothrin

Although phenothrin is probably too photosensitive for any widespread agricultural use, studies have nevertheless been carried out on the fate of this compound on beans and rice plants under glasshouse conditions (Nambu et al.,

1980). As might be expected for a photosensitive compound, the initial degradation process involves a very rapid, photo-initiated, ozonolysis of the isobutenyl double bond. Such a phototransformation process has been established for other pyrethroids containing chrysanthemic acid (see Figure 5.2 and Chen & Casida, 1969). The ozonide formed is also unstable, rapidly decomposing to an aldehyde, which in turn is oxidised to a carboxylic acid. Ester cleavage, hydroxylation at the 2'- and 4'-positions of the alcohol moiety, and photo- induced interconversion of the *cis*- and *trans*-isomers were also shown to occur, but only as fairly minor processes. These phototransformations and metabolic reactions are summarised in Figure 5.31.

It was also established in this study that phenothrin is not translocated from its site of application to other parts of the plant, and only minimal amounts of degradation products (1−3% of the treatment given) are translocated. Similarly, bean plants, grown in soil containing 1 mg/kg of phenothrin, do not translocate phenothrin from the soil into the aerial parts of the plant.

Permethrin

Thorough investigations have been completed of the metabolism of permethrin by bean plants and cotton plants (Ohkawa et al., 1977; Gaughan & Casida, 1978). In the case of the bean plants both topical application to leaves and stem injection were used to treat the plants. After treatment the plants were maintained in a glasshouse where the insecticide was found to degrade with a half-life of 7−9 days. With the cotton plants, treatment was by topical application to leaves only, but in this experiment some plants were then maintained in a glasshouse while others were kept under "field" conditions. No significant differences in the types of metabolic pathways were detected for the two plant species with either field or glasshouse conditions.

As with mammals and insects, the major primary metabolic process in plants was ester cleavage. However, in contrast to mammals and insects, plants do not readily oxidise the 3-phenoxybenzyl alcohol thus generated to a carboxylic acid, such oxidation being undetected in cotton plants and a very minor reaction in beans. Hydroxylation of the intact pyrethroid, or its cleavage products, at the alcohol moiety 2'- and 4'-positions and of the geminal dimethyl group are other reactions detected in plants. Attack on the geminal dimethyl is the most significant of these hydroxylation reactions. Photo-initiated interconversion of the *cis*- and *trans*-isomers was also found to occur. These metabolic pathways for permethrin in plants are summarised in Figure 5.32.

In these studies with cotton and bean plants, very little translocation of permethrin, or its metabolites, was detected in the plants following either topical application or stem injection. However, it has been shown that plants (sugar beet, wheat, lettuce and cotton) grown in a soil of very high sand content (sand 82%, clay 10%, silt 8% and organic matter 1.4%) may translocate low

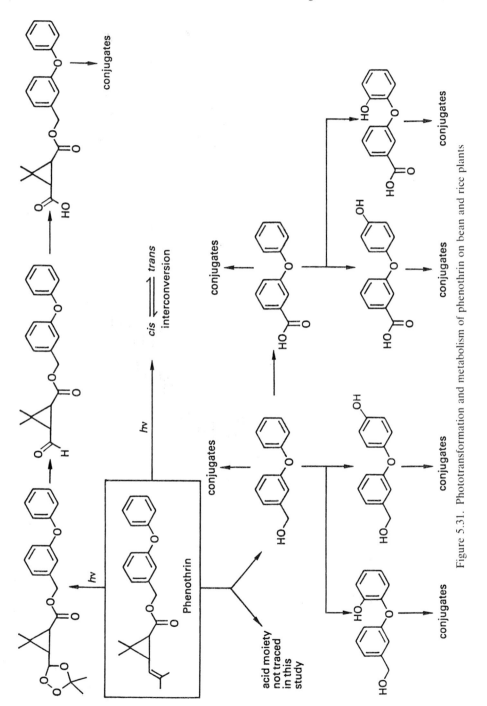

Figure 5.31. Phototransformation and metabolism of phenothrin on bean and rice plants

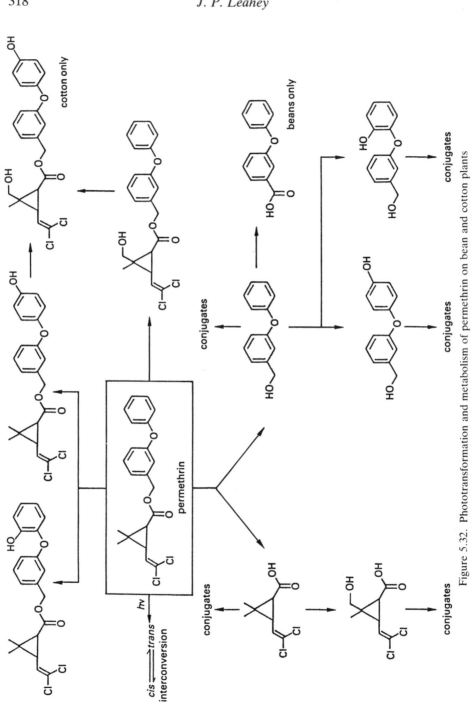

Figure 5.32. Phototransformation and metabolism of permethrin on bean and cotton plants

levels of degradation products from the soil into the root and aerial parts of these plants (Leahey & Carpenter, 1980). Major constituents of these translocated compounds were identified as conjugates of 3-(2,2-dichloro-vinyl)-2,2-dimethylcyclopropanecarboxylic acid and of 3-(2,2-dichloro-vinyl)-1-methylcyclopropane-1,2-dicarboxylic acid.

Cypermethrin and deltamethrin

The degradation of deltamethrin after topical application to cotton leaves has been studied by Ruzo & Casida (1979). Similar metabolic pathways (summarised in Figure 5.33) were detected with plants maintained under either glasshouse or field conditions, but a more rapid rate of degradation occurred in the field (glasshouse half-life ~8 days, exact field half-life not specified). Ester cleavage is the major primary metabolic reaction, with hydroxylation at various positions on the molecule, as indicated in Figure 5.33, also occurring, but only as fairly minor processes. Ester cleavage generates α-cyano-3-phenoxybenzyl alcohol. In mammals this compound is extremely unstable, readily losing hydrogen cyanide to yield 3-phenoxybenzaldehyde, which in turn is rapidly oxidised to 3-phenoxybenzoic acid. Reduction to 3- phenoxybenzyl alcohol also occurs as a minor process in some mammals. Such a metabolic pathway also occurs in cotton plants, but additionally, conjugates of α-cyano-3-phenoxybenzyl alcohol are formed. These conjugates presumably form very rapidly in the plant before loss of hydrogen cyanide can occur, and they are especially significant in plants maintained under field conditions.

Metabolic pathways analogous to those established for deltamethrin would also be expected for cypermethrin, and studies with lettuce and cabbage certainly confirm that ester cleavage is the major primary reaction (Wright et al., 1980; Roberts, 1981). Minor metabolic pathways, however, were not traced in detail in these studies. Instead the authors concentrated on the characterisation of the conjugates formed with the products of ester cleavage, since these are the major products formed in plants. Mono- and disaccharide derivatives, including β-D-glucopyranose, glucosylarabinose and glucosyl-xylose esters, were characterised as probable conjugates of the acid moiety of cypermethrin. Additional *in vitro* studies using 3-phenoxybenzoic acid (a major metabolite derived from the alcohol fragment) and abscised leaves from a variety of plants indicated that similar mono- and disaccharides could also result from the alcohol moiety of cypermethrin (More et al., 1978).

Fenvalerate

The metabolism of fenvalerate, under glasshouse conditions, has been studied after topical application to the leaves of kidney bean plants (Ohkawa et al., 1980b). The half-life for fenvalerate in this experiment was approximately two weeks, although the authors noted that much more rapid degradation

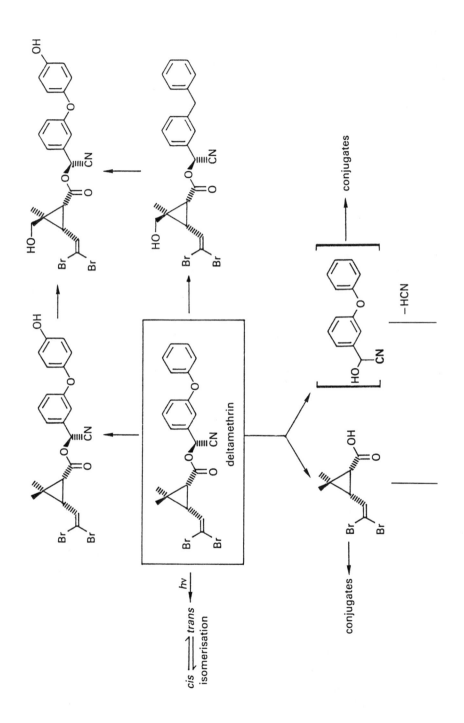

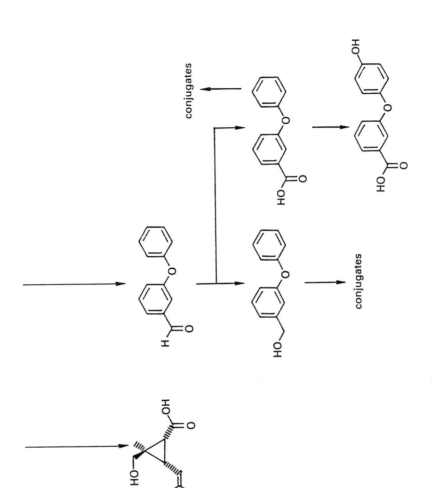

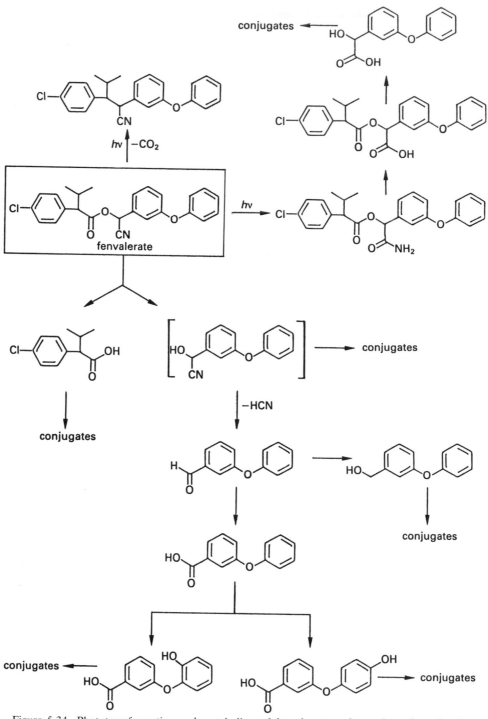

Figure 5.34. Phototransformation and metabolism of fenvalerate on bean plants (postulated intermediates shown in square brackets)

occurred under field conditions. The metabolic pathways elucidated in this study are summarised in Figure 5.34.

Ester cleavage was the major metabolic process detected, with further metabolism of the α-cyano-3-phenoxybenzyl alcohol released following the pathway reported for this group in studies with deltamethrin (see Figure 5.33). Hydroxylation at the 2'- and 4'-positions of the alcohol moiety also occurred. In addition, decarboxylation of fenvalerate to yield 3-(4-chlorophenyl)-4-methyl-2-(3-phenoxybenzyl)-valeronitrile and hydrolysis of the cyanide group to an amide and then a carboxylic acid were detected as minor processes. Such decarboxylation and hydrolysis reactions, which are probably photo-initiated (see Figure 5.10), have not been detected on plants with most other α-cyano-3-phenoxybenzyl alcohol derived pyrethroids, although the analogous amide has been reported by Roberts (1981) as a fairly minor degradation product of cypermethrin on apple leaves.

Fluvalinate

A range of plants (cotton, tomato, tobacco, lettuce and cabbage) have been treated with fluvalinate and the plants have been maintained under glasshouse conditions (Quistad et al., 1982b). Analysis of various parts of these plants (e.g. leaves, fruit, bolls) at a range of time intervals indicated a half-life of 4–6 weeks for fluvalinate on these crops. Similar pathways of degradation, summarised in Figure 5.35, were detected with all of the plants. Ester cleavage is the major metabolic process, and the acid fragment released is then mainly conjugated with glucose and probably with other sugars. Hydroxylation of the 4'-position of the terminal aromatic ring was a minor pathway, as was the formation of 1-chloro-4-(trifluoromethyl) aniline. This latter compound was detected as a residue on the plants, but was also lost from the plant surface by volatilisation. The fate of the alcohol fragment, after ester cleavage, was not traced in these studies, but its fate can readily be predicted from studies with other α-cyano-3-phenoxybenzyl alcohol derived pyrethroids (i.e. deltamethrin, cypermethrin and fenvalerate).

Tralomethrin and tralocythrin

A study to compare the degradation, on cotton and bean leaves, of tralomethrin and tralocythrin with that of deltamethrin and cypermethrin has been carried out by Cole et al. (1982b). Plants were treated topically with these four compounds and were then exposed to daylight. Rapid, photo-induced debromination was the major degradation process for tralomethrin and tralocythrin, thus generating deltamethrin and cypermethrin (see Figure 5.11). Further metabolism paralleling that occurring on the plants treated with deltamethrin and cypermethrin was then established. Only one degradation pathway was detected which did not generate products identical to those

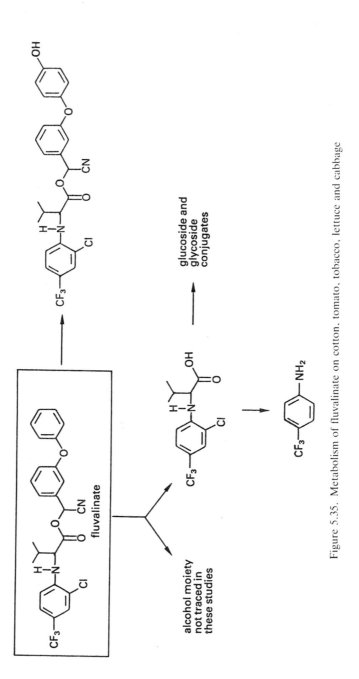

Figure 5.35. Metabolism of fluvalinate on cotton, tomato, tobacco, lettuce and cabbage

derived directly from deltamethrin and cypermethrin. This pathway, which involves loss of hydrogen bromide rather than a bromine molecule (see Figure 5.11), was, however, a very minor process, accounting for 2 to 3% of the degradation of tralomethrin and less than 1% of the degradation of tralocythrin.

General discussion

With the exception of the photolabile compound, phenothrin, the pyrethroids studied have been shown to have half-lives on plants, under glasshouse conditions, ranging from 1 to 6 weeks. Where comparisons have been made, degradation under field conditions has been found to be rather more rapid. In the glasshouse or the field the initial metabolic processes are identical to those elucidated in mammals, that is ester cleavage and hydroxylation of alkyl carbons and aromatic rings. However, oxidation of primary alcohols, formed by such initial metabolism, to carboxylic acids is a much less important process in plants than in mammals.

The metabolites formed on plants are generally readily conjugated. These conjugates, where they have ben identified, are usually with sugars or amino acids. However, much less characterisation of the conjugates formed has been achieved with plants, compared to animals. This is not surprising, since much smaller amounts of metabolic products are available from plant studies, making the identification of conjugates much more difficult.

With most of the pyrethroids discussed in this section an evaluation of the translocation of these compounds from treated leaves to other parts of the plant, or from the soil into the plant, has been carried out. In all cases it was shown that the parent pyrethroid was not translocated, although minimal levels of metabolites were found to move through the plant, both from treated leaves and from the soil.

In addition to metabolism, photo-initiated reactions are also possible on the surface of treated plants. In the case of phenothrin, phototransformation is in fact the dominant degradation process. Even for the more photostable pyrethroids, photoisomerisation has been detected on leaf surfaces, and minor photo-initiated reactions, that is hydrolysis of the cyanide group and decarboxylation, have been detected with fenvalerate and to a lesser extent with cypermethrin. Photodegradation studies with pyrethroids (see early part of this chapter) have also shown that photoinitiated ester cleavage can occur. It is therefore possible that ester cleavage, which is a major process for pyrethroids on plants, is a photo-induced as well as a metabolic reaction. Such photodegradation may well explain the faster degradation rates of pyrethroids on plants maintained in direct sunlight as opposed to under glasshouse conditions.

Degradation in soil and natural waters

Any compound that is used to protect plants will inevitably also find its way into the soil in which the plants are grown. Therefore, for the pyrethroids developed for agricultural use, an evaluation of the fate of these compounds in soil is of great importance. It is therefore not surprising that for most pyrethroids of agricultural importance fairly comprehensive data have been published on their degradation in soil.

Good agricultural practice will usually ensure that plants and soil are the only parts of the environment directly exposed to pyrethroids. However, spray drift, especially from aerial applications, may also result in pyrethroids entering rivers, lakes and other natural water. Information on the fate of pyrethroids in natural water is therefore also of interest. However, rather less information is available on this aspect of the environmental fate of pyrethroids.

Phenothrin

The degradation of phenothrin has been studied in two soil types (a light clay and a sandy loam) under both aerobic and anaerobic (flooded) conditions (Nambu et al., 1980). These studies utilised [^{14}C]-phenothrin radiolabelled in the benzylic carbon. Under aerobic conditions degradation is very rapid (half-life 1−2 days), with ester cleavage, ether cleavage, and hydroxylation and oxidation reactions yielding a range of compounds (see Figure 5.36), all of which are further degraded via loss of the radiolabelled carbon as carbon dioxide. Unextractable residues also form rapidly in the soil, reaching a maximum level of up to 55% of the radioactivity applied within 30 days. This unextractable radioactivity then gradually decreases, probably via conversion to [^{14}C]-carbon dioxide.

Under anaerobic conditions degradation is much slower (half-life 2−8 weeks). The pathways of degradation are, however, qualitatively similar to those detected under aerobic conditions. The major quantitative difference between anaerobic and aerobic degradation is a much slowed conversion of the radiolabelled carbon to [^{14}C]-carbon dioxide in anaerobic soils.

Permethrin

A number of workers have been involved in investigations of the degradation of permethrin in soil (Kaufmann et al., 1977; Hill et al., 1978; Kaneko et al., 1978; Jordan et al., 1982), so that data with a wide variety of soil types have been obtained. In addition, these data have been generated utilising [^{14}C]-permethrin radiolabelled at five different positions (see Figure 5.37), so that the fate of virtually all of the significant subunits of this molecule have been traced. In all soil types, under aerobic conditions, degradation was fairly rapid (half-life 5−55 days), with conversion to carbon dioxide as the major ultimate

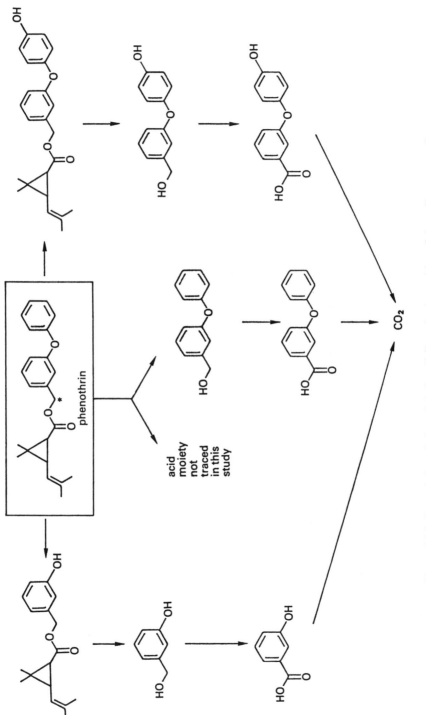

Figure 5.36. Degradation of phenothrin in soil (* marks the position of the radiotracer used)

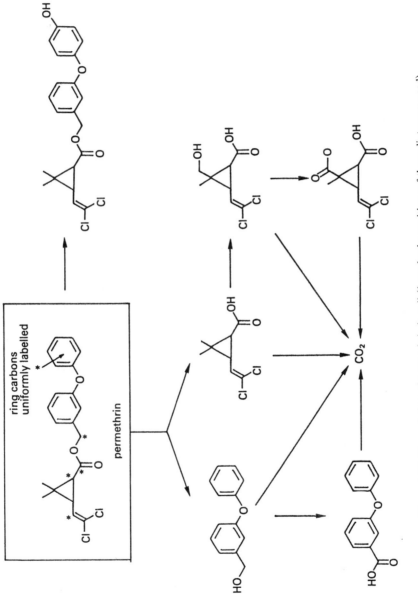

Figure 5.37. Degradation of permethrin in soil (* marks the positions of the radiotracers used)

fate of all the radiolabelled carbons used. With the large number of labelled positions utilised this is very strong evidence that complete mineralisation is the ultimate fate of this molecule.

With all soils and all positions of radiolabelling the formation of unextractable residues is a major occurrence. These unextractable residues appear to reach a maximum level fairly quickly, generally in 5–10 weeks. Where studies have continued for up to 51 weeks (Hill et al., 1978) it was established that with all positions of radiolabelling degradation of this bound material continues, ultimately releasing [^{14}C]-carbon dioxide. Low levels of various metabolites, resulting from ester cleavage, oxidation and hydroxylation, have been detected in the various soils. These results allow the construction of the degradation pathways which lead to the mineralisation of this molecule (see Figure 5.37). Under anaerobic conditions (either flooded or in an atmosphere of nitrogen) similar degradation processes seem to occur. However, the rate of ultimate conversion to [^{14}C]-carbon dioxide is rather slower than under aerobic conditions.

The fate of permethrin after addition to river and pond waters maintained statically above their natural sediments has also been studied (Rawn et al., 1982; Hill et al., 1979). In these studies loss of permethrin from the water, by absorption on to the sediments ensured that <2% of the applied permethrin remained in the aqueous phase after 7 days. In another study, where lake water containing permethrin was shaken with the lake sediment, 95% of the permethrin was absorbed on to the sediment within 1 minute (Sharom & Solomon, 1981). Thus absorption on to sediments is a very important process for the removal of permethrin from natural waters. Degradation also occurs in natural water/sediment mixtures, with ester cleavage as the major degradation process. The acid moiety thus released is then partially retained by the sediment, but is also detected in solution in the aqueous phase. The 3-phenoxybenzyl alcohol fragment is further oxidised to 3-phenoxybenzoic acid, and these two degradation products mainly remain absorbed on to the sediment. In the work of Hill et al. (1979) slow evolution of [^{14}C]-carbon dioxide was detected from river water/sediment mixtures. This study utilised [^{14}C]-permethrin radiolabelled in the cyclopropane ring. However, data from flooded soils (Hill et al., 1978) indicate that slow mineralisation of all other parts of this molecule will also occur in natural waters and sediments.

Cypermethrin

The degradation of cypermethrin, under aerobic conditions, has been studied in a range of soils by Roberts & Standen (1977a, 1981). Variable rates of degradation were detected in these soils, with half-lives ranging from 1 to 10 weeks. Ester cleavage is the major initial degradation process and the α-cyano-3-phenoxybenzyl alcohol thus released is readily converted to 3-phenoxybenzoic acid, via loss of hydrogen cyanide and oxidation. Hence the major initial degradation products from cypermethrin are identical to those from permethrin (see Figure 6.37). Further degradation of these fragments then follows the pathways established for permethrin, with eventual conversion of the radiolabelled carbons used (see Figure 5.38) to $[^{14}C]$-carbon dioxide. Generally such complete degradation is a major process, but in one soil studied (a clay loam from Spain: pH 7.7, clay 66%, silt 32%, sand 1.9% and organic matter 1.83%) most of the applied radioactivity was recovered from the soil after 52 weeks. Thus mineralisation of the molecule in this soil is slow. 3-Phenoxybenzoic acid accounted for up to 43% of the residue in this soil after 52 weeks (the cyclopropane acid was not traced in this particular soil) and unextracted material accounted for much of the remaining residue. However, such persistence of 3-phenoxybenzoic acid in soil under aerobic conditions is a very unusual occurrence and does not happen with cypermethrin in other soils, nor in any of the soils in which other pyrethroids generating 3-phenoxybenzoic acid have been studied.

Under anaerobic conditions, in a flooded sandy loam soil (Roberts & Standen, 1977a), degradation was similar to that in the Spanish clay soil, with rapid formation of 3-phenoxybenzoic acid being detected, but with little further degradation of this compound.

The formation of unextractable residues occurred in all soils under aerobic and anaerobic conditions; such "binding" is, however, more important under aerobic conditions. In soils maintained aerobically for periods up to 52 weeks these bound residues were shown to slowly mineralise. This is in agreement with the results obtained with permethrin.

A number of minor degradation pathways, involving hydrolysis of the cyanide group, hydroxylation and oxidation of the geminal dimethyl group and hydroxylation of the alcohol moiety 4′-position, have also been elucidated. These degradation pathways are summarised in Figure 5.38.

Studies in which cypermethrin (radiolabelled as shown in Figure 5.38) was added to river water/sediment mixtures have also been carried out (Arnold et al., 1980). As with permethrin, rapid absorption of the pyrethroid on to the sediment was found to occur, followed by rapid degradation (half-life ~5 days). Ester cleavage was the major initial degradation process and the cyclopropane acid moiety thus released was detected both absorbed to the sediment and in the aqueous phase. This acid is then slowly further degraded to yield carbon dioxide (~3% of the radioactivity added in 15 weeks). The α-cyano-3-phenoxybenzyl alcohol moiety degrades very rapidly to 3-phenoxy-benzaldehyde, so that significant amounts of this compound were detected, absorbed to the sediment, after 2 weeks. This compound is then degraded via 3-phenoxybenzoic acid to eventually yield carbon dioxide (30–50% of the radioactivity added in 15 weeks).

Fenpropathrin

Roberts & Standen (1977b) have studied the degradation of fenpropathrin in

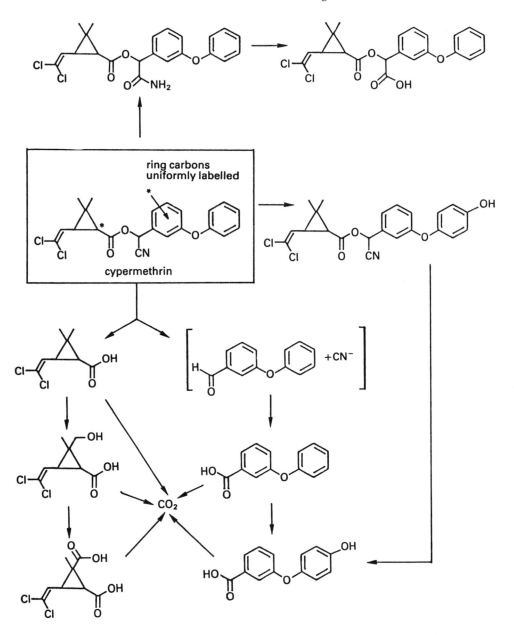

Figure 5.38. Degradation of cypermethrin in soil (postulated intermediates shown in square brackets, * marks the positions of the radiotracers used)

three of the soils they have used in their investigations of cypermethrin (see above). Under aerobic conditions fenpropathrin was less readily degraded in these soils than was cypermethrin, having a half-life ranging from 4 to 16

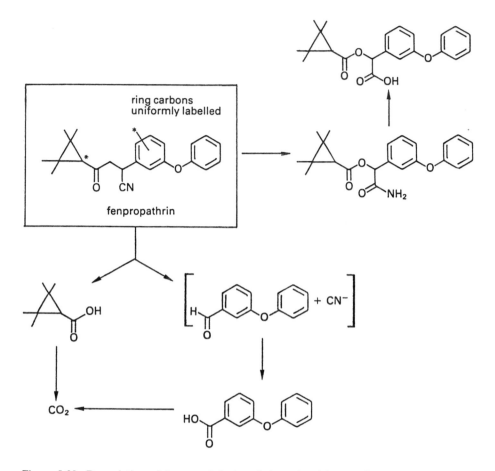

Figure 5.39. Degradation of fenpropathrin in soil (postulated intermediates shown in square brackets; * marks the positions of the radiotracers used)

weeks. Degradation was mainly initiated by ester cleavage, with further degradation of the fragments released eventually yielding [^{14}C]-carbon dioxide from the radiolabelled carbons utilised (see Figure 5.39 for radiolabel positions). Hydrolysis of the cyanide group was also detected, but, in contrast to cypermethrin, hydroxylation of the acid and alcohol moieties did not occur.

Under anaerobic conditions degradation was still initiated by ester cleavage, but further degradation of the 2,2,3,3-tetramethylcyclopropanecarboxylic acid and 3-phenoxybenzoic acid generated was very much slower than under aerobic conditions.

The degradation pathways for fenpropathrin in soil are summarised in Figure 5.39.

Fenvalerate

Ohkawa et al. (1978) have studied the degradation of fenvalerate in four different soils, using [^{14}C]-fenvalerate radiolabelled in the carbonyl group of the acid moiety and the cyanide group of the alcohol moiety. These positions of radiolabelling limit the amount of information on degradation pathways which can be obtained, since ester cleavage will generate fragments from which the labelled carbon can easily be lost. This is especially true for the cyanide label.

Under aerobic conditions, variable rates of degradation (half-lives 15–90 days) were detected in the different soils studied. The formation of bound material and the evolution of [^{14}C]-carbon dioxide were the major processes which occurred in all the soils. Such ready evolution of carbon dioxide is strong evidence, considering the ^{14}C-label positions used, that ester cleavage is the major initial degradation process. A range of degradation products, resulting from hydrolysis of the cyanide group, ester and ether cleavage, and hydroxylation of the alcohol moiety, were detected at minor levels. These degradation processes are summarised in Figure 5.40.

In soils maintained anaerobically, degradation is slower and evolution of [^{14}C]-carbon dioxide from the carbonyl-labelled [^{14}C]-fenvalerate becomes insignificant, although for the cyanide-labelled [^{14}C]-fenvalerate evolution of [^{14}C]-carbon dioxide is still the major process detected. With the exception of ether cleavage, the degradation pathways under anaerobic conditions are similar to those elucidated for aerobic degradation.

The fate of the isomers of fenvalerate that have an (*S*)-configuration in the acid moiety has also been studied in a model ecosystem (Ohkawa et al., 1980b). In this study samples of [^{14}C]-(*S*)-fenvalerate radiolabelled in the carbonyl group, the cyanide group and the benzyl carbon, were each added to a sandy loam soil in the bottom of glass aquaria. The aquaria were then filled with water and allowed to equilibrate, with continuous aeration, for 7 days. Fish (carp), daphnia, snails, algae, protozoa and rotifers were then added and the ecosystems thus generated were then maintained for periods up to 30 days.

After 30 days the radioactivity initially added to the ecosystems was distributed as follows: soil 90–93%, water 0.94–1.1%, snails 0.35–0.4%, fish 0.10–0.24%, algae and daphnia 0.1%, with unchanged (*S*)-fenvalerate accounting for 67–82%, 33–40%, 61%, 27–31% and 67–94% respectively of this radioactivity. A complex range of degradation products were also identified in various parts of the ecosystem, and these are summarised in Figure 5.41.

Fluvalinate

The aerobic and anaerobic degradation of fluvalinate, [^{14}C]-radiolabelled in the trifluoromethyl group, has been studied in three soils by Staiger & Quistad (1983). Rapid aerobic degradation (half-life 6–28 days) was detected in all

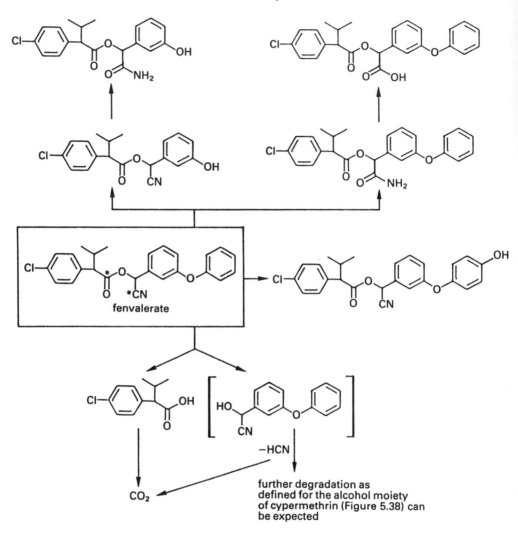

Figure 5.40. Degradation of fenvalerate in soil (postulated intermediates shown in square brackets; * marks the positions of the radiotracers used)

three soils, with ester cleavage as the initial degradation process. The acid moiety thus generated was then further degraded to 1-chloro-4-(trifluoro-methyl)aniline. This latter compound is a major degradation product, accounting for 37–51% of the applied radioactivity after 8 weeks. It is also readily lost from the soil by volatilisation, so that 22–47% of the applied radioactivity was evolved as this compound during 8 weeks. Radiolabelled carbon dioxide is also evolved from the soils, but this only accounts for 3–9% of the applied radioactivity during the same period.

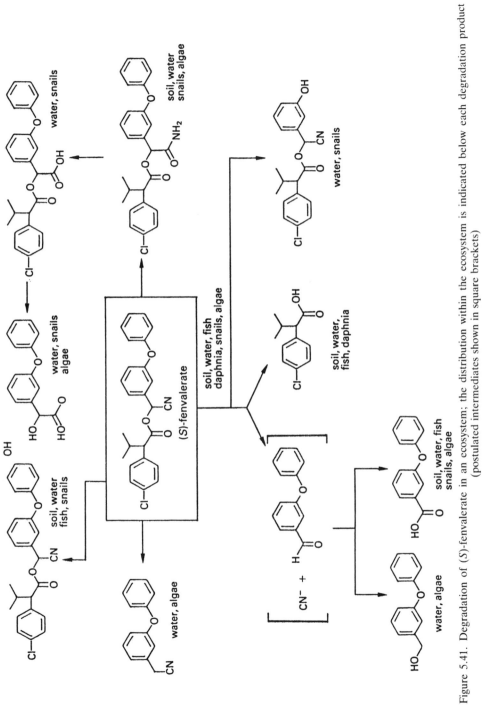

Figure 5.41. Degradation of (*S*)-fenvalerate in an ecosystem; the distribution within the ecosystem is indicated below each degradation product (postulated intermediates shown in square brackets)

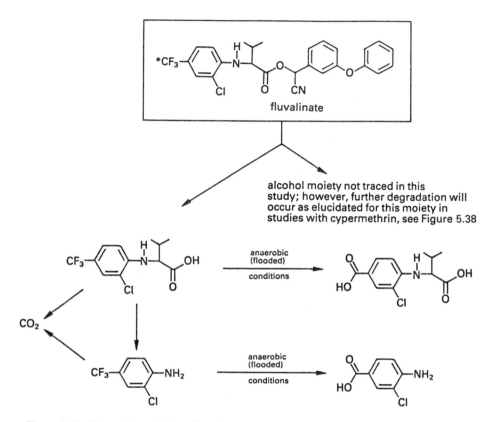

Figure 5.42. Degradation of fluvanilate in soil (* marks the position of the radiotracer used)

Under anaerobic (flooded) conditions degradation is only slightly slower (half-life 15 days). The degradation pathways established for aerobic degradation also occurred anaerobically, and in addition two further metabolites were detected. These metabolites, which were detected in the flood water, resulted from hydrolysis of the trifluoromethyl group to a carboxylic acid.

The aerobic and anaerobic degradation pathways detected for fluvalinate are summarised in Figure 5.42. It is interesting that hydrolysis of the cyanide group, as detected with fenvalerate, fenpropathrin and cypermethrin, does not occur with this compound. Hydroxylation at the alcohol moiety 4′-position is also not detected.

General discussion

Ready degradation in soils maintained under aerobic conditions is characteristic of all of the pyrethroids used in agriculture. The actual rate of degradation varies from compound to compound and also with the type of soil investigated, so that a wide range of half-lives, ranging from 1 day to 16 weeks, have been reported.

With the exception of fluvalinate, degradation to yield $^{14}CO_2$ from those parts of the molecule which had been radiolabelled is the ultimate fate in soil of all the pyrethroids investigated. In the case of permethrin the comprehensive range of radiolabelling positions used shows that virtually the whole molecule is completely mineralised. The close similarity of permethrin to most of the other pyrethroids studied strongly indicates that they too will be mineralised completely. Fluvalinate is the exception to this, in that a volatile degradation product is generated from its acid moiety, which evaporates from the soil more quickly than it is further degraded to carbon dioxide.

Under anaerobic conditions the degradation of pyrethroids is slower, especially in terms of breakdown to yield carbon dioxide. However, the pathways of degradation are, with a few minor exceptions, similar to those detected under aerobic conditions. In the few cases studied where a pyrethroid has been added to a natural water/sediment mixture, rapid loss of the pyrethroid from the aqueous phase by absorption to the sediment has been shown to occur. The pyrethroid is then slowly degraded eventually to yield carbon dioxide.

In most of the studies referenced in this chapter (see also Kaufmann et al., 1981) an evaluation of the potential for leaching of pyrethroids in soil has been carried out. In all cases it was found that the pyrethroids are virtually immobile in soil, which is not surprising considering the highly hydrophobic nature of these compounds. Such immobility, coupled with ready degradation, means that the main point of entry of pyrethroids into the environment, that is, soil, will first of all irreversibly bind and will then completely degrade these insecticides. Thus pyrethroids are highly unlikely to move from soil into any other parts of the general environment.

References

Abe, Y., Tsuda, K. & Fujita, Y., 1972, Studies on pyrethroidal compounds. *Botyu-Kagaku*, **37**(3), 102–11.

Ando, T., Ruzo, L. O., Engel, J. L. & Casida, J. E., 1983, 3-(3,3-dihalo-2- propenyl) analogues of allethrin and related pyrethroids: synthesis, biological activity, and photostability. *J. Agric. Food Chem.*, **31**(2), 250–3.

Arnold, D. J., Rapley, J. H., Vincent, J. & Moore, D., 1980, ICI, Plant Protection Division, unpublished results.

Bigley, W. S. & Plapp, F. W., 1978, Metabolism of *cis*- and *trans*-[^{14}C]permethrin by the tobacco budworm and the bollworm. *J. Agric. Food Chem.* **26**(5), 1128–34.

Bullivant, M. J. & Pattenden, G., 1973, Photolyis of bio-allethrin. *Tet. Lett.*, **38**, 3679–80.

Bullivant, M. J. & Pattenden, G., 1976, Photodecomposition of natural pyrethrins and related compounds. *Pestic. Sci.*, **7**, 231–5.

Casida, J. E., Kimmel, E. C., Elliott, M. & Janes, N. F., 1971, Oxidative metabolism of pyrethrins in mammals. *Nature, Lond.*, **230**, 326–7.

Chang, C. K. & Jordan, T. W., 1982, Penetration and metabolism of topically applied permethrin and cypermethrin in pyrethroid-tolerant *Wiseana cervinata* larvae. *Pestic. Biochem. Physiol.*, **17**, 196-204.

Chen, Y. L. & Casida, J. E., 1969, Photodecomposition of pyrethrin I, allethrin, phthalthrin, and dimethrin, modifications in the acid moiety. *J. Agric. Food.Chem.*, **17**(2), 208–15.

Cole, L. M., Ruzo, L. O., Wood, E. J. & Casida, J. E., 1982a, Pyrethroid metabolism: comparative fate in rats of tralomethrin, tralocythrin, deltamethrin and (1*R*, α*S*)-*cis*-cypermethrin. *J. Agric. Food Chem.*, **30**(4), 631–6.

Cole, L. M., Casida, J. E. & Ruzo, L. O., 1982b, Comparative degradation of the pyrethroids tralomethrin, tralocythrin, deltamethrin and cypermethrin on cotton and bean foliage. *J. Agric. Food Chem.* **30**(5), 916–20.

Crawford, M. J. & Hutson, D. H., 1977, The metabolism of the pyrethroid insecticide (±)-α-cyano-3-phenoxybenzyl-2,2,3,3-tetramethylcyclopropanecarboxylate, WL 41706, in the rat. *Pestic. Sci.*, **8**, 579–99.

Crawford, M. J., Croucher, A. & Hutson, D. H.,1981, The metabolism of the pyrethroid insecticide cypermethrin in rats: excreted metabolites. *Pestic. Sci.*, **12**, 399–411.

Crayford, J. V. & Hutson, D. H., 1980a, The metabolism of 3-phenoxybenzoic acid and its glucoside conjugate in rats. *Xenobiotica*, **10**(5), 355–64.

Crayford, J. V. & Hutson, D. H. 1980b. Xenobiotic triglyceride formation. *Xenobiotica*, **10**(5), 349–54.

Elliott, M., Janes, N. F., Kimmel, E. C. & Casida, J. E., 1972, Metabolic fate of pyrethrin I, pyrethrin II and allethrin administered orally to rats. *J. Agric. Food Chem.*, **20**(2), 300–13.

Elliott, M., Farnham, A. W., Janes, N. F., Needham, P. H. & Pulman, D. A., 1973a, Potent pyrethroid insecticides from modified cyclopropane acids. *Nature, Lond.*, **244**, 456–7.

Elliott, M., Farnham, A. W., Janes, N. F., Needham, P. H., Pulman, D. A. & Stevenson, J. H., 1973b, A photostable pyrethroid. *Nature, Lond.*, **246**, 169–70.

Elliott, M., Janes, N. F., Pulman, D. A., Gaughan, L. C., Unai, T. & Casida, J. E., 1976, Radiosynthesis and metabolism in rats of the 1*R* isomers of the insecticide permethrin. *J. Agric. Food Chem.*, **24**(2), 270–6.

Farnham, A.W., 1971, Changes in cross-resistance patterns of houseflies selected with natural pyrethrins or resmethrin (5-benzyl-3-furylmethyl (±)-*cis-trans*-chrysanthe-mate). *Pestic. Sci.*, **2**, 138–43.

Farnham, A. W., 1973, Genetics of resistance of pyrethroid selected houseflies, *Musca domestica* L. *Pestic. Sci.*, **4**, 513–20.

Gaughan, L. C., Unai, T. & Casida, J. E., 1977, Permethrin metabolism in rats. *J. Agric. Food Chem.*, **25**(1), 9-17.

Gaughan, L. C. & Casida, J. E., 1978, Degradation of *trans*- and *cis*-permethrin on cotton and bean plants. *J. Agric. Food Chem.*, **26**(3), 525–8.

Gaughan, L. C., Ackerman, M. E., Unai, T. & Casida, J. E., 1978a, Distribution and metabolism of *trans*- and *cis*- permethrin in lactating jersey cows. *J. Agric. Chem.*, **26**(3), 613–18.

Gaughan, L. C., Robinson, R. A. & Casida, J. E., 1978b, Distribution and metabolic fate of *trans*- and *cis*-permethrin in laying hens. *J. Agric. Food Chem.*, **26**(6), 1374–80.

Glickman, A. H., Shono, T., Casida, J. E. & Lech, J. J., 1979, *In vitro* metabolism of permethrin isomers by carp and rainbow trout liver microsomes. *J. Agric. Food Chem.*, **27**(5), 1038–41.

Glickman, A. H., Hamid, A. A. R., Rickert, D. E. & Lech, J. J., 1981, Elimination and metabolism of permethrin isomers in rainbow trout. *Toxicol. Appl. Pharmacol.,* **57**, 88–98.

Glickman, A. H. & Lech, J. J., 1981, Hydrolysis of permethrin, a pyrethroid insecticide, by rainbow trout and mouse tissues *in vitro*: a comparative study. *Toxicol. Appl. Pharmacol.,* **60**, 186–92.

Glickman, A. H., Weitmann, S. D. & Lech, J. J., 1982, Differential toxicity of *trans*-permethrin in rainbow trout and mice. *Toxicol. Appl. Pharmacol.,* **66**, 153–161.

Harrison, M. P., Moss, S. R. & Fowkes, A. G., 1983, The metabolism of the pyrethroid insecticide cyhalothrin (ICI 146 814) in the rat, dog and cow, analysis of intact metabolite conjugates by ^{13}C-NMR spectroscopy and fast atom bombardment mass spectroscopy. Poster presentation at the First International Symposium on Foreign Compound Metabolism, Sponsored by International Society for the Study of Xenobiotics.

Hill, I. R., Arnold, D. J. & Cleverley, B. A., 1978, ICI, Plant Protection Division, unpublished results.

Hill, I. R., Harvey, B. R. & Weissler, M. S., 1979, ICI, Plant Protection Division, unpublished results.

Holden, J. S., 1979, Absorption and metabolism of permethrin and cypermethrin in the cockroach and the cotton-leafworm larvae. *Pestic. Sci.,* **10**, 295–307.

Holmstead, R. L. & Casida, J. E., 1977, Photochemical reaction of pyrethroid insecticides. In *Synthetic Pyrethroids,* edited by M. Elliott, ACS Symposium Series No. 42 (Washington, DC: American Chemical Society), pp. 137–46.

Holmstead, R. L., Casida, J. E., Ruzo, L. O. & Fullmer, D. G., 1978a, Pyrethroid photodecomposition: permethrin. *J. Agric. Food Chem.,* **26**(3), 590–5.

Holmstead, R. L., Fullmer, D. G. & Ruzo, L. O., 1978b, Pyrethroid photodecomposition: pydrin. *J. Agric. Food Chem.,* **26**(4), 954–9.

Hunt, L. M. & Gilbert, B. N., 1977, Distribution and excretion rates of ^{14}C-labelled permethrin isomers administered orally to four lactating goats for 10 days. *J. Agric. Food Chem.,* **25**(3), 673–6.

Hutson, D. H. & Casida, J. E., 1978, Taurine conjugation in metabolism of 3-phenoxybenzoic acid and the pyrethroid insecticide cypermethrin in mouse. *Xenobiotica,* **8**(9), 565–71.

Hutson, D. H., Gaughan, L. C & Casida, J. E., 1981, Metabolism of the *cis-* and *trans*-isomers of cypermethrin in mice. *Pestic. Sci.,* **12**, 385–98.

Ishaaya, I. & Casida, J. E., 1981, Pyrethroid esterase(s) may contribute to natural pyrethroid tolerance of the larvae of the common green lacewing. *Environ. Entomol.,* **10**, 681–4.

Ivie, G. W. & Hunt, L. M., 1980, Metabolites of *cis-* and *trans*-permethrin in lactating goats. *J. Agric. Food Chem.,* **28**(6), 1131–8.

Jao, L. T. & Casida, J. E., 1974, Insect pyrethroid-hydrolysing esterases. *Pestic. Biochem. Physiol.,* **4**, 465–72.

Jordan, E. G., Kaufman, D. D. & Kayser, A. J., 1982, The effect of soil temperature on the degradation of *cis, trans*-permethrin in soil. *J. Environ. Sci. Health,* **B17**(1), 1–17.

Kaneko, H., Ohkawa, H., & Miyamoto, J., 1978, Degradation and movement of permethrin isomers in soil. *J. Pestic. Sci.,* **3**, 43–51.

Kaneko, H., Ohkawa, H. & Miyamoto, J., 1981, Comparative metabolism of fenvalerate and the [2S,αS]-isomer in rats and mice. *J. Pestic. Sci.,* **6**, 317–26.

Kaufmann, D. D., Haynes, S. C., Jordan, E. G. & Kayser, A. J., 1977, Permethrin degradation in soil and microbial cultures. In *Synthetic pyrethroids*, edited by M. Elliott, ACS Symposium Series No. 42, (Washington DC: American Chemical Society), pp. 147–61.

Kaufmann, D. D., Russell, B. A., Helling, C. S. & Kayser, A. J., 1981, Movement of cypermethrin, decamethrin, permethrin, and their degradation products in soil. *J. Agric. Food Chem.,* **29**(2), 239–45.

Kimmel, E. C., Casida, J. E. & Ruzo, L. O., 1982, Identification of mutagenic photoproducts of the pyrethroids allethrin and terallethrin. *J. Agric. Food Chem.,* **30**(4), 623–6.

Leahey, J. P., Bewick, D. W., Saunders, R., Curl, E. A. & Milner, S. D., 1980, ICI, Plant Protection Division, unpublished results.

Leahey, J. P. & Carpenter, P. K., 1980, The uptake of metabolites of permethrin by plants grown in soil treated with [^{14}C]permethrin. *Pestic. Sci.,* **11**, 279–89.

Liu, M., Tzeng, Y. & Sun, C., 1981, Diamondback moth resistance to several synthetic pyrethoids. *J. Econ. Entomol.,* **74**, 393–6.

Masri, M. S., Jones, F. T., Lundin, R. E., Bailey, G. F. & DeEds, F., 1964, Metabolic fate of two chrysanthemic acid esters: barthrin and dimethrin. *Toxicol. Appl. Pharmacol.,* **6**, 711–15.

Mihara, K., Ohkawa, H. & Miyamoto, J. (1981). Metabolism of terallethrin in rats. *J. Pestic. Sci.,* **6**, 211–22.

Mikami, N., Takahashi, N., Hayashi, K. & Miyamoto, J., 1980, Photodegradation of fenvalerate (sumicidin) in water and on soil surface. *J. Pestic. Sci.,* **5**, 225–36.

Miskus, R. P. & Andrews, T. L., 1972, Stabilisation of thin films of pyrethrins and allethrin. *J. Agric. Food Chem.,* **20**(2), 313–5.

Miyamoto, J., Sato, Y., Yamamoto, K., Endo, M. & Suzuki, S., 1968, Biochemical studies on the mode of action of pyrethroidal insecticides. Part I, metabolic fate of phthalthrin in mammals. *Agr. Biol. Chem.,* **32**(5), 628–640.

Miyamoto, J., Nishida, T. & Ueda, K., 1971, Metabolic fate of resmethrin, 5-benzyl-3-furylmethyl *dl-trans*-chrysanthemate in the rat. *Pestic. Biochem. Physiol.,* **1**, 293–306.

Miyamoto, J. & Suzuki, T., 1973, Metabolism of tetramethrin in houseflies *in vivo*. *Pestic. Biochem. Physiol.,* **3**, 30–41.

Miyamoto, J., Suzuki, T. & Nakae, C., 1974, Metabolism of phenothrin or 3-phenoxybenzyl *d-trans*-chrysanthemate in mammals. *Pestic. Biochem. Physiol.,* **4**, 438–50.

More, J. E., Roberts, T. R. & Wright, A. N., 1978, Studies of the metabolism of 3-phenoxybenzoic acid in plants. *Pestic. Biochem. Physiol.,* **9**, 268–80.

Nakanishi, M., Kato, Y, Furuta, T. & Miura, S., 1971, Metabolic fate of proparthrin. *Botyu-Kagaku,* **36**, 116–21.

Nambu, K. Ohkawa, H. & Miyamoto, J., 1980, Metabolic fate of phenothrin in plants and soils. *J. Pestic. Sci.,* **5**, 177–97.

Ohkawa, H., Kaneko, H. & Miyamoto, J., 1977, Metabolism of permethrin in bean plants. *J. Pestic. Sci.,* **2**, 67–76.

Ohkawa, H., Nambu, K. & Miyamoto, J., 1978, Metabolic fate of fenvalerate (sumicidin) in soil and by soil microorganisms. *J. Pestic. Sci.,* **3**, 129–41.

Ohkawa, H., Kaneko, H., Tsuji, H. & Miyamoto, J., 1979, Metabolism of fenvalerate (sumicidin) in rats. *J. Pestic. Sci.,* **4**, 143–55.

Ohkawa, H., Kikuchi, R. & Miyamoto, J., 1980a, Bioaccumulation and biodegradation of the (*S*)-acid isomer of fenvalerate (sumicidin) in an aquatic model ecosystem. *J. Pestic. Sci.,* **5**, 11–22.

Ohkawa, H., Nambu, K. & Miyamoto, J., 1980b, Metabolic fate of fenvalerate (sumicidin) in bean plants. *J. Pestic. Sci.,* **5**, 215–23.

Ohsawa, K. & Casida, J. E., 1979, Photochemistry of the potent knockdown pyrethroid kadethrin. *J. Agric. Food Chem.,* **27**(5),1112–20.

Ohsawa, K. & Casida, J. E., 1980, Metabolism in rats of the potent knockdown pyrethroid kadethrin. *J. Agric. Food Chem.* **28**(2), 250–5.

Quistad, G. B., Staiger, L. E., Jamieson, G. C & Schooley, D. A., 1982a, Metabolism of fluvalinate by a lactating dairy cow. *J. Agric. Food Chem.,* **30**(5), 895–901.

Quistad, G. B., Staiger, L. E., Mulholland, K. M. & Schooley, D. A., 1982b. Plant metabolism of fluvalinate (α-cyano-3-phenoxybenzyl-2-[2-chloro-4-(trifluoromethyl)anilino]-3-methylbutanoate). *J. Agric. Food Chem.*, **30**(5), 888–95.

Quistad, G. B., Staiger, L. E., Jamieson, G. C. & Schooley, D. A., 1983, Fluvalinate metabolism by rats. *J. Agric. Food Chem.*, **31**(3), 589–96.

Quistad, G. B. & Selim, S., 1983, Fluvalinate metabolism by rhesus monkeys. *J. Agric. Food Chem.*, **31**(3), 596–9.

Rawn, G. P., Webster, G. R. B. & Muir, D. C. G., 1982, Fate of permethrin in model outdoor ponds. *J. Environ. Sci. Health*, **B17**(5), 463–86.

Roberts, T. R. & Standen, M. E., 1977a, Degradation of the pyrethroid cypermethrin NRDC 149 (±)-α-cyano-3-phenoxybenzyl (±)-*cis*, *trans*-3-(2,2-dichlorovinyl)-2,2-dimethlcyclopropanecarboxylate and the respective cis-(NRDC 160) and trans-(NRDC 159) isomers in soil. *Pestic. Sci.*, **8**, 305–19.

Roberts, T. R. & Standen, M. E., 1977b, Degradation of the pyrethroid insecticide WL 41706, (±)-α-cyano-3-phenoxybenzyl-2,2,3,3-tetramethylcyclopropanecarboxylate, in soils. *Pestic. Sci.*, **8**, 600–10.

Roberts, T. R. & Standen, M. E., 1981, Further studies of the degradation of the pyrethroid insecticide cypermethrin in soils. *Pestic. Sci.*, **12**, 285–96.

Roberts, T. R., 1981, The metabolism of the synthetic pyrethroids in plants and soils, In *Progress in Pesticide Biochemistry*, edited by T. R. Roberts & D. H. Hutson, Vol. 1, (Chichester: John Wiley and Sons), p. 127.

Ruzo, L. O., Holmstead, R. L. & Casida, J. E., 1977, Pyrethroid photochemistry: decamethrin. *J. Agric. Food Chem.*, **25**(6), 1385–94.

Ruzo, L. O., Unai, T. & Casida, J. E., 1978, Decamethrin metabolism in rats. *J. Agric. Food Chem.*, **26**(4), 918–24.

Ruzo, L. O. & Casida, J. E., 1979, Degradation of decamethrin on cotton plants. *J. Agric. Food Chem.*, **27**(3) , 572–5.

Ruzo, L. O., Engel, J. L. Casida, J. E., 1979, Decamethrin metabolites from oxidative, hydrolytic and conjugative reactions in mice. *J. Agric. Food Chem.*, **27**(4), 725–31.

Ruzo, L. O., Gaughan, L. C. & Casida, J. E., 1980, Pyrethroid photochemistry: S-bioallethrin. *J. Agric. Food Chem.*, **28**(2), 246–9.

Ruzo, L. O., Gaughan, L. C. & Casida, J. E., 1981, Metabolism and degradation of the pyrethroids tralomethrin and tralocythrin in insects. *Pestic. Biochem. Physiol.*, **15**, 137–42.

Ruzo, L. O. & Casida, J. E., 1981, Pyrethroid photochemistry: (*S*)-α-cyano-3-phenoxybenzyl *cis*-(1*R*, 3*R*, 1'*R* or *S*)-3-(1',2'-dibromo-2',2'-dihaloethyl)-2,2-dimethylcyclopropanecarboxylates. *J. Agric. Food Chem.*, **29**(4), 702–6.

Ruzo, L. O., Smith, I. H. & Casida, J. E., 1982, Pyrethroid photochemistry: photooxidation reactions of the chrysanthemates phenothrin and tetramethrin. *J. Agric. Food Chem.*, **30**(1), 110–15.

Sasaki, T., Eguchi, S. & Ohno, M., 1970, Studies on chrysanthemic acid. IV. Photochemical behaviour of chrysanthemic acid and its derivatives. *J. Org. Chem.*, **20**(2), 35, 790–3.

Schnitzerling, H. J., Nolan, J. & Hughes, S., 1983, Toxicology and metabolism of some synthetic pyrethroids in larvae of susceptible and resistant strains of the cattle tick *Boophilus microplus* (can). *Pestic. Sci.*, **14**, 64–72.

Sharom, M. S. & Solomon, K. R., 1981, Adsorption–desorption, degradation and distribution of permethrin in aqueous systems. *J. Agric. Food Chem.*, **29**(6), 1122–5.

Shono, T., Unai, T. & Casida, J. E., 1978, Metabolism of permethrin isomers in American cockroach adults, housefly adults and cabbage looper larvae. *Pestic. Biochem. Physiol.*, **9**, 96–106.

Shono, T. & Casida, J. E., 1978, Species-specificity in enzymatic oxidation of pyrethroid insecticides: 3-phenoxybenzyl and α-cyano-3-phenoxybenzyl-3-(2,2,dihalovinyl)-2,2-dimethylcyclopropanecarboxylates. *J. Pestic. Sci.,* **3,** 165–8.

Shono, T., Ohsawa, K. & Casida, J. E., 1979, Metabolism of *trans*-and *cis*- permethrin, *trans*- and *cis*-cypermethrin and decamethrin by microsomal enzymes. *J. Agric. Food Chem.* **27**(2), 316–25.

Smith, I. H. & Casida, J. E., 1981, Epoxychrysanthemic acid as an intermediate in metabolic decarboxylation of chrysanthemate insecticides. *Tet. Lett.,* **22,** 203–6.

Soderlund, D. M. & Casida, J. E, 1977, Effects of pyrethroid structure on rates of hydrolysis and oxidation by mouse liver microsomal enzymes. *Pestic. Biochem. Physiol.,* **7,** 391–401.

Soderlund, D. M., Sanborn, J. R. & Lee, P. W., 1983, Metabolism of pyrethrins and pyrethroids in insects. In *Progress in Pesticide Biochemistry,* Vol. 3, Edited by D. H. Hutson & T. R. Roberts (Chichester: John Wiley & Sons), pp. 401–435.

Staiger, L. E., Quistad, G. B., Duddy, S. K. & Schooley, D. A., 1982, Metabolism of fluvalinate by laying hens. *J. Agric. Food Chem.* **30**(5), 901–6.

Staiger, L. E. & Quistad, G. B., 1983, Degradation and movement of fluvalinate in soil. *J. Agric. Food Chem.,* **31**(3), 599–603.

Suzuki, T. & Miyamoto, J., 1974, Metabolism of tetramethrin in houseflies and rats *in vitro. Pestic. Biochem. Physiol.,* **4,** 86–97.

Suzuki, T., Ohno, N. & Miyamoto, J., 1976, New metabolites of (+)-*cis*-fenothrin, 3-phenoxybenzyl (+)-*cis*-chrysanthemumate, in rats. *J. Pestic. Sci.,* **1,** 151–2.

Tattersfield, F., 1932, The loss of toxicity of pyrethrum dusts on exposure to light and air. *J. Agric. Sci.,* **22,** 396–417.

Ueda, K. & Matsui, M., 1971, Studies on chrysanthemic acid. XXI. Photochemical isomerisation of chrysanthemic acid and its derivatives. *Tetrahedron,* **27,** 2771–4.

Ueda, K., Gaughan, L. C. & Casida, J. E., 1974, Photodecomposition of resmethrin and related pyrethroids. *J. Agric. Food Chem.,* **22**(2), 212–19.

Ueda, K., Gaughan, L. C. & Casida, J. E., 1975a, Metabolism of (+)-*trans*- and (+)-*cis*-resmethrin in rats. *J. Agric. Food Chem.,* **23**(1), 106–15.

Ueda, K., Gaughan, L. C. & Casida, J. E., 1975b, Metabolism of four resmethrin isomers by liver microsomes. *Pestic. Biochem. Physiol.,* **5,** 280–94.

Williams, R. T., 1959, *Detoxication Mechanisms* (London: Chapman & Hall), p. 30.

Wood, J. L. & Cooley, S. L., 1956, Detoxification of cyanide by cystine. *J. Biol. Chem.,* **218,** 449–57.

Wright, A. N., Roberts, T. R., Dutton, A. J. & Doig, M. V., 1980, The metabolism of cypermethrin in plants: the conjugation of the cyclopropyl moiety. *Pestic. Biochem. Physiol.,* **13,** 71–80.

Yamamoto, I. & Casida, J. E., 1966, O-Demethyl pyrethrin II analogs from oxidation of pyrethrin I, allethrin, dimethrin and phthalthrin by a housefly enzyme system. *J. Econ. Entomol.,* **59**(6), 1542–3.

Yamamoto, I., Kimmel, E. C. & Casida, J. E., 1969, Oxidative metabolism of pyrethroids in houseflies. *J. Agric. Food Chem.,* **17**(6), 1227–36.

Yamamoto, I., Elliott, M. & Casida, J. E., 1971, The metabolic fate of pyrethrin I, pyrethrin II and allethrin. *Bull. World Health Org.,* **44,** 347–8.

Yamamoto, I. 1973, Mode of action of synergists in enhancing the insecticidal activity of pyrethrum and pyrethroids. In *Pyrethrum the Natural Insecticide,* edited by J. E. Casida (New York & London: Academic Press), pp. 200–2.

6. Agricultural, public health and animal health usage

J. J. Hervé

Introduction

The launching of a new family of insecticides on the world agricultural market is always an event, especially if the compounds are the broad-spectrum type like organophosphorus compounds or organochlorine compounds. However, in 1976, the advent of light-stable analogues of the natural pyrethrins opened a new chapter in the history of plant protection. At the time of writing, seven years after permethrin was launched, it can be said that these compounds have not disappointed the agricultural world, and that their success bears witness to this. A few figures will illustrate the great impact of these new compounds on the world market, starting with the user-value of pyrethroids:

Million US $
1976	10	
1977	80	
1978	220	(introduction of pyrethroids to US market)
1979	260	
1980	300	
1981	390	(extension to new crops in US and European markets and first commercial use in China)
1982	460	
1983	630	(introduction to new markets like India and Russia and large scale use in China)

In 1976, organophosphates represented a total of 40% of the world foliar insecticide market, organochlorine compounds 30%, carbamates 25%, miscellaneous 5%. In 1983, organophosphates represented about 35–40% of the world foliar insecticide market, organochlorines 15%, carbamates 20%, pyrethroids 20–25%, miscellaneous 5%.

Table 6.1. Crops use of pyrethroids in 1982. Wood-Mackenzie (1983)

	User value (million US $)	Percentage area treated
Cotton	68·5	65
Vegetables	8	10
Orchards	5	3
Maize	4	3
Soybean	3	4
Forests	2	6
Vines	2	2
Coffee	1·5	2
Other crops	6	5

Table 6.2. Percentage breakdown of usage of pyrethroids according to continent. Wood-Mackenzie (1983)

	1981	1983
USA/Canada	27·5	16·5
Latin America	24·0	16·5
Africa	18·0	9·5
Asia/Australia	15·5	34·5
Europe	15·0	23·0

It is interesting to see how pyrethroids were used in 1982 and what part of the pyrethroid market each crop represented (Table 6.1).

At present pyrethroids are not widely used in rice crops for three basic reasons:

1. The potential to induce growth of *Nilaparvata lugens* (brown plant hopper) populations following application (resurgence).
2. The toxicity to fish.
3. Poor efficacy against early attacks of rice stem-borers.

A number of studies in progress today are progressively eliminating each of these obstacles, and it is obvious that pyrethroids will be sold extensively for use on rice in the next five years. This will undoubtedly bring about a profound change in the areas treated with pyrethroids as will the use (initially on cotton) of these compounds on a large scale in China, Russia and India, which began in 1982 (Table 6.2).

A graph showing the growth of the areas treated with pyrethroids from 1976 to 1982, and projected to 1986, best demonstrates the great success achieved by this new family (Figure 6.1). The key to this success has been the versatility, and especially the safety in use of the pyrethroids, both for the user and the environment.

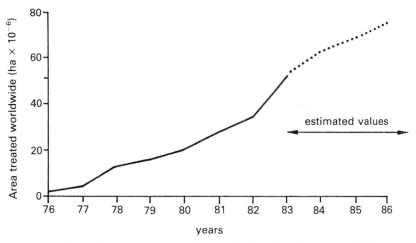

Figure 6.1. Growth of areas treated with pyrethroids. Wood-Mackenzie (1984a)

Wood-Mackenzie estimated that in 1983 pyrethroid users spent about 630 million dollars to treat about 52 million hectares. It is evident that cotton remains by far the leading crop treated, but since 1981 there has been an increasingly rapid diversification of use, hence the projected increase in treated areas from 1983 to 1986.

In terms of products, several pyrethroids are currently marketed, but four products, those which established the "pyrethroid image", still represented most of the market (98%) in 1982/3.

The volumes of pyrethroids used in 1982 and 1983 were, according to Wood-Mackenzie (1983):

	1982	*1983*
Fenvalerate	32%	30%
Deltamethrin	29%	35%
Cypermethrin	21%	22%
Permethrin	16%	10%
Others	2%	3%

These figures are expressed in deltamethrin equivalents.

In fact – and this is a basic point in the present history of pyrethroids – current products are substantially equivalent in terms of efficacy provided that the following ratio of application rates on the crops is observed:

Deltamethrin	1
Cypermethrin	4
Fenvalerate	7
Permethrin	10
Flucythrinate	4
Cyfluthrin	3

These ratios, derived by Wood-Mackenzie, are relatively robust; however, it is more sensible in biological terms to suggest a range of ratios, such a range is tabulated below.

Deltamethrin	1
Cypermethrin	4 to 5
Fenvalerate	4 to 9
Permethrin	8 to 12
Flucythrinate	4 to 5
Cyfluthrin	2·5 to 3·5

These ratios have been estimated at the 95% control level under pest pressure, that is an infestation that in the absence of treatment is likely to induce a crop loss unacceptable to the grower.

Pyrethroids are not used solely in agriculture – although non-photostable pyrethroids already held a large part of other markets, photostable pyrethroids are gradually becoming established. In 1980 non-agricultural uses of all insecticides, which represented a value of 675 million dollars (ex-manufacturer), were broken down as follows:

	Million US $
Veterinary usage (ectoparasites, external treatments)	200
Public health	100
Household and industrial usage	250
Stored grain	100
Treatment of animal houses	25
Total	675

Within these markets pyrethroids generated a turn-over of 100 million dollars, ex-manufacturer, but it is estimated that in 1985 this will rise to 300 million dollars (1980 value) of which photostable pyrethroids will command a 60% share; the remainder will be non-photostable pyrethroids.

In terms of active ingredient, non-agricultural uses could reach 20% of the total usage of pyrethroids by 1985, according to Cox (1981), which presupposes a rapid growth of this sector. Thus in this chapter the authors will address both agricultural and non-agricultural usage of these compounds.

Agricultural usage of photostable synthetic pyrethroids

The photostable pyrethroids have proven to be broad-spectrum insecticides, effective against a great many insect species. In this chapter the effectiveness of these compounds will be examined in relation to relevant insect families, rather than the types of crop on which they may be used. In general, only the main insect families of agricultural importance will be taken into consideration, but it will soon appear obvious that, for the major commercial products, if a particular pyrethroid is active against an insect, others will also be effective at a rate which can be derived from the ratios given by Wood-Mackenzie in the introduction to this chapter.

Thus, it is fairly easy to find the approximate effective dose for most pyrethroids for any given pest/crop combination on the basis of a known reference.

Synthetic pyrethroids are generally recognised as having the following main characteristics:

- Knock-down effect
- Excitation and flushing effects
- Repellent effect
- Relative absence of vapour activity
- Highly lipophilic tendencies
- Absence of systemic effect

These characteristics must be well understood in devising the best control strategies using these compounds. Thus those specialised in the use of pyrethroids will understand that it is as important to know how these compounds should be used as it is to know the dose rate for each insect and crop. In the following section use strategies for various pests and crops will be discussed with reference to selected experimental trials.

Lepidoptera

Our discussion will begin with the Lepidoptera, against which pyrethroids are highly effective, especially in cotton, the leading market for the current products.

The *Heliothis* genus

Heliothis is a major pest throughout the world, and a particular problem in cotton. Three species of this genus are particularly difficult to control namely:

- *Heliothis armigera*   (old world cotton bollworm)
- *Heliothis virescens*   (tobacco budworm)
- *Heliothis zea*   (cotton bollworm)

Strangely enough, these pests are sometimes neglected while attention is directed towards other insects that are well controlled but in fact have less of an impact on yield than *Heliothis*. Such is the case in Egypt, where the campaign against *Spodoptera* has taken priority over the control of *Heliothis*.

Many authors have evaluated pyrethroids against *Heliothis*. As early as 1975, Davis et al. published results for *H. virescens* and *H. zea* showing permethrin to be far more effective than parathion. In 1976, Angelini & Couilloud observed that *RU 22950*, (RS) α-cyano-(3-phenoxyphenyl)methyl-

(1R-cis)-3(2,2-dibromoethenyl)-2,2-dimethylcyclopropane carboxylate, was highly effective at a rate of only 12·5 g ai/ha against *H. armigera* in cotton on the Ivory Coast, although it had no effect on mites (*Hemitarsonemus latus* and *Tetranychus urticae*). In 1977, however, All et al. in the US were already studying the possibility of combining pyrethroids with organophosphates, and noted the potentiation of organophosphate action achieved by adding small doses of pyrethroid (in a ratio of 1:10) leading to incredibly high yields in 1975, despite heavy *Heliothis* infestation, as shown by the following figures:

- Methyl parathion + permethrin   671.2 kg/ha
- Methyl parathion   344.6 kg/ha
- Control   171.1 kg/ha

In Nigeria, Caswell & Raheja (1977) also reported a decrease in damage to cotton bolls which was quite amazing at that time: only 4 to 6% of the bolls treated with deltamethrin and permethrin showed feeding damage compared with 20 to 40% of those treated by conventional means. Similarly Davis et al. (1977) demonstrated the tremendous potential of pyrethroids for the control of *H. virescens* in a trial in cotton in which 93% of the squares in untreated plots were damaged (Table 6.3). Comparative curves showing remarkable results were also presented by Ruscoe (1979) (Figures 6.2 and 6.3). In addition a number of authors (e.g. Pfrimmer 1979, Hervé 1982) observed improved protection of early-setting bolls at the base of cotton plants following pyrethroid use, which facilitated early harvesting.

Despite the excellent performance of pyrethroids when used alone, work on pyrethroid mixtures has continued with three basic objectives:

1. To limit the cost of treatment which was relatively high when the pyrethroids were first introduced, by lowering the dosages of the individual products to be mixed.

2. To find compounds with complementary modes of action, particularly to compensate for the absence of any practical ovicidal effect and the lack of vapour activity.

3. To attempt to delay the development of resistance by combining products with different modes of action, taking care to avoid antagonistic mixtures, for example El Okda et al. (1979) with *Spodoptera littoralis*.

Plapp (1979) first drew attention to the advantages of combining pyrethroids with formamidines. This combination has been particularly effective in providing ovicidal activity and pyrethroid synergy against resistant *Heliothis*. However, full rates of both components are needed to produce the desired effects. Plapp noted that permethrin and fenvalerate were synergised by formamidines to a greater extent than was deltamethrin.

Another factor contributing to the success of synthetic pyrethroids for *Heliothis* control is their ability to act upon strains that are resistant to organo-chlorines or organophosphates. This fact was reported by Sukhoruchenko

Table 6.3. Effect of insecticides on damage by *Heliothis* spp. (80 to 94% tobacco budworms) to squares and bolls of cotton. J. W. Davis et al. (1977)

Treatment[a]	Rate (kg a.i./ha)	Damaged (%)		Undamaged[b] squares/ha
		Squares	Bolls	
Permethrin	0·056	70 e	80 fg	6·630 d
	0·10	48 cd	48 de	58·667 cd
	0·43	8 a	12 ab	241·185 a
	0·112	37 cd	36 cd	44·815 d
	0·22	18 ab	21 abc	141·556 b
Cypermethrin	0·056	35 cd	41 b	34·222 d
	0·11	18 ab	13 ab	128·740 b
FMC-45497 (NRDC 160)[c]	0·022	66 e	68 ef	21·185 d
	0·056	25 bc	33 bd	60·296 c
	0·11	16 ab	17 abc	164·593 b
Deltamethrin	0·0056	43 cd	58 de	17·926 d
	0·011	26 bc	31 bcd	112·444 bc
	0·022	12 a	10 a	164·593 b
FMC-35171 (NRDC 148)[c]	0·056	67 ce	63 de	27·703 d
	0·11	41 cd	37 cd	101·037 bc
Methyl parathion	1·682	62 e	67 ef	2·444 d
Check		93 f	93 g	0 d

Means followed by the same letter are not significantly different at the 5% level of probability
[a] Treated 3 times; applied July 29 or 30, Aug. 2 or 3, and Aug. 5 or 6.
[b] Mean of 2 tests.
[c] Experimental pyrethroids.

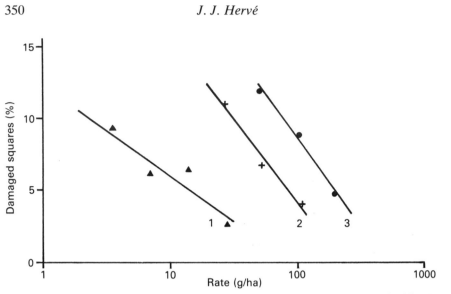

Figure 6.2. Effect of pyrethroids against *Heliothis* spp. in cotton. Ruscoe (1979). 1 = deltamethrin; 2 = cypermethrin; 3 = permethrin

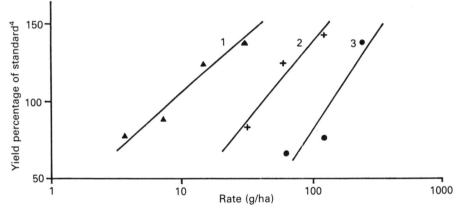

Figure 6.3. Effect of pyrethroids on cotton yield. Ruscoe (1979). 1 = deltamethrin; 2 = cypermethrin; 3 = permethrin; 4 = EPN + methyl parathion, 560 + 560 g/ha

et al. (1981) in Tadjikistan (USSR) in trials on *H. armigera*. In the same trials the number of plant-sucking pests (*Aphis, Acyrthosiphon*) and mites (*T. urticae*) were shown to increase. Limitations of the use of pyrethroids for controlling plant-sucking pests have been pointed out by several authors, for example Angelini & Couilloud (1976), Cauquil & Guillaumont (1979), Delattre (1978), and Lhoste & Piedallu (1977), who advocated combinations using triazophos, chlorpyrifos and monocrotophos to achieve broad-spectrum control. Today these combinations are widely used as preformulated products.

Many authors consider that the use of mixtures delays the build up of resistance and prevents it from reaching unacceptable levels. A recent study by

McDonald et al. (1983) showed that the best way to keep *Musca domestica* from developing resistance was to use a combination of dichlorphos and permethrin, instead of alternating one after the other, or using either one separately. This is contrary to the observations of other entomologists who favour alternation strategies. The problem is likely to remain unresolved until further studies on field-resistant insects can be undertaken.

One of the aspects of *Heliothis* control developed by several authors is the fact that fewer applications of pyrethroid are necessary than in pest control treatments using standard products. This finding was supported by Bastos et al. (1979) and by Piedallu & Roa (unpublished results). Although this is a positive advantage for pyrethroids this argument has not always been put forward by promoters of cotton plant protection for a number of reasons which deserve explanation.

In several examples of cotton pest control using pyrethroids, cotton has been treated incorrectly because the scouting methods used for these products were based on the conventional methods used for organophosphates, showing a lack of understanding of the mode of action of the pyrethroids. In other cases, spraying problems in aerial treatment have been mistakenly attributed to product quality, and consideration has not been given to what may have been an inappropriate spray regime.

Pyrethroids are active against all larval stages of *Heliothis* and most commercial treatment programmes are based on scouting results for eggs and first instar larvae. But pyrethroids are not effective as ovicides, and first instar larvae penetrate cotton squares shortly after hatching where they are not available for control with non-mobile insecticides. Thus a strategy which can provide good control but uses fewer spray treatments can be based on scouting for first to fourth instar larvae only as shown in Figure 6.4.

An alternative and highly effective strategy was proposed by Morton (1979) and Morton et al. (1981) in which sprays were related to plant growth stage and not applied in response to insect damage levels or numbers, thus providing a "toxic carpet" to kill first instar *Heliothis* soon after egg-hatch.

Both the infestation level and the occurrence of pest species with different levels of susceptibility are important factors in determining pyrethroid rates. This was clearly demonstrated on the Ivory Coast by Angelini et al. (1982), who showed that 10 g/ha of deltamethrin was equivalent to 30 g/ha of cypermethrin and 50 g/ha of fenvalerate against light to moderate attack of *Pectinophora gossypiella* in cotton. In instances where *Heliothis armigera* was also present, higher rates of 15 g/ha of deltamethrin or 75 g/ha of fenvalerate were needed to produce excellent results; cypermethrin at 60 g/ha was less effective in this trial (Figure 6.5). In general, however, cypermethrin at 30 g/ha can provide acceptable protection against low level attacks of *Heliothis*. Relatively high doses of pyrethroids are required for the control of *H. virescens* (tobacco budworm) and *H. zea* (cotton bollworm) in US cotton. In three tests located in Mississippi, Pitts & Pieters (1981) compared a number of synthetic

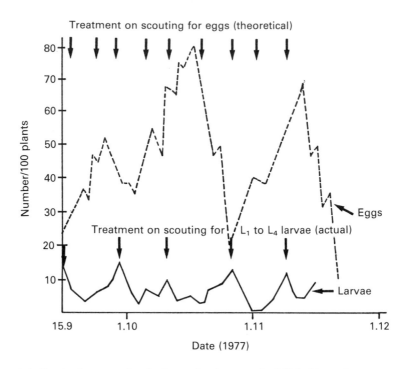

Figure 6.4. Results from scouting for larvae for the control of *Heliothis armigera* on cotton in Guatemala, using deltamethrin at 11 g a.i./ha. (Piedallu & Roa, 1982)

pyrethroids, all demonstrating excellent control of heavy *Heliothis* infestations (Table 6.4).

An important factor which may explain the variation in dosage from one location to another is the influence of temperature on pyrethroid activity. This was particularly well demonstrated on *H. virescens* by Whitney & Wettshein (1979) (Table 6.5). Sparks et al. (1982) confirmed that for most insects studied, pyrethroids have a negative temperature coefficient. However in the case of *H. virescens* these workers obtained results which actually indicated that fenvalerate has a neutral temperature coefficient and deltamethrin a positive coefficient. Experience shows that for relatively high levels of *Heliothis* infestation in cotton, effective control is obtained using 10 to 15 g/ha of deltamethrin, 50 to 70 g/ha of cypermethrin, 100 to 150 g/ha of permethrin, 75 to 110 g/ha of flucythrinate or 50 to 75 g/ha of fluvalinate, for an average protection of two weeks. Similar doses were recommended by Wolfenbarger & Harding (1982) for *Heliothis spp.* control in Texas (USA).

Heliothis is polyphagous and may be found on several different plants such as corn, sorghum, tomatoes, tobacco, eggplants and others. While damage to these plants may be considerable, it is much easier to control than damage to cotton plants. In general, the doses used on these crops are 20 to 25% lower than those recommended for cotton.

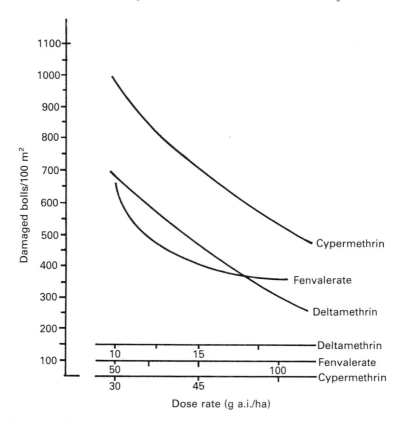

Figure 6.5. The dose response of pyrethroids controlling insects on the Ivory Coast. Angelini et al. (1982)

To conclude this brief review of the use of pyrethroids against *Heliothis* spp. it may be stated that all major pyrethroids are effective in controlling these pests, the only differences being in the recommended dose. The dose of each synthetic pyrethroid depends on certain local conditions which may be summarised as follows:

– Insect pressure
– Environmental conditions
– Application methods

While *Heliothis* remains a considerable problem, other Lepidoptera continue to draw attention. Most of these pests are defoliating caterpillars which, like *Heliothis*, are polyphagous; and it has been shown that their susceptibility to insecticides can be greatly influenced by the nature of their diet. Various authors have conducted studies on *Spodoptera littoralis, S. exigua* and *S. frugiperda;* species that often develop resistance, or that require much

J. J. Hervé

Table 6.4. Comparison of pyrethroids for the control of *Heliothis sp.* in cotton. Pitts & Pieters (1981)

Treatment[a] and lb a.i./acre	Percent damaged squares[b,c]	Live larvae/100 squares	Yield (lb seed cotton/acre)
Trial no. 1			
Fenvalerate 2·4 EC 0·1	4 a	0·8 a	2797 a
Permethrin 3·2 EC 0·1	8 a	3·6 bcd	2333 bc
Permethrin 2· EC 0·1	10 a	4·8 d	2058 cd
Untreated	68 d	7·2 e	191 g
Trial no. 2			
Cypermethrin 3 EC 0·06	8 a	0·4 a	2305 a
Fenvalerate 2·4 EC 0·1	10 a	0·4 a	2287 a
Cyfluthrin 1·67 EC 0·088	10 a	0·8 a	2235 a
Cypermethrin 3 EC 0·12	8 a	0·4 a	2187 a
Flucythrinate 2·5 EC 0·04	10 a	0·4 a	2126 a
Flucythrinate 2·5 EC 0·025	11 a	1·6 a	2097 a
Fluvalinate 2·0 EC 0·1	11 a	0·8 a	
Untreated	56 b	6·8 b	151 b
Trial no. 3			
Fenvalerate 3·4 EC 0·1	13 a	0·4 a	2296 a
Tralomethrin ·3 EC 0·02	17 a	1·2 a	2256 a
Untreated	61 e	4·8 bc	96 e

[a] Insecticides applied: July 14, 18, 24, 28 Aug. 1, 6, 11, 15, 20, 25, 29 and Sept. 3.
[b] Means for 5 dates: July 23, 31, Aug. 5, 14, 19.
[c] Means followed by the same letter are not significantly different ($P = 0.05$) using Duncan's multiple range test.

Table 6.5. Effects of post-treatment temperature on pyrethroid toxicity to *Heliothis virescens*. Whitney & Wettshein (1979)

Insecticides	LD 50 (ng/3rd instar larva)		
	7 °C	27 °C	38 °C
Flucythrinate	7·2	13·2	15·2
Deltamethrin	1·4	3·6	3·1
Fenvalerate	13·2	20·1	31·0
Permethrin	14·1	31·0	108·4

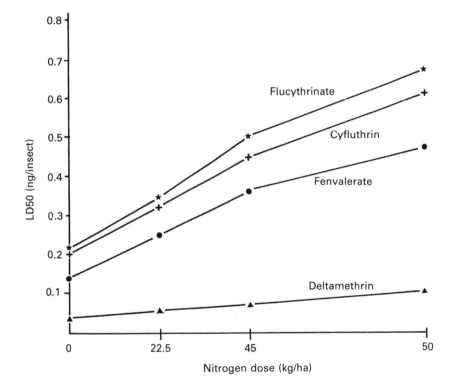

Figure 6.6. Effect of rate of nitrogenous fertilisation on the LD50 value of some pyrethroids against fourth instar larvae of cotton leafworm. Hervé et al. (1983)

higher-than-average doses for control than other Lepidoptera. Hervé et al. (1983) showed that temperature, photoperiod, diet and nitrogenous fertilisation all affected the insecticidal susceptibility of these pests. For instance, when nitrogen doses (especially those applied as foliar spray) were increased, the required dose of pyrethroid also increased 3- to 4-fold (Figure 6.6). Similarly,

Spodoptera reared on tomatoes are much more susceptible to a wide range of pyrethroids than insects raised on cotton, which, in turn, are more susceptible than those reared on cow peas (Table 6.6). An explanation of the varying susceptibility of different strains was proposed by Ishaaya et al. (1983) who demonstrated that profenophos acts as a synergist to the pyrethroids inhibiting the esterases of the insects. These esterases vary greatly from one strain to another, especially when their diets are different. Testing is currently in progress to assess the combined effects on insecticidal susceptibility of the interaction between temperature, photoperiod, nitrogenous fertilisers and diet.

Spodoptera littoralis is considered to be the most serious pest infestation in Egypt, and as a result it has been the object of numerous studies. As early as 1977, Ford et al. were conducting experiments using ten different synthetic pyrethroids on two different strains, one resistant to and the other susceptible to the organophosphate, parathion. The authors were able to find no evidence of cross-resistance between the two chemical groups. El Okda et al. (1978) compared the toxicity levels of 4 synthetic pyrethroids, 8 organophosphates, 2 organochlorines, and 1 carbamate. Their conclusions showed that the substances most toxic to *Spodoptera littoralis* were the pyrethroids. However the tests also showed that cypermethrin and fenvalerate actually inhibited the action of the tested non-pyrethroid compounds. All et al. (1977) reported findings on a strain with resistance to permethrin which also proved to have a cross-resistance to cypermethrin and fenvalerate. The use of high dose rates may have been an important factor in selecting the potential for resistance. The type of test used in Egypt to determine pyrethroid efficacy can result in recommended doses that are perhaps excessive. This results from the fact that Egyptian official tests only consider toxic effects caused by ingestion or residual contact. This is potentially worrying since, as Atkins & Kellum in 1978, Ruscoe in 1977 and Hervé in 1982 reported, pyrethroids also have an extremely repellent effect on insects. Saad et al. (1981) confirmed this repellent effect in experiments which demonstrated that adults avoided treated leaves when ovipositing (Table 6.7).

El-Guindy et al. (1981) reported that in Egypt strains resistant to all the major groups of insecticides could be found, and that studies were underway to find mixtures that might overcome resistance. These authors demonstrated that synergistic effects observed in susceptible strains could also be found in resistant strains, especially when using combinations of cypermethrin and chlordimeform or cypermethrin and endrin or methomyl. Synergy was not found when fenvalerate was combined with chlordimeform. However, Zidan et al. (1981) considered that combinations of synthetic pyrethroids and organophosphates were always antagonistic against resistant insects. These results were contested by Saad et al. (1981) who demonstrated that equal parts of chlorpyrifos and cypermethrin gave an additive effect, although delta-methrin and phosfolan were antagonistic, as were fenvalerate and chlorpyrifos.

Table 6.6. Variation in susceptibility of cotton leaf worm larvae to pyrethroid insecticides when reared for three generations on different host plants in Egypt. LD50 (ng/insect). Hervé et al. (1983)

	Deltamethrin	Fenvalerate	Cyfloxylate	Flucythrinate
Cow pea	0·228	1·120	1·380	1·500
Soybean	0·185	0·930	1·100	1·200
Cotton	0·073	0·364	0·460	0·500
Castor bean	0·052	0·273	0·320	0·360
Clover	0·048	0·250	0·280	0·340
Grape vine	0·043	0·232	0·260	0·300
Tomato	0·032	0·150	0·190	0·210
Ratio of maximum to minimum rate	7·125	7·41	7·26	7·14
Ratio relative to cotton	1	4·9	6·3	6·84

Table 6.7. Repellency and toxicity of cyfluthrinate and other insecticides to adult *Spodoptera littoralis*[a]. Saad et al. (1981)

Treatment	Dose (g a.i./ha)	Percentage of egg masses laid on treated plants	Total no. egg masses/20 moths	Moth mortality (%) at day				
				1	2	3	4	5
Cyfluthrinate 30% EC	54	6	4·25	50	76	93	94	100
	71	9	2·75	51	80	94	95	100
	107	0	2·25	45	60	75	96	100
	143	0	2·00	49	59	85	96	100
Fenvalerate 20% EC	286	40	17·50	58	80	95	100	100
Cypermethrin 30% EC	143	40	10·00	45	65	95	100	100
Deltamethrin 2·5% EC	45	7	8·00	63	98	98	100	100
Phospholan 25% EC	893	22	4·50	34	50	800	93	100
Mephosfolan 75% EC	1071	20	2·50	56	88	94	100	100
Chlorpyrifos 48% EC	1143	36	11·00	45	75	90	100	100
D.C. 702[b] WP	1143 + 71	44	12·50	23	60	95	100	90
Untreated	—	—	17·00	33	62	77	90	100

[a] Free choice cage test with 20 moths on two treated and untreated cotton plants in the same cage.
[b] 38·4% chlorpyrifos + 2·4% diflubenzuron.

Flowering start date

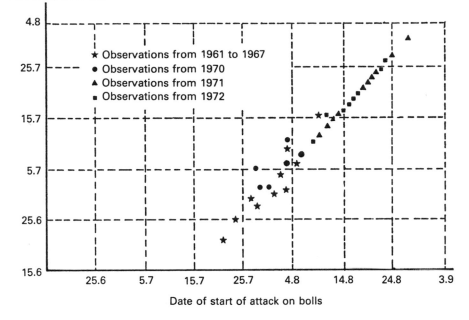

Figure 6.7. Correlation between start of flowering on cotton and start of *P.gossypiella* attack on bolls in Morocco. Le Rumeur (1973)

They also found synergism between equal parts of two non-pyrethroid compounds, chlorpyrifos and diflubenzuron, whereas diflubenzuron and cypermethrin proved to be antagonistic. It is thus difficult to state firmly which specific pyrethroid combinations should be used to control resistant *Spodoptera littoralis*. Despite the potential for resistance, the value of pyrethroids is recognised and these products are still used throughout the Nile Valley in Egypt, albeit limited to a single "spray round" and not for use in mixtures.

In addition to *Heliothis* and *Spodoptera,* there are several species of Lepidoptera that cause considerable damage to cotton plants. The third species to be taken into consideration is *Pectinophora gossypiella* (the pink bollworm). This insect is found on all continents. This dangerous cotton pest attacks buds and bolls, damaging and soiling fibres, and gnawing on the seeds. This insect is well known for its ability to develop resistance, a characteristic which was often reported before synthetic pyrethroids were introduced on the market. Extensive studies carried out in Morocco by Le Rumeur (1973) revealed that attacks are highly synchronised with the development of cotton bolls and occur in parallel with the flowering process of the cotton plants (Figure 6.7). The peak of the boll attacks is reached approximately four weeks after flowering. The effective use of pyrethroids requires that two applications be made at very precise moments: 3 weeks after flowering and 10–12 days after the first application. While a definite relationship exists between the development of the plant and the maturation of the pest in Morocco, this is certainly not the case in all other countries. It is therefore essential to study the life cycle of *Pectinophora* in order to plan rational and efficient control programmes, bearing in mind that a large number of failures are caused by taking action against pink bollworm infestations when it is already too late, and the larvae are protected within the cotton bolls.

Butter et al. (1982), working in India, found permethrin and fenvalerate to be much more effective than the standard treatment (carbaryl, applied 4 times at 10-day intervals at a rate of 1·25 kg/ha) for pink bollworm control. They observed that high doses (100 g/ha) were more effective and resulted in greater yield than low doses (50 g/ha), even though assessment of the results was complicated by simultaneous attack of *Earias* sp and *H. armigera*. In other cases, Angelini & Couilloud (1976) working in the Ivory Coast where pink bollworm infestations are much lighter than in India, found it was easier to control *P. gossypiella* than *H. armigera*. They recommended doses of 10 g/ha of deltamethrin, 30 g/ha of cypermethrin and 50 g/ha of fenvalerate, as compared with 15, 70 and 75 g/ha respectively against *H. armigera*. Tests conducted by Keerthisinghe (1982) in Sri Lanka concurred with the results of Angelini & Couilloud. Keerthisinghe recommended using 100 g/ha, which although the lowest pyrethroid dose used during the experiment, was much more effective than the standard treatment (monocrotophos, applied 8 to 10 times at doses of 500 g/ha) in reducing boll damage.

In India, where *P. gossypiella* is a very serious problem, Jayaswal & Saini (1981) recognised that pyrethroids were much more effective than carbaryl applied at 1000 g/ha, but they found it was necessary to use high doses. They compared 75 and 100 g/ha rates for both fenvalerate and permethrin, 40 and 80 g/ha for cypermethrin, and 10 and 20 g/ha for deltamethrin. The best result was achieved with 20 g/ha of deltamethrin which provided a yield increase of 4·1 bushels per hectare over the carbaryl treatment. In California, an interesting technique was proposed by Butler & Las (1983) which consisted of combining an adhesive (320 g/ha) with gossyplure (1·41 g/ha) (the sex pheromone of the pink bollworm) and a very low dose of permethrin (3·8 g/ha) to enhance the effects of the pheromone by killing male moths. This treatment was far safer to beneficial insects than standard insecticide applications. However, it is unfortunate that the results did not demonstrate the real benefit of these "alternative" measures against *P. gossypiella* by comparing the resulting yields with those obtained using standard chemical treatment.

In the field there are large numbers of other Lepidoptera species that are dangerous to the cotton plant, such as *Earias* spp., *Alabama*, *Bucculatrix* and *Diaropsis*. These pests are not especially troublesome however, since they are kept well under control by the treatments used on the three species discussed above. For example, the doses of active ingredient used against *Heliothis*, or

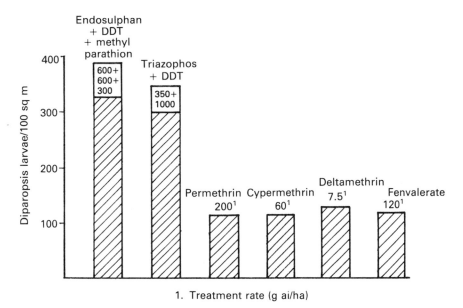

1. Treatment rate (g ai/ha)

Figure 6.8. Effect of pyrethroids (8 applications, 14-day intervals) against red bollworm (*Diparopsis watersi*) in Chad. Ruscoe (1979)

even smaller amounts, are sufficient, as shown in the results presented by Ruscoe (1979), to control the red bollworm (*Diparopsis watersi*) (Figure 6.8). However, there remains one species that is difficult to control in West Africa: *Sylepta*, which rolls itself up in plant leaves; this problem was disclosed by Delattre in 1978. At present conventional systemic or fumigant insecticides are needed to control this insect.

Lepidoptera attack not only cotton but also other important agricultural crops. In South America, two Lepidoptera, in addition to *Heliothis* and *Spodoptera*, are a serious threat to soybean crops, namely *Anticarsia gemmatalis* and *Pseudoplusia includens*. Piedallu & Roa (1982) demonstrated that effective doses of deltamethrin varied between 3 and 7 g/ha, depending on the degree of infestation and the density of the vegetation. As a reference, the treatment threshold level recommended by the EMBRAPA in Brazil is 40 larvae (1·5 cm long) per hectare obtained through taking 6 counts over 10 ha, or 30% defoliation before flowering and 15% after flowering. Table 6.8 gives some examples of the results obtained in Brazil, showing that late-season infestation decreased sharply in the untreated controls. This is due primarily to rain, but more importantly to the often explosive multiplication of the entomophagous fungus *Nomurea*. Research work is being conducted on insecticide combinations which should make it possible, as was the case with endosulfan, to optimise the effectiveness of pyrethroids against these two pests. By increasing the dose of each component, such combinations provide

Table 6.8. Results against *Anticarsia gemmatalis* and *Pseudoplusia includens* on soya in Brazil. Piedallu & Roa (1982)

Insecticide	g a.i./ ha	Trial I[a]				Trial II[b]			
		Numb. larvae count[c]	% reduction of infestation			Numb. larvae count[c]	% reduction of infestation[d]		
		D 0	D + 3	D + 7	D + 14	D 0	D + 2	D + 7	D + 16
Deltamethrin 2·5%	2	67·3	63·6	79·2	75·3	81·6	87·2	89·9	75·7
	2·5	—				77·6	90·9	91·7	81·2
	3	—				75·2	85·7	93·2	78·9
	4	65·3	85·2	94·3	90·0	—			
Deltamethrin + endosulfan 0·8 + 32%	1·5 + 59	—				75·2	95·2	96·7	81·6
Deltamethrin + endosulfan 0·8 + 32%	2 + 80	60·8	88·7	97·9	89·0	—			
Endosulfan 35%	175	63·3	97·7	98·7	82·8	99·3	96·6	99·3	95·3
Control	—	73·3	(71·7)	(39)	(28·8)	89·3	(91·1)	(47·7)	(13·5)

[a] 1 treatment only – Trial I 50 days after sowing.
[b] 1 treatment only – Trial II 53 days after sowing.
[c] Official EMBRAPA method.
[d] Percentage infestation calculated using the formula of Henderson & Tilton.

effective control of other pests (such as Hemiptera) which are sometimes present at the same time as *Anticarsia* and *Pseudoplusia*.

Synthetic pyrethroids have also proven to be very effective against another important Lepidoptera, the European corn borer (*Ostrinia nubilalis*). In this case, successful control depends on correct timing, since the treatment must be applied before the larvae have penetrated the plant stems. The importance of precise timing was demonstrated by Cranshaw & Radcliffe (1983) in Minnesota where they treated a bean crop with flucythrinate on 19, 23 and 26 July. As their results show, it is imperative that synthetic pyrethroids are applied as a preventative measure, unlike organophosphates, such as acephate or parathion, which may be used as a curative treatment because of their vapour pressure effect (Table 6.9).

As early as 1979, Hofmaster & Francis reported that all pyrethroids were effective against first generation European corn borer when used at rates of 0·01 lb/acre for deltamethrin and 0·1 lb/acre for fenvalerate or permethrin. Damage to corn and potato crops was reported by these authors to have fallen by 95% with respect to the untreated control plots. Witkowski, working alone (1979) and then in collaboration, Koziol & Witkowski (1982), demonstrated the advantage of using pyrethroids in combination with parathion or chlorpyrifos to control European corn borer (Table 6.10). The possibility of controlling European corn borer through spraying rather than granules applied when male panicles appear is of great interest to farmers, and opens the way to aerial treatments of the second generation. In some situations, however, certain species of aphid have a tendency to multiply on corn crops after application of a liquid pyrethroid formulation. This implies that crops must be observed closely for 20 days after treatment. No clear explanation has been found for this phenomenon which should be studied using a biochemical rather than a population dynamics approach. A solution to this problem may lie in the use of combinations, such as those suggested by Koziol & Witkowski. The successful control of another borer, *Sesamia*, is determined by the timing of application. So far, experiments using pyrethroids on corn against *Sesamia* have reaped more losses than gains. This also applies for the control of first generation rice borers (*Chilo suppressalis, Tryporyza*) although treatments applied on the second generation, after plant flowering, have proven successful, especially in the case of *Chilo suppressalis*.

Some studies conducted on cabbage also deserve mention. They concern the three major Lepidoptera that attack cabbage crops: *Plutella*, *Pieris* and *Trichoplusia ni*. The first study in question was carried out in Maryland in 1980 and reported in 1981 by Linduska & Bagley as summarised in Table 6.11. Results were obtained from cabbages heavily infested by *Pieris rapae* (imported cabbageworm) and *Trichoplusia ni* (cabbage looper). The high injury ratings of the untreated heads (50 to 100%, means that the cabbages cannot be marketed) proved the high effectiveness of the tested synthetic pyrethroids, even at low doses.

Table 6.9. The efficacy of insecticides used to control an artificial infestation[a] of *Ostrinia nubilalis* in relation to the date of application to a bean crop.

Treatments[b]	Dose (a.i./lb/acre)	Treatment date	No. tunnels		Living[c] Borers
			Stems	Pods	
Acephate	0·5	July 23	16	4	7 a
Flucythrinate	0·08	July 19	10	3	8 a
Flucythrinate	0·04	July 19	19	3	9 a
Flucythrinate	0·08	July 23	16	11	16 ab
Flucythrinate	0·04	July 23	27	10	18 ab
Flucythrinate	0·08	July 26	39	16	27 ab
Flucythrinate	0·04	July 26	49	20	27 ab
Fenvalerate	0·10	July 23	67	13	39 b
Untreated	—	—	112	39	65 c

[a] 10 plants from the centre of each row were artifically infested on 23 July with 2 egg masses.
[b] Insecticides were applied with a CO_2 powered sprayer delivering 20 gal/acre at ca 25 psi.
[c] Counts are totals for 3 replicates, number followed by the same letter are not significantly different ($P = 0.05$).

Table 6.10. Dosage-mortality and synergism data for permethrin in combination with methyl parathion, chlorpyrifos and malathion, and for the insecticides alone, against European corn borer larvae. Koziol & Witkowski (1982)

Insecticide mixture	Ratio	LD50 (μg/g larval wt)[a]	95% fiducial limits	Slope	Joint toxicity coefficient ± SE[b]	Synergistic activity[c]
Permethrin	1:0	9.47	8.94–9.99	2.88	—	—
Methyl-parathion	1:0	25.0	23.8–26.6	2.20	—	—
Methyl-parathion:permethrin	9:1	1.84a	1.77–1.91	2.98	1.06 ± 0.05 *	11.5
Methyl-parathion:permethrin	4:1	0.37b	0.32–0.42	2.63	1.69 ± 0.08 *	48.9
Methyl-parathion:permethrin	2.3:1	0.34b	0.30–0.37	3.02	1.70 ± 0.07 *	50.3
Methyl-parathion:permethrin	1.5:1	0.28b	0.24–0.33	2.89	1.71 ± 0.07 *	51.8
Chlorpyrifos	1:0	51.3	48.7–53.9	3.06	—	—
Chlorpyrifos:permethrin	9:1	1.78a	1.52–2.04	3.72	1.29 ± 0.03 *	19.9
Chlorpyrifos:permethrin	4:1	2.04a	1.66–2.43	3.21	1.13 ± 0.05 *	13.5
Chlorpyrifos:permethrin	2.3:1	1.84a	1.46–2.23	2.94	1.09 ± 0.03 *	12.2
Chlorpyrifos:permethrin	1.5:1	1.91a	1.49–2.33	3.12	0.98 ± 0.06 *	9.6
Malathion	1:0	1015.8	811.9–1219.7	1.34	—	—
Malathion:permethrin	9:1	565.8	548.0–581.6	2.08	-0.81	—
Malathion:permethrin	4:1	750.0	742.8–759.2	1.49	-1.21	—
Malathion:permethrin	2.3:1	109.2	108.5–109.9	1.87	-0.56	—
Malathion:permethrin	1.5:1	67.1	66.6–67.6	2.58	-0.45	—

[a] LD50 values for a given insecticide:permethrin group followed by a common letter are not significantly different at the 5% level of probability based on the 95% fiducial limits.

[b] >0 = synergism, <0 = lack of synergism. Joint toxicity coefficients followed by asterisks are significantly greater than 0 at the 5% level of probability by the t-test.

[c] Synergistic activity = LD50 (predicted)/LD50 (observed).

Table 6.11. Efficacy of various insecticides against major cabbage pests. Linduska & Bagley (1981)

Treatment and dose (lb a.i./acre)	Mean foliage injury rating[a] 10 plants × 4 reps = 40 plants
	Sep. 9, 80[b]
Untreated check	4·90 d
Methomyl (1·8 L) 0·45	2·40 ab
Methomyl (1·8 L) 0·9	2·40 ab
Oxamyl (2·0 L) 0·5	4·30 cd
Oxamyl (2·0 L) 0·25 + Methomyl (1·8 L) 0·225	3·33 bd
Thiodicarb (4·18 EC) 0·225	3·13 ac
Thiodicarb (4·18 EC) 0·45	2·38 ab
Larvin 500 (4·18 EC) 0·75	2·48 ab
Flucythrinate (2·5 EC) 0·03	1·30 a
Tralomethrin (0·3 EC) 0·0108	2·08 ab
Tralomethrin (0·3 EC) 0·0134	1·50 ab
Tralomethrin (0·3 EC) 0·0156	1·38 a
Tralomethrin (0·3 EC) 0·0269	1·25 a
Cypermethrin (3·0 EC) 0·03	1·73 ab
Cypermethrin (3·0 EC) 0·06	1·28 a
Permethrin (2·0 EC) 0·1	1·20 a
Fenvalerate (2·4 EC) 0·1	1·40 a
Permethrin (3·2 OC) 0·1	1·43 ab
Acephate (75 S) 0·75	1·28 a
Methomyl (1·8 L) 0·5 + fenvalerate (2·4 EC) 0·5	1·50 ab
Methomyl (1·8 L) + acephate (75 S) 0·25	2·75 ac
Cyfluthrin (1·66 EC) 0·01125	1·88 ab
Cyfluthrin (1·66 EC) 0·0225	1·48 ab
Cyfluthrin (1·66 EC) 0·04437	1·50 ab
Methamidophos (4 WM) 0·75	1·50 ab

[a] Injury rating: 1 = 0–3% damaged leaves, 2 = 4–10% damaged leaves, 3 = 11–25% damaged leaves, 4 = 26–50% damaged leaves, 5 = 51–100% damaged leaves.
[b] Any two numbers in the same column followed by the same letter are not significantly different as determined by Duncan's multiple range test at the 5% level of significance.

An excellent study on *Pieris brassicae* (cabbage white butterfly) was conducted by Tan (1982) who tested the effects of pyrethroids when applied at sublethal rates. Treatments using cypermethrin and permethrin caused an estimated 50% decrease in the feeding damage caused by *Pieris*. The results in Tables 6.12 and 6.13 clearly illustrate the secondary anti-feeding effects of low doses of pyrethroids. An interesting associated phenomenon is the delay of pupation, perhaps through malnutrition or other side effects not as yet identified. Such results support the relationship between dose and the mode of action of pyrethroids proposed by Ruscoe (1977) and re-iterated by Hervé (1982), illustrated in Figure 6.9. As suggested by Tan, these sublethal effects

Table 6.12. Calculated antifeeding and mortality probit regression equation; ED50 and LD50 values of cypermethrin and permethrin, as treated leaf-discs, against fifth-instar larvae of *Pieris brassicae*. Tan (1982)

Pyrethroid	Response	Regression equation	x^2	D.f.	ED50 or LD50 (p.p.m.)	95% fiducial limits	
						Upper	Lower
Cypermethrin	Antifeeding	$y = 5.59 + 2.20\,x$	5.46	3	0.54	0.64	0.46
	Mortality	$y = 1.29 + 6.68\,x$	2.37	3	3.60	3.77	3.40
Permethrin	Antifeeding	$y = 5.69 + 2.18\,x$	0.21	3	0.48	0.50	0.46
	Mortality	$y = 0.79 + 10.34\,x$	1.29	3	2.55	2.66	2.45

ED50 is the dose required to induce 50% reduction in leaf consumption.
LD50 is the dose required to kill 50% of the test species.

Table 6.13. Maximum weight attained by larvae and pupae: pupation delay and total leaf area consumed when 1-day-old fifth-instar larvae of *Pieris brassicae* were exposed to pyrethroid-treated leaf-discs (1 p.p.m.) throughout development. Tan (1982)

Compound	No. larva	Mortality (4–7 days) No. larva	Maximum weight attained (mg)		Total leaf area consumed per larva (mm^2)	Time of pupation after last moult (days)
			Larva	Pupa		
			Mean ± SD	Mean ± SD	Mean ± SD	Mean ± SD
Control	9	0	406·5 ± 39·4	362·5 ± 43·7	4634 ± 444	4·9 ± 0·3
Cypermethrin	10	0	353·1 ± 30·6 ($P = 0.004$)	303·1 ± 21·7 ($P = 0.001$)	3954 ± 434 ($P < 0.001$)	9·7 ± 0·8 ($P < 0.001$)
Permethrin	10	3	325·4 ± 28·3 ($P < 0.001$)	267·8 ± 32·8 ($P < 0.001$)	3969 ± 345 ($P < 0.005$)	8·0 ± 0·8 ($P < 0.001$)

P values in parentheses are the probabilities for the comparison of the means with the respective control.

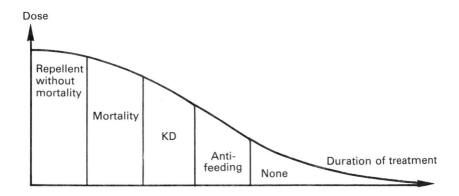

Figure 6.9. Theoretical representation of effects of one application on several species. Hervé (1982), adapted from Ruscoe (1977). KD, knockdown

may explain why pyrethroids have been so successful compared with more conventional insecticides over the last few years.

Plutella xylostella (the diamond-backed moth) can be difficult to control. This pest was studied in 1983 by Ho et al., who tested two different strains: one from Singapore which was resistant (R) and the other from Malaysia which was susceptible (S). The two strains were taken from glasshouse *Brassica chinensis* plants; only the adult pests had been raised on Holoway's artificial medium. Deltamethrin and cypermethrin, often used alone or in combination with several other types of insecticide, were the synthetic pyrethroids used in testing. Their results are presented in Table 6.14 and show the high levels of resistance reached. The authors showed that a slight synergistic effect (6 times) was produced between deltamethrin and piperonyl butoxide (PB) on 5 different strains when a deltamethrin/PB ratio of 10:1 was applied. This effect is much stronger (182 times) on R strains which leads one to believe that microsomal oxidases play an important role in this process (Table 6.15). As Ho mentions, however, it is important to draw attention to the fact that *P. xylostella* is prone to insecticide resistance because of selection pressure created by the climatic conditions of South-east Asia, which can foster up to 30 generations per year needing continuous treatment. It is also important to note that these pests have no diapause, that crop rotation is not practised, and that at the time of the experiment these strains had been subjected to insecticide sprayings almost every 3 days for the past year.

Other market-garden crops often damaged by Lepidoptera are tomatoes and eggplants which are commonly attacked by *H. armigera* and *H. zea*. These pests are controlled by using doses of insecticide that are generally 10 to 20% lower than those used on cotton plants. Other pests, including *Lyriomyza sativae* (vegetable leaf miner), *Spodoptera exigua* (beet armyworm) and *Keiferia lycopersicella* (tomato pinworm) are also present on these plants. Tests

Table 6.14. Data from probit analysis for topical application of deltamethrin and cypermethrin to two different strains of *Plutella xylostella* (at 48 h). Ho et al. (1983)

Insecticide	Strain	Slope of probit line (b ± SE)	LD50 ± SE (µg a.i./larva)	95% confidence limit		LD95 (µg a.i./larva)	RF[a] (LD 50)
				Lower	Upper		
Deltamethrin	S	1·21 ± 0·08	0·0014 ± 0·0011	0·0011	0·0019	0·034	
	R	0·69 ± 0·09	2·301 ± 0·0116	1·052	4·699	501·2	1644
Cypermethrin	S	1·35 ± 0·13	0·0046 ± 0·0001	0·0036	0·0058	0·68	
	R	1·50 ± 0·21	169·82 ± 1·05	125·89	229·09	2188	36917

[a] RF, Resistance Factor = $\dfrac{\text{LD50 of R strain}}{\text{LD50 of S strain}}$

Table 6.15. Effect of the synergist piperonyl butoxide on the activity of deltamethrin against 4th instar larvae of *P. xylostella* (at 48 h). Ho et al. (1983)

Strain	Treatment	LD50 (μg/larva)	SF	LD95 (μg/larva)	SF[a]
S	Deltamethrin alone	0·0014	—	0·034	—
	D:PB (1:10)	0·00095	1·5	0·0055	6·2
R	Deltamethrin alone	2·30	—	501·2	—
	D:PB (1:1)	0·24	9·6	5·50	91
	D:PB (1:10)	0·072	32	2·75	182

[a] Synergy factor = $\dfrac{\text{LD50 (or LD95) of deltamethrin}}{\text{LD50 (or LD95) of deltamethrin + synergist}}$

(Taking PB to be non-insecticidal *per se*).

involving these pests were conducted in California in 1981 by Van Steenvyk & Hayashi (results are shown in Table 6.16). External injury was due essentially to the beet armyworm. The difference between the total percentage of damage and the damage caused by both the armyworm and the pinworm is accounted for by *Heliothis zea*. It appears evident that all the pyrethroids (except for ZR 3210, fluvalinate) were highly effective against every type of insect tested, which is all the more remarkable considering the rates used. Nonetheless, the authors did note a definite rise in the *Auchops lycopersici* (mite) population, a phenomenon which had already been observed on cotton plants and is generally attributed to the low level of effectiveness of these compounds on phytophagous mites.

In India, excellent results have been obtained using pyrethroids against *Leucinodes orbonalis* (Brinjal shoot and fruit borer) on eggplant crops. Kuppuswany & Balsubramanian (1980) and Basha et al. (1982) rated deltamethrin as the most effective insecticide at doses as low as 0.005%. Tests showed sizeable increases in crop yield, rising from 5 271 kg on untreated plots to 19 360 kg on the treated ones. However, these authors also reported a rise in the *T. cinnabarinus* (mite) population, except when fenpropathrin was used. But this was also accompanied by a notable decrease in the number of certain Cicadellidae which were better controlled using pyrethroids than by using methomyl or Neem oil extracts.

A review of Lepidoptera control programmes would not be complete without a word on arboriculture and viticulture. Excellent results in orchards were reported by Pastre et al. (1978) and Ruscoe (1979) as is illustrated in Figure 6.10. An interesting report summarising the potential of first generation pyrethroid use in orchards was presented by Vanwetswinkel & Seutin (1978). Tables 6.17 and 6.18 show the results obtained on several types of Lepidoptera such as *Clysia*, *Operophtera* and *Adoxophyes*. One of the conclusions drawn by

Table 6.16. Control of lepidopterous larvae and vegetable leafminer on tomatoes. Van Steenvyk & Hayashi (1981)

Treatment and dose (lb a.i./acre)	Percentage damage as a seasonal average by:			
	Lepidoptera			Leafminer
	Total damage	Pinworm	External	Pupae/3 days
Fenvalerate 2·4 EC 0·1	0·9 a	0·2 a	0·5 a	0·6 a
Permethrin 3·2 EC 0·2	1·3 ab	0·5 ab	1·2 a	1·3 a
Permethrin 3·2 EC 0·1	2·2 ab	0·6 ab	1·5 ab	1·9 a
Permethrin 2·0 EC 0·2	2·6 ab	0·6 ab	1·8 ab	0·8 a
Methomyl 1·8 L 0·45	2·8 ab	1·0 abc	1·3 a	31·5 b
Fenvalerate 2·4 EC 0·2	3·0 ab	0·2 a	2·3 abc	1·1 a
Permethrin 2·0 EC 0·1	3·4 b	0·6 ab	1·6 ab	0·9 a
Permethrin 3·2 EC 0·05	3·7 b	1·7 bcd	1·5 ab	1·7 a
Fluvalinate 2·0 EC 0·1	9·0 c	1·9 bcd	5·6 cd	5·3 a
Fluvalinate 2·0 EC 0·2	9·2 c	0·6 ab	7·2 d	3·3 a
Dipel[a] 16·000 IU/mg WP 1·0	10·7 c	3·1 cd	4·8 bcd	4·4 a
Fluvalinate 2·0 EC 0·05	10·8 c	1·5 abcd	8·8 d	3·9 a
Sandoz virus 4041 (Autographa californica) 0·44	11·6 c	3·4 d	7·0 d	3·8 a
Untreated	18·9 d	2·6 cd	15·0 e	4·5 a

[a] Bacillus thuringiensis

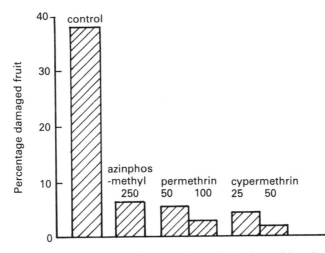

Figure 6.10. Effect of pyrethroid sprays (5 applications, 17-day intervals) against codling moth *Cydia pomonella* in Spain. Ruscoe (1979). Treatment rates are given in parts per million

Table 6.17. Efficacy of various insecticides against second and third larval instars of *Adoxophyes*. Vanwetswinkel & Seutin (1978)

| Products | Rates (in g or ml per 100 l) | | Mortality 8 DAT (%) ($n = 80\text{–}100$) |
	Commercial formulation	a.i.	
Deltamethrin SE 25 g/l	10	0·25	100
	20	0·50	100
Fenvalerate SE 300 g/l	10	3·0	99
	20	6·0	100
Fenvalerate SE 250 g/l	10	4·0	100
	20	8·0	100
Permethrin SE 250 g/l	10	2·5	100
	20	5·0	100
Azinphos-methyl PM 25%	150	37·5	100

these authors was that it would be possible to reduce the number of treatments presently used by introducing pyrethroids into the pest control programmes. This philosophy was contested in the USA in 1978 by Croft & Hoyt, who felt that by using synthetic pyrethroids in fruit pest control programmes especially in apple integrated pest management, natural balances between insect groups would be disturbed. Interesting results were presented in 1981 in Pennsylvania, USA by Hull on two varieties of Golden apple and York Imperial apple. The

Table 6.18. Efficacy of various insecticides against *Operopthtera brumata* L. Vanwetswinkel &
Seutin (1978)

Products	Rates		No. of caterpillars remaining on 200 leaf clusters (15 days after treatment)
	Commercial formulation (ml/100 l)	a.i. (g/100 l)	
Deltamethrin SE 25 g/l	33	0·8	0
Deltamethrin SE 25 g/l	67	1·8	0
Permethrin SE 250 g/l	10	2·5	0
Fenvalerate SE 300 g/l	25	7·5	0
Fenvalerate SE 400 g/l	19	7·5	0
Azinphos-methyl PM 25%	200	50·0	0
Methidathion SE 400 g/l	100	40·0	0
Untreated	—	—	51

insecticides were applied on May 21, June 5 and 19, July 6 and 20, August 4 and 8, and September 2 on a basis of 8 gallons of water per tree (or until run-off). The results shown in Table 6.19 demonstrate the high degree of efficacy of the pyrethroids against *Cydia pomonella*, *Laspeyresia* and *Grapholita molesta* when the level of infestation was fairly high. The gain in marketable fruits was considerable and in addition other minor pest infestations were also controlled.

In the area of viticulture, a certain number of authors have demonstrated an interest in the use of synthetic pyrethroids, but warn of potential problems with mite resurgence. Excellent results have nonetheless been obtained against *Lobesia botrana*, *Clysia ambiguella* and *Sparganothis pilleriana*. Similar results were obtained by Ishaaya et al. in 1983 with other pyrethroids (Table 6.20).

Finally, mention should also be made of the excellent effectiveness of low doses of pyrethroids against cutworms, (see Table 6.21). Harris et al. (1978) working in Ontario Canada, found that by treating the surface of the soil it was possible to eliminate *Euxoa messoria scades* and *Agrotis ipsilon* from tobacco crops. It would also be possible to use pyrethroids against *A. segetum* and *A. ipsilon* on corn crops in the USA with perhaps the possibility of controlling coleopterous pests, the corn root worms *(Diabrotica longicornis*: Northern corn rootworm, *Diabrotica virgifera*: Western corn rootworm, *Diabrotica undecimpunctata*: Southern corn rootworm) at the same time, if the farmers were to apply the treatment 30 or 40 days after sowing.

Coleoptera

Although Lepidoptera represent the largest proportion of the market for

Table 6.19. Efficacy of some insecticides against lepidopterous species on apples. Hull (1981)

Treatment and rate/100 gal (lb a.i.)	Injuries/100 apples					Uninfected (%)
	CM + OFM		RBL	OBL	TABM	
	Stings	Entries				
Flucythrinate 2·5 EC 38 ml (0·025)	0·55	5·41	0·15	0·14	0·79	93·09
Flucythrinate 2·5 EC 76 ml (0·05)	1·05	3·03	0·21	0·15	0·27	95·33
FMC-54617ª 0·8 EC 24 ml (0·005)	1·80	14·10	0·87	0·68	1·01	81·49
FMC-54617ª0·8 EC 48 ml (0·001)	2·13	17·40	0·52	0·48	1·64	79·24
FMC-54800ª 0·8 EC 24 ml (0·005)	2·08	10·59	1·09	0·64	2·78	83·26
FMC-54800ª 0·8 EC 48 ml (0·01)	1·05	5·52	0·52	0·32	1·41	88·77
Cypermethrin 2·5 EC 19 ml (0·0125)	0·23	5·79	0·51	0·17	0·87	92·59
Cypermethrin 2·5 EC 38 ml (0·025)	1·06	5·34	0·20	0·20	0·60	92·33
Thiodicarb 500 g/l 228 ml (0·25)	1·04	4·28	0·49	0·00	0·00	94·18
Thiodicarb 500 g/l 456 ml (0·5)	0·57	1·24	0·09	0·36	0·34	97·38
Methomyl 1.8 L 474 ml (0·225)	1·45	8·65	0·84	0·69	2·35	86·82
Methomyl 1.8 L 474 ml (0·225) + oxamyl 2L 237 ml (0·125)	1·44	5·84	1·13	0·66	1·77	89·65
Tralomethrin 0·3 EC 52 ml (0·00416)	0·54	6·93	0·56	0·31	0·61	91·16
Cyfluthrin 200 g/l 28·4 ml (0·142)	0·08	2·10	0·22	0·37	0·56	96·68
Azinphos-methyl 50 W 114 g (0·125)	1·67	6·89	1·20	0·55	2·15	87·42
Azinphos-methyl 50 W 114 g (0·125) + methyl parathion 2F 237 ml (0·125)	0·93	3·30	0·28	0·30	1·77	93·40
Untreated control	1·97	41·14	2·67	2·92	3·13	45·91

ª Experimental pyrethroids.

CM = *Cydia pompnella* = codling moth; OMF = *Grapholita molesta* = oriental fruit moth; RBL = *Argyrotaenia velutina* = redbanded leafroller; OBL = *Choristoneura rosaceana* = oblique-banded leafroller; TABM = *Platynotaidacusalis* = tufted apple budmoth.

Table 6.20. Effect of synthetic pyrethroids on larvae of the grape-berry moth, *Lobesia botrana*, under laboratory conditions.[a] Ishaaya et al. (1983)

Compound	Larval mortality (%)		
	0·005% a.i.	0·01% a.i.	0·015% a.i.
Cypermethrin	87 ± 7 a	92 ± 3 a	92 ± 4 a
Fenvalerate	56 ± 8 b	61 ± 10 b	88 ± 5 a
Fenpropathrin	53 ± 8 b	45 ± 4 b	95 ± 3 a
Flucythrinate	39 ± 8 b	50 ± 7 b	74 ± 10 a

[a] Six replicates of ten larvae were used for each treatment. Data, after Abbott's correction for natural mortality, are expressed as means with their SE values.
Means followed by different letters within the same column differ significantly at $P = 0·05$ according to Duncan's multiple range test.

Table 6.21. The control of *Agrotis segetum* on lettuce. Piedallu & Roa (1982)

Insecticide	Dose (g a.i./ha)	Mean infestation (%)		
		Trial I 7 DAT	Trial II 10 DAT	Trial III 6 DAT
Deltamethrin	7·5	2·5	0·0	0·0
Endosulfan +	400 +			
dimethoate	600	0·0	0·0	0·0
Control	—	47·5	62·5	52·5

synthetic pyrethroids, it is possible and advisable to use these compounds to control other pest families that are a danger to crops.

As destructive as *Heliothis, Anthonomus grandis* is the most dangerous cotton pest in the US and Central America. This fearful weevil is not well controlled by present-day pyrethroids which must be used at high rates and at frequent intervals. The first group to define the exact conditions of application for pyrethroids in cotton boll weevil control was Burris et al., (1981) Their test findings are summarised in Table 6.22. However, the possibility of using pyrethroids against boll weevils was first suggested in 1975 in Texas by Davis et al., whose findings on permethrin are shown in Table 6.23.

The major factor inhibiting this use of pyrethroids is the economic aspect which is unfavourable when compared with standard parathion treatments. This may become less important as prices for pyrethroids in the USA continue to decline. The increase of pyrethroid treatments against *A. grandis* may not be an entirely favourable event, however, since increased usage against *Heliothis* spp. may accelerate the build-up of resistance. The development of pyrethroid

Table 6.22. The efficacy of various insecticides against cotton bollweevil. Burris et al. (1981)

Treatment[a] and dose (lb a.i./acre)		Percent damaged squares[b] bollweevil	Yield[b] (lb seed cotton/ acre)
Flucythrinate	2·5 EC ·04	16·4 ab	2424 b
Fluvalinate	2 E 0·1	26·0 c	2227 b
Fenvalerate	2·4 EC 0·1	11·7 a	2552 b
Permethrin	3·2 EC 0·1	12·4 a	2410 b
Tralomethrin	3 EC 0·02	13·4 a	3047 ab
Azinphos-methyl	2 F 0·25	14·7 a	2577 b
Azinphos-methyl	2 F 0·125	25·8 c	2352 b
Azinphos-methyl	2 L 0·25	17·5 ab	2565 b
EPN + methyl parathion	0·75 + 0·75	13·5 a	2679 ab
RH 0994	0·4 EC 0·75	22·5 ab	2345 b
Untreated		54·1 d	1078 c

[a] Applications were made on July 1, 5, 10, 15, 21, 29 and Aug. 6, 11, 16, 22.
[b] Means with the same letter are not significantly different at the 0·05 level according to Duncan's multiple range test.

Table 6.23. Effectiveness of insecticides applied aerially for control of the bollweevil and *Heliothis* spp. (predominantly tobacco budworms) on cotton. Davies et al. (1977)

Treatment	Rate (kg a.i./ha)[a]	Percentage damaged squares by	
		Bollweevil	*Heliothis* spp.
Permethrin	0·10	7·1 a[b]	0·8 a
Permethrin	0·11	11·7 a	0·6 a
Permethrin	0·21	8·4 a	0·4 a
Permethrin	0·43	2·3 a	0·2 a
Methyl parathion	1·68	16·7 a	1·1 a
Untreated control		60·7 b	4·8 b

[a] Treated 9 times; applied June 8, 12, 16, 21, 25 and 30, July 5, 14 and 19.
[b] Means in a column followed by the same letter are not significantly different at the 5% level of probability.

treatments for cotton pest control in the USA will be interesting to observe over the next few years, since extension services and industry will be faced with the challenge of providing sufficient guidance to farmers to avoid resistance problems.

The order Coleoptera includes many rape pests which may be adequately controlled through very low doses of pyrethroid treatments. Today there is a growing market for pyrethroids in Europe and Canada, especially since several authors such as Atkins & Kellum (1978), Bocquet et al. (1982), Gerig (1982), Benedek (1982), Pettinga (1982), Wilkinson (1983) and others, have proven that pyrethroids are not toxic to bees and other pollinating insects under the actual conditions of application. With the knowledge that pyrethroids are harmless to bees in practice, these substances have become outstanding weapons for integrated pest management in rape crops.

Today's techniques for protecting rape crops include attracting pests with pheromones and then spraying with a pyrethroid. This method is remarkably efficient at any temperature and requires very small doses of the active ingredient. For example, insects killed with a dose of only 5 g/ha of deltamethrin include *Ceutorrhynchus assimimilis, Meligethes aeneus, Psylliodes chrysocephala* and *Phyllotreta* spp., Piedallu & Roa (1982).

Another beetle that poses serious problems in Europe and North America is *Leptinotarsa decemlineata* (Colorado potato beetle, CPB) attacking potatoes, tomatoes and eggplants. Doses vary from one crop to another, but significant resistance build-up to pyrethroids has been reported for this pest in potato crops on the east coast of the USA (Virginia, New York State, etc.). This case of resistance is actually the only one to have been satisfactorily corrected by adding piperonyl butoxide (PB) to the treatments, which provides a typical example of metabolic resistance due to oxidases. Linduska (1982), working in Maryland, demonstrated that pyrethroids are very effective without the addition of PB for the control of CPB on tomatoes, as indicated in Table 6.24. Treatments were applied on 23 June, 2 and 7 July, and 4 August.

Pyrethroids have been used to good effect against the curculionid, *Bothynoderes punctiventris*. These insects are very active pests in beet plantations. While it is impossible to reach the larvae, which feed by digging tunnels through the root, it is possible to control the adults. Useful results have been obtained in Hungary and Turkey, as indicated by the tests published by the Sugar Institute of Ankara (Table 6.25).

Synthetic pyrethroids are also effective against Chrysomelidae and Coccinellidae, and are especially deadly for two quite dangerous pests, *Epilachna varivestris* and *Epitrix hirtipennis* (tobacco flea beetle) as was proven by Pless & Shamiyed (1981). These authors, working in Tennessee, tested the effectiveness of fluvalinate on tobacco crops infested with *E. hirtipennis* by applying the treatment 4 times beginning on 11 June (Table 6.26).

On the Mexican bean beetle (Table 6.27), McClanahan (1981) showed unusual differences in the effectiveness of synthetic pytheroids, cypermethrin proving to be the most active compound. Good results were also observed in Tunisia in 1978 at the model farm in Fretissa and in Morocco where 7.5 g/ha doses of deltamethrin were used against *Cassida nobilis* on sugar beet.

Table 6.24. Foliar sprays to control the Colorado potato beetle on tomatoes. Linduska (1982)

Treatment and dose (lb a.i./acre)	Colorado potato beetle larvae/10ft row 15 July	% defoliation 28 August	Yield (tons/acre) 9 September
Untreated control	81·50	72·50	8·5
Azinphos-methyl 50 WP 0·5	8·25	55·00	13·9
Fenpropathrin 2·4 EC 0·05	21·50	46·25	9·8
Fenpropathrin 2·4 EC 0·1	15·50	20·00	17·6
Fenpropathrin 2·4 EC 0·2	17·50	23·75	16·7
Flucythrinate 2·5 EC 0·05	16·75	11·25	15·5
Flucythrinate 2·5 EC 0·1	8·50	7·50	21·1
Permethrin 3 EC 0·1	1·25	8·75	19·1
Oxamyl 2L 0·5	12·00	18·75	20·1
Oxamyl 2L 1·0	0·25	15·00	17·7
Oxamyl 2L 0·5 + permethrin 3 EC 0·05	1·00	7·50	19·7
Oxamyl 2L 0·5 + methoxychlor 2L 1·0	0·25	45·00	14·5
Cypermethrin 2·5 EC 0·5	3·00	10·00	21·7
Cypermethrin 2·5 EC 0·075	3·25	8·75	17·0
Cypermethrin 2·5 EC 0·1	0·50	5·00	18·1
Permethrin 2 EC 0·2	31·50	10·00	17·2
Tralomethrin 0·3 EC 0·0156	16·75	13·75	17·9

Table 6.25. Insecticide efficacy against *Bothynoderes punctiventris* on sugar-beet in Ankara. Piedallu & Roa (1982)

Insecticide	(g a.i./ha)	Mortality of adults relative to untreated control (%)
Deltamethrin	12·5	96
Endosulfan	700	87
Chlorpyriphos	960	91

Treatment: 24.5.77.

Table 6.26. Control of *E. hirtipennis* on barley tobacco. Shamiyeh & Pless (1981)

Treatment and dose (lb a.i./acre)	Live flea beetles/10 plants[a]		Flea beetle holes/10 leaves[a]	
	July 2	July 17	July 10	July 28
Fluvalinate 2 E 0·02	33·00 a	4·67 a	316·67 a	42·67 a
Orthene 75 S 0·75	79·33 b	13·67 a	319·33 a	116·00 a
Untreated	206·33 c	79·00 b	544·00 b	388·00 b

[a] Means followed by the same letter are not significantly different ($P = 0.05$) by Duncan's new multiple range test.

Table 6.27. Toxicity of selected insecticides to first instar Mexican bean beetles. McClanahan (1981)

Insecticide	Dosage (mg a.i./litre)				
	LD50	95% confidence limits	LD95	95% confidence limits	Probit line slope
Cypermethrin (PP383)	0·091	0·081–0·100	0·277	0·232–0·358	3·40
Cypermethrin (Ripcord®)	0·129	0·115–0·143	0·450	0·376–0·573	3·04
Fenvalerate	0·210	0·194–0·229	0·816	0·686–1·01	2·80
Deltamethrin	0·379	0·322–0·439	1·34	1·05–1·91	3·00
Fenpropathrin	0·888	0·776–1·01	2·37	1·96–3·18	3·86
Carbofuran	1·18	1·08–1·27	3·24	2·84–3·84	3·74
Permethrin (Pounce®)	1·29	1·12–1·58	5·21	3·30–14·3	2·71
Permethrin (Ambush®)	2·63	2·41–2·92	9·50	7·56–12·9	2·94
Methomyl	3·23	2·73–3·60	5·81	4·98–7·88	6·45
Carbaryl	4·28	3·88–4·68	7·66	6·79–9·13	6·51
Parathion	5·74	5·05–6·53	14·3	11·4–21·1	4·14
Triazophos	8·15	7·37–8·94	13·1	11·5–16·2	8·04
Azinphosmethyl	8·28	7·62–9·01	25·7	22·0–31·4	3·35
Phosalone	9·08	8·08–10·2	21·6	18·1–27·8	4·37
Phosmet	9·89	8·02–11·9	40·0	29·2–67·4	2·71
Methamidophos	13·2	11·8–14·7	36·1	30·5–45·6	3·77
Diazinon	15·1	13·0–17·3	41·1	33·3–56·1	3·78
Endosulfan	27·3	24·6–30·1	51·3	44·0–65·2	6·00
Profenofos	28·7	26·0–31·5	73·7	63·5–90·1	4·02
Dimethoate	37·2	33·5–41·2	73·1	62·6–91·8	5·61
Rotenone	23·3	19·9–26·8	75·7	60·2–106	3·21
Acephate	28·6	25·5–32·0	94·0	76·7–125	3·18
Malathion	36·5	32·6–40·6	113·0	93·1–150	3·34

As was mentioned in the previous section, pyrethroids may also have an important role in the future for the control of the devastating corn rootworms (*Diabrotica* spp.).

Diptera

This section is mainly concerned with the various flies found on fruits, legumes and cereals. A few examples will serve to illustrate the potential of synthetic pyrethroids, but it is important to emphasise two points. First, synthetic pyrethroid treatments must be used on a preventive schedule since, unlike for example dimethoate, they have poor curative ability. Second, synthetic pyrethroids cannot be used on fly species such as *Hylemyia* sp., *Phorbia* sp., and *Psila rosae* which are terricolous in the larval stage. Finally, the section on hygiene, endemic diseases and veterinary uses will present results concerning *Musca, Drosophila,* and *simulids.*

The principle pest control treatments for the three major fly species *Ceratitis capitata, Rhagoletis cerasi,* and *Dacus oleae* are preventive. This is usually carried out by treating every two or three rows of trees with an insecticide mixture that also contains a protein hydrolysate bait (based on fish, corn, etc.). Forsythe (1982) demonstrated the high rate of effectiveness of synthetic pyrethroids against *Rhagoletis* on apples (Table 6.28). These results were confirmed in tests conducted near the Napa Valley in California in 1981 by Hislop et al., who used fenvalerate and permethrin at 1·0 lb ai/100 gal in combination with a protein hydrolysate bait (1 qt/100 gal.) against *Rhagoletis completa* (the corn husk fly). They also showed that results were much poorer than those obtained with conventional products when the bait was not used. The control of *Dacus oleae* has been demonstrated by Pastre (1984) in North Africa, where treatments of deltamethrin were applied to every other row of olive trees giving excellent control (Table 6.29).

Hemiptera

This group contains a certain number of economically important insects such as aphids, leaf hoppers and whitefly which are dangerous vectors of virus diseases and mycoplasmas. These pests are divided into two main groups: (*a*) *Homoptera*, which includes the Aphididae, Coccidae, Cicadellidae, Psyllidae, Aleurodidae and Jassidae; and (*b*) *Heteroptera*, which includes certain particularly damaging genera such as *Aelia, Eurygaster, Nezara, Piezodorus, Euschistus, Lygus Dysdercus, Salbergella, Distantiella* and *Helopeltis.*

The Aphididae are highly susceptible to synthetic pyrethroids apart from a few exceptions such as the woolly aphid (*Eriasoma lanigerum*) and aphids that

Table 6.28. Effectiveness of some chemicals against apple maggot.
Forsythe (1981)

Material and oz formulation/100 gal (superscript shows date of spraying)	% apples injured
Flucythrinate 2·5 EC 1·3[1,2]	1·0
Flucythrinate 2·5 EC 2·6[1,2]	0·2
Flucythrinate 2·5 EC 5·2[1,2]	0·6
Methomyl 1·8 L 11[1,2]	7·2
Cypermethrin 2·5 EC 0·9[3,1,2]	1·4
FMC 54617* 0·8 EC 2·7[3] then FMC 54617 0·8 EC 0·53[1,2]	13·1
FMC 54617 0·8 EC 5·3[3] then FMC 54617 0·8 EC 1·1[1,2]	10·6
FMC 54800* 0·8 EC 2·7[3] then FMC 54800 0·8 EC 0·53[1,2]	16·9
FMC 54800 0·8 EC 5·3[3] then FMC 54800 0·8 EC 1·1[1,2]	10·1
Azinphos methyl 50 W 8[1,2]	4·0
Permethrin 3·2 EC 1·3[1,2]	2·9
Triflumuron 25 W 12[1,2]	
Triflumuron 25 W 6·8 + Azinphos methyl 50 W 4[1,2]	3·9
Phosmet 50 W 24[3] then phosmet 50 W 16[1,2]	1·3
Untreated	16·0
Untreated	17·5

* Experimental pyrethroids.
[1] 28–30 May.
[2] 12 and 20 August.
[3] 9 May.

Table 6.29. The control of *Dacus oleae* in olives in North Africa. Pastre (1984)

Treatment	Rate (g a.i./ha)[a]	Damaged olives (%)	
		TEST I[b]	TEST II[c]
Deltamethrin	6·25	1·5	1·0
Control	—	25·6	24·5

[a] Alternate rows of trees sprayed by air.
[b] Date of treatment: 20.6.81; date of observation 3.8.81.
[c] Date of treatment: 11.7.82–15.9.82–18.10.82–11.11.82; date of observation: 15.12.82.
In Test I, the farmer refused to continue the experiment because of the extended damage caused to the control plots. The second application was therefore an overall treatment and testing was not pursued.

Table 6.30. Proportions of *N. tabacum* seedlings infected with PVY by aphids in flight chamber experiments. Rice et al. (1983)

	Deltamethrin	Control	SED[a]
Treated PVY source plants			
2 h experiment	0·006	0·104	
n = 480	(−3·98)[b]	(−2·46)	0·237
16 h experiment	0·021	0·263	
n = 240	(−2·00)	(−0·55)	0·226
Treated healthy seedlings			
16 h experiment	0·117	0·371	
n = 240	(−1·09)	(−0·40)	0·145

[a] SED, Significant experimental difference.
[b] Bracketed figures logit-transformed.

cannot be reached through spraying, such as those hidden in cabbages. Another interesting facet of synthetic pyrethroids, which is partly due to their secondary effects, is the opportunity they provide for improved control of plant virus diseases transmitted by sucking insects. This activity may be attributed to three causes: first, the knockdown effect of synthetic pyrethroids on established populations; second, their repellent effect; third, the disturbance caused which is often strong enough to prevent the insect from probing the leaf cuticle. Interesting results have been obtained on three important crops, namely, potatoes, barley, and sugarbeet. Gibson et al. (1982) and Rice et al. (1983) demonstrated the effectiveness of deltamethrin against PVY-type non-persistent viruses on tobacco and potato crops. The results obtained in the flight chamber were more convincing than those obtained in the field (Table 6.30). In the latter case, however, only three treatments were applied, on 18 May, 19 June, and 12 August. Potato foliage was not burnt until 12 September. Plant growth had not been taken into account when calculating application dates and therefore there was no insecticide on the newly grown leaves. It should also be noted that leaves on the seed potato plants, planted on 6 April, were burned relatively late, especially for this variety whose foliage is generally destroyed at the end of July. Nonetheless, significant differences showing reduced incidence of disease in the treated plot are shown in Table 6.31. It is also interesting to note that in studies carried out by the same authors, deltamethrin has been shown to be effective against organophosphate and pyrethroid-resistant *Myzus*, although to a lesser degree than on susceptible insects. This study also noted changes in the behaviour of the winged forms of this pest which were perhaps related to a knockdown effect.

In a rather different, but nonetheless important area, promising results have been obtained on sugarbeet (Bocquet et al., 1983). These authors were able to show that when aphids carrying virus yellows disease were introduced to sugarbeet leaves one hour after treatment with deltamethrin, virus transmission

Table 6.31. Plants showing PVY foliar symptoms or infected tubers in field plots sprayed or unsprayed (with deltamethrin). Rice et al. (1983)

	Plots with infector plants		Plots without infector plants		
	Deltamethrin	Control	Deltamethrin	Control	SED[a]
Plants with diseased foliage	69	134	6	20	0·125
	$(-1·17)^b$	$(-0·48)$	$(-1·86)$	$(-1·44)$	
Plants with infected tubers (%)	81	94	59	70	0·246
	$(-0·79)$	$(-1·38)$	$(-0·22)$	$(-0·47)$	

[a] SED, significant experimental difference.
[b] Numbers in brackets have been logit-transformed.

Table 6.32. Insecticide efficacy on sugar beet. Bocquet et al. (1983)

Infestation	Treatment	Dosage (g a.i./ha)	Gross yield (t/ha)	Sugar content	Theoretical sugar yield (t/ha)
Virus	Deltamethrin	25	54·50	17·33	9·50
infested	Deltamethrin +	25 +	54·67	17·47	9·57
aphids	Heptenophos	400			
introduced	Pirimicarb	500	53·13	17·33	9·20
1 h after	Heptenophos	400	51·33	16·87	8·67
treatment	Untreated	—	46·73	16·70	7·68
	Least significant difference P = 0·05	Not significant	0·45	1·19	
	Coefficient of variation	7·69	1·40	7·05	

was suppressed which led to yield improvements (Table 6.32). Today fenvalerate, cypermethrin and deltamethrin are authorised in many European countries not only for protecting crops, but also for improving yield and producing healthy seeds. This final objective will be an important application for synthetic pyrethroids in the future, since by optimising knockdown effect, disturbance, and repellent effects it is possible to prevent transmission of persistent and non-persistent viruses. Today's synthetic pyrethroids are already showing great potential in this area and a number of applications that are of great interest to farmers have already been developed. In similar vein, an interesting comment was made by Ishaaya et al. in 1983 on pyrethroid mode of

Table 6.33. Effectiveness of various insecticides against adult *Bemisia tabaci* in Yemen. Polleh (1982)

Treatment	Rate (g a.i./ha)	Percentage of control (Abbotts Formula)	
		1 DAT	7 DAT
Mevinphos	350[a]	98	45
Deltamethrin	13	72	15
Cypermethrin	52	63	30
Fenvalerate	104	64	31

[a] 700 l of spray mixture per hectare. Average of 6 tests on different crops.

action against the flour beetle *Tribolium castaneum*, "the delay in pupation and emergence at sub-lethal pyrethroid doses is associated with a drastic reduction in larval growth, possibly due to an antifeeding effect". It indeed appears that yet another secondary effect of synthetic pyrethroids must be studied in greater depth in order to make possible even more rational use of these compounds.

One Homoptera genus poorly controlled by synthetic pyrethroids is *Bemisia*. Poor results are not due to a lack of intrinsic activity, but to a problem involving contact between the compound and sessile larvae, the treatment having proven to be very effective on adults and mobile larvae. It is therefore possible to observe variable results on *Bemisia*, depending on which stages are predominant during testing, especially since synthetic pyrethroids are not ovicides; Table 6.33, summarising tests conducted in Yemen by Pollehn (1982), shows the weak persistence of synthetic pyrethroid action against this pest. In this case effects were limited even on adults. Since adults are often hidden under leaves, they are difficult to control, and sessile larvae rapidly become new adults given the very short biological cycle of these pests in hot climates. In another test conducted by Gerhardt (1982) in Arizona, synthetic pyrethroids were shown to lack ovicidal activity and were only relatively effective against adults and nymphs in lettuce crops (Table 6.34). It is obvious that existing pyrethroids are not well suited for the control of whitefly so that, if they contribute to controlling other simultaneous pest infestations, it is preferable to combine them with other insecticides having some vapour activity, such as mevinphos or dimethoate.

Another important group of pests in this family is the scale insects; here again, results are quite variable, depending on the species. Although it is possible to obtain a high rate of effectiveness when scales are mobile, it is difficult to achieve good results with substances that have no vapour action since sessile scales are covered by a waxy protective coating. Nonetheless, good results are regularly achieved on *Quadraspidiotus perniciosus* (the San Jose scale) as demonstrated by Cobb & Poe (1981) in Virginia (Table 6.35).

Table 6.34. Effectiveness of certain synthetic pyrethroids against *Bemisia tabaci* on lettuce in the USA. Gerhardt (1981)

Treatment[a]	Rate (lb a.i./acre)	Number/4 square inches					
		October 7		October 15		November 4	
		Eggs	Nymphs	Eggs	Nymphs	Eggs	Nymphs
Fenpropathrin 2·4 EC	0·05	1004	0	790	100	331	84
Fenpropathrin 2·4 EC	0·1	749	0	700	80	241	34
Fenpropathrin 2·4 EC	0·2	679	0	566	59	213	11
Cypermethrin 3·0 EC	0·06	1292	0	1628	75	586	148
Cypermethrin 3·0 EC	0·12	1216	0	1320	96	643	97
Permethrin 2·0 EC	0·1	882	0	1236	39	276	105
Permethrin 2·0 EC	0·2	1010	0	1688	51	161	66
Cyfluthrinate 2·5 EC	0·025	1750	0	2160	70	839	211
Cyfluthrinate 2·5 EC	0·05	1379	0	1977	63	727	127
Cyfluthrinate 2·5 EC	0·1	1080	0	1705	39	487	112
Methomyl 1·8 L	0·9	2461	0	2658	188	697	257
Fluvalinate 2·0 EC	0·05	—	—	—	—	769	266
Untreated control		1134	0	2338	118	1449	511

[a] Treatments applied October 1, 7, 14, 20, 27; November 3 at 35 gallons/acre.

Table 6.35. Protection of apples from damage by San Jose scale. Cobb & Poe (1980)

Treatment	Formulation		Rate (fl. oz a.i./ 100 gallons)	Damaged fruit (%)
Fenvalerate	2·4	EC	1	4·4
Fenvalerate	2·4	EC	2	3·2
Permethrin	2·0	EC	3·2	4·0
Permethrin	2·0	EC	6·4	0
Cyfluthrin	1·66	EC	1·5	0
Untreated	—		—	16

Table 6.36. Pyrethroids for the control of *Coccus* in olives. Pastre & Hervé (1983)

Treatment	Dose (g a.i./hl)	Larvae, percentage mortality = x			
		Bollulos[a]–II–8.08.83		Bollulos[a]–III–8.08.83	
		x	Arc sin x	x	Arc sin x
Deltamethrin	0·75	98·9	87·079	95·7	80·620
Deltamethrin	1·5	97·2	83·437	90·5	73·927
Fenvalerate	4·5	96·8	79·940	74·0	55·443
Fenvalerate	9	96·6	82·309	91·8	75·342
Cypermethrin	3	96·7	81·454	76·9	57·347
Cypermethrin	6	92·6	74·542	86·8	70·224
Phosmet	50	82·8	66·076	88·7	70·642
Untreated	—	59·4	50·643	37·7	46·448
LSD[b] for P = 0·05			10·453		11·20
Coefficient of Variation			2·49		3·70

[a] Location.
[b] Least significant difference.
Spray: Bollulos–II–7.07.83 at 500 litres/ha; Bollulos–III–8.07.83 at 500 litres/ha.

The best results were obtained with high doses, especially in the case of permethrin. Even more promising results were obtained by Pastre & Hervé (1983) in tests conducted in Spain against *Coccus* spp. on olive trees (Table 6.36).

A plant-sucking pest which is particularly dangerous to pear orchards, especially since it is resistant to several organophosphates, is *Psylla piri*. Synthetic pyrethroids have been known to be very effective on *Psylla*, but tolerance build-up has been reported in several orchards in Europe and in the USA. It would therefore be wise to develop programmes that alternate or combine products with different modes of action before this phenomenon develops into widespread resistance. Nonetheless, in most treatments, synthetic pyrethroids are quite effective, for example against *Psylla piri* and *Psylla mali* in Kentville, Nova Scotia as demonstrated by Sanford (1980) (Table 6.37).

Another Homoptera group, the Cicadellidae, is one of the most common vectors of virus. The ability of synthetic pyrethroids to give control varies from one species to another, and although effective against adult pests, pyrethroids have only a limited effect on nymphs and eggs. Thus immediate results are good but infestations re-appear fairly quickly, as in the case of whitefly, for example control of *Jacobiasca lybica* (Bourdouxhe & Collingwood 1982). Similar results were obtained on the potato leaf hopper (*Empoasca abrupta*) by Vyman et al. (1981) at the University of Wisconsin (USA). Pyrethroid efficacy was strongly dependent on the dose of active ingredient used (Table 3.38).

At present, synthetic pyrethroids as well as several organophosphates are ineffective for the control of rice leaf hoppers, especially *Nilaparvata lugens* (brown plant hopper), which proliferate after the first treatment. This phenomenon was manifest immediately after pyrethroid treatments were applied to rice at the Institute of Rice Research (IRRI) in the Philippines, and has since been confirmed in several other countries. The flare up of rice plant hoppers is not solely due to an imbalance between pests and their predators, but is caused by the complex action of synthetic pyrethroids and organophosphates on plant hoppers and the rice plant, which modifies the life span and the ovogenesis of female brown plant hoppers. Cypermethrin is, however, used successfully in the Philippines to control *Nephotettix cinciceps* (green leaf hopper), which is particularly susceptible to pyrethroids, in areas where it is the predominant pest of rice. In addition, satisfactory results were achieved on another genus in this group, *Empoasca* sp., as demonstrated by Piedallu & Roa (1982) in the Punjab (Figure 6.11). These results were confirmed by Piedallu against *E. flavescens* on vines in France using doses of deltamethrin as low as 1.25 g a.i./hl.

Leaving the Homoptera group, discussion will now focus on the Heteroptera, which includes bugs and mirids. Results against mirids, especially those of the cocoa tree, are not very good, although significant yield increases have been observed on treated plantations (Hervé & Delabarre, 1979). This, however, is probably due to the effectiveness of synthetic pyrethroids against other plant-sucking pests than against genera such as *Salbergella* and *Distantiella* (Coulibaly & Decazy, 1981). Satisfactory results have been achieved in Brazil, the Philippines and Indonesia against *Helopeltis* using 7·5 g of deltamethrin on cocoa trees.

Although the above-mentioned results have shown a poor control of mirids by pyrethroids, surprisingly good activity has been observed against the predatory bug *Daraeocoris fasciolus*. This result, obtained by Sanford (1980), emphasises the important role played by food substrate on the susceptibility of a given species. Even so, the findings of Sanford in Canada (Table 6.39) show that

Table 6.37. Efficacy of various insecticides against *Psylla mali*. Sanford (1980)

| Treatment[a] | Rate/100 l | No. of pests on 20 injured clusters | | | No. of apple plant-sucking adults in visual counts of 3 min per plot | | |
| | | Winter moth June 10 | Pale apple moth June 10 | Leafroller June 14 | Apple sucker[b] | | |
					June 10	June 14	July 14
Permethrin 25 WP	12·5 g	0	0	0	10	0	41
Permethrin 25 WP	25·0 g	2	4	0	14	0	51
Fenvalerate 30 EC	6·6 ml	0	2	2	13	0	11
Fenvalerate 30 EC	13·2 ml	1	4	0	6	14	4
Cypermethrin 12·5 WP	12·8 g	0	3	2	15	0	21
Cypermethrin 12·5 WP	25·6 g	0	1	0	4	2	2
Deltamethrin 2·5 EC	17·6 ml	1	0	0	0	8	4
Deltamethrin 2·5 EC	11·7 ml	0	2	0	2	2	9
Deltamethrin 2·5 EC	5·9 ml	2	1	0	1	1	11
Flucythrinate 30 EC	75 ml	0	3	2	3	12	2
Flucythrinate 30 EC	100 ml	0	1	4	2	24	0
Methomyl 24 EC	155 ml	0	0	2	0	8	69
Methomyl 24 EC	580 ml	0	0	0	0	0	39
Carbaryl 50 WP	100 g	2	4	4	15	14	31
Untreated		7	5	20	116	34	80
Check		8	10	44	67	24	75

[a] Single tree plots treated June 7 (hand gun to run-off).
[b] Counts on 10 cluster samples from north and south sides of tree.

Table 6.38. Potato leafhopper populations on potatoes following applications of synthetic pyrethroids. Vyman et al. (1981)

Treatment		Rate (lb a.i./acre[a])	Mean no. of potato leafhoppers/200 sweeps			
			3- and 2-day counts	5-day counts	8-day counts
Cyfluthrin	1·67 E	0·05	13 a	20 a	33 a[b]
Cyfluthrin	1·67 E	0·025	19 a	21 a	80 a
Flucythrinate	2·5 E	0·08	12 a	21 a	50 a
Flucythrinate	2·5 E	0·04	12 a	23 a	93 ab
Flucythrinate	2·5 E	0·025	19 a	65 ab	146 abcd
Cypermethrin	2·5 E	0·05	20 a	61 ab	145 abcd
Cypermethrin	2·5 E	0·025	20 a	69 ab	227 def
Tralomethrin	0·3 E	0·0156	34 a	68 ab	114 abc
Tralomethrin	0·3 E	0·0134	20 a	68 ab	190 bcde
Tralomethrin	0·3 E	0·0108	25 a	81 abc	187 bcde
Fenvalerate	2·4 E	0·1	37 ab	49 a	111 abc
Permethrin	2·0 E	0·1	57 ab	256 e	420 gh
Permethrin	3·2 E	0·1	24 a	117 abcd	249 def

[a] Applied July 19 and 29.
[b] Means in a column followed by the same letter are not significantly different ($P = 0.05$).

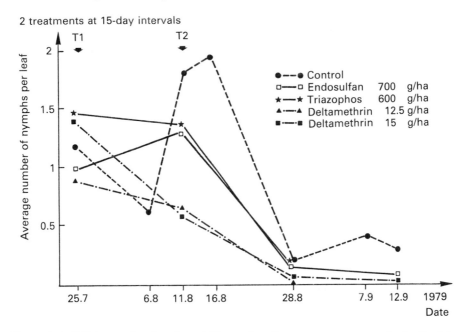

Figure 6.11. Efficiency of various insecticides against *Empoasca* sp. on cotton in the Punjab, 1979

synthetic pyrethroids are also effective against bug-type pests such as *Campylomma verbasci* (Mullein bug) and the apple bug, whereas the surviving predatory insects in the untreated plots did not prevent a build-up of the pest populations.

This leads to the problem of bugs in general. Some species of the Pentatomidae feed on cereals when the plants have reached the milky stage. The bug species which cause the greatest damage in the Mediterranean area are *Aelia* spp., *Eurygaster* spp. and *Nezara* sp.; many others are active elsewhere. All these Heteroptera, both in their larval and imago stages, are easily controlled by spraying synthetic pyrethroids at relatively low rates, such as 6·0 to 10 g a.i./ha of deltamethrin. However, *Euschistus heros*, found in soybean, is not well controlled with synthetic pyrethroids; the required doses are not economical and application requires special techniques such as fogging and ULV.

Results on other species of bug are excellent. For example, Brettel & Chipfuda (1980) in Zimbabwe achieved excellent results against *Dysdercus fasciatus* on cotton crops as shown in Table 6.40. Synthetic pyrethroids have a high intrinsic activity against this group which is illustrated by the LD50 values (g/bug) reported by Leigh & Wynholds (Table 6.41) from the University of California. Adult *Lygus* collected from alfalfa in the Shafter area (California) were used for this study, since alfalfa is the major source of those bugs which invade local cotton crops.

Table 6.39. Effect of certain pyrethroids and other chemicals against predatory mirids, Apple brown bug and Mullein bug on apple trees. Sanford (1980)

Treatment		Rate/100 l	No. of insects collected on beating trays			
			Predator *Deraeocoris fasciolus*	Pest *Campylomma verbasci*	Predator *Anthocoris musculus*	Pest *Atractotomus mali*
Permethrin	25 WP	12·5 g	0	0	0	0
Fenvalerate	30 EC	6·6 ml	0	0	0	0
Cypermethrin	12·5 WP	12·8 g	0	0	0	0
Deltamethrin	2·5 EC	5·9 ml	0	0	0	0
Methomyl	24 EC	155 ml	0	0	0	0
Chlorpyrifos	50 WP	120 g	0	0	0	0
Flucythrinate	30 EC	75 ml	0	0	0	0
Pirimicarb	50 WP	50 g	0	0	4	0
Cyfluthrin	200 EC	7 ml	0	0	0	0
Insect growth regulator	25 WP	80 g	9	5	6	54
Carbaryl	50 WP	100 g	0	2	1	14
Untreated			7	9	16	160

Table 6.40. Effect of insecticides on stainers, applied by hand ULV in Zimbabwe. Brettell & Chipfuda (1980)

	Rate of application (g/100 l)	Percentage of damaged strawberries[a]
Permethrin	6·5	1·4 a[b]
Fenvalerate	6·6	1·6 a
Cypermethrin	4·37	2·1 a
Deltamethrin	0·55	2·6 a
Azinphos-methyl	250	11·9 b
Untreated		15·8 c

[a] Average of three repeated applications on a row 3 m long.
[b] Treatments with letters in common are not significantly different ($P = 0.05$, Duncan test).

Table 6.41. Topical toxicity of insecticides to *Lygus* bugs. Leigh & Wynholds (1980)

Compound	LD50 g/bug at 24 h corrected by Abbott's formula
Acephate	0·134
Cypermethrin	0·00275
Fluvalinate	0·0197
cis-Permethrin	0·0164
trans-Permethrin	0·0181
40/60 Permethrin	0·0175
52/48 Permethrin	0·0151

Paradis & Simarg (1980) in Quebec showed the excellent effectiveness of synthetic pyrethroids against another bug, *Lygus lineolaris*, on strawberry plants.

Orthoptera

Orthoptera exhibit a rather variable susceptibility towards synthetic pyrethroids. The results obtained in Senegal by Ouattara in 1976 on grasshoppers are given as an example in Table 6.42. In 1980 MacCuaig published results on the desert locust (*Schistocerca gregaria*). These data are summarised in Table 6.43. MacCuaig noted that synthetic pyrethroids acted quickly, but that the locusts also recovered from the poisoning symptoms. It was also necessary to spray synthetic pyrethroids at relatively high concentrations, making their use uneconomical. This author observed the same results

Table 6.42. Susceptibility of some grasshoppers to deltamethrin, in Senegal. Ouattara (1976)

Insecticide	Dose (g a.i./ha)	Percentage mortality after about 24 h		
		Hieroglyphus daganensis	*Cataloipus* sp.	*Kraussaria angulifera*
Deltamethrin	5	90–100	30–50	30–50
	10	100	80–100	80–100
	20	100	100	100
Propoxur	100	80–90	60–70	60–70

Table 6.43. Toxicity of varoius insecticides to desert locusts compared with fenitrothion and dieldrin. MacCuaig (1980)

Insecticide	LD50 (µg/g)	LD99 (µg/g)	Time for completion of action (days)
Bioresmethrin, adults	4–10	19–>100	*c.* 4
Cismethrin, adults	1·9–2·8	—	*c.* 4
Cypermethrin			
Fifth instar nymphs	*c.* 5–10	—	—
Adults	*c.* 22	—	1–2
Deltamethrin			
Fifth instar nymphs	1·8	16·2	4
Adults	*c.* 0·6	6·9	4–6
Fenpropathrin			
Fifth instar nymphs	*c.* 10	—	—
Adults	*c.* 1·5	—	2–3
Fenvalerate			
Fifth instar nymphs	*c.* 3	—	2
Adults	*c.* 1·5	—	2–4
Permethrin			
Fifth instar nymphs	12·0	33	2–3
Adults	3·5	7·8	1
Dieldrin			
Fifth instar nymphs	1–3·5	3–9	4–10
Adults	4–7	9–25	7–21
Fenitrothion			
Fifth instar nymphs	6–8	12–30	0·5–1
Adults	4–8	9–35	1–6

on the red locust *Nomadacris septemfasciata*, using permethrin at doses of 100–200 g a.i./ha, confirming that it is difficult to control locusts using synthetic pyrethroids. On other grasshopper species results are quite different, and synthetic pyrethroids may be used at economical doses which completely kill the pests, whereas other locust species can only be knocked down.

Table 6.44. Efficacy of various insecticides used to control *Thrips tabaci* (Thysanoptera) on an onion crop. Bourdouxhe & Collingwood (1982)

Active ingredients	Dosage (g a.i./ha)	Larvae and adults counted on 5 plants per plot	
		Before treatment (20.05.81)[a]	After treatment (11.06.81)[b]
Deltamethrin	15	196·7 a	17·0 d
Bromophos	360	198·0 a	44·0 ab
Cypermethrin	50	231·7 a	47·67 ab
Acephate	500	158·3 a	58·33 abc
Ethyl chlorpyriphos	480	181·0 a	64·67 bcd
Quinalphos	375	220·0 a	84·30 bcd
Pyridaphenthion	800	177·6 a	86·30 cd
Fenvalerate	75	251·6 a	102·00 cdc
Diazinon	600	219·3 a	102·00 cdc
Endosulfan	700	220·0 a	108·30 de
Dimethoate	400	231·0 a	138·67 e
Control	–	198·3 a	469·00 f

[a] Coefficient of variation = 31%.
[b] Means followed by the same letter not significantly different. LSD ($P = 0·05$) = 45·29; Coefficient of variation = 24%.

Thysanoptera

Successful thrips control can be achieved using synthetic pyrethroids. Several tests conducted on many different crops are particularly revealing, for example Table 6.44 (Bourdouxhe & Collingwood, 1981). Thrips, like aphids and *Bemisia*, are virus vectors, hence there is an advantage to the use of synthetic pyrethroids for their control. This was demonstrated by Slawinski (1981) in Poland in tests conducted on tobacco plants (Table 6.45).

Similarly excellent results were also achieved in greenhouse testing conducted in the Netherlands (Hervé, unpublished results) on *Chaetanaphothrips* using a 1·25 g/hl dose of deltamethrin.

Non-agricultural usage

It is certain that until recently the future of pyrethroids was seen to be in agriculture, but it is now equally certain that there will be a great increase in their use for:

- – Veterinary applications
- – The protection of stored foodstuffs
- – The control of great endemics and public health

Table 6.45. Comparison of a synthetic pyrethroid with other compounds for the control of tobacco viruses (V3). Slawinski (1983)

Dosage (a.i.)	No. of thrips/leaf					Virus infected plants (%)					Yield (t/ha)
	25.6[a]	6.7	16.7	25.8	5.8	25.6	6.7	16.7	27.7	5.8	
Deltamethrin 2·5 g/hl	0·1	1·5	1·2	3·5	6·5	0·8	2·3	4·3	14·0	26·3	2·46
Carbofuran 1190 g/ha	0	1·4	1·6	3·1	9·5	1·4	2·3	8·1	16·3	40·1	2·27
Methamidophos 600 g/hl	0	0·8	1·2	3·0	5·8	1·5	2·0	4·5	12·6	36·1	2·26
Untreated plot	0·3	2·1	3·3	7·6	9·2	3·7	6·8	12·7	23·8	46·8	2·02

[a] Date of assessment.

and other fields. Each of these will be examined in a little more detail. It is evident that our knowledge in these fields is less advanced than in agriculture, although considerable information is available, especially on the first four synthetic pyrethroids to be developed, namely, deltamethrin, permethrin, fenvalerate and cypermethrin.

Veterinary applications

Elliot et al. had in 1978 already shown the great potential of pyrethroids for the control of flies, tsetse flies and ticks (Table 6.46). They compared the efficacy of a large number of products with a reference product, bioresmethrin, a non-photostable pyrethroid. It was clear from the comparisons at the time that pyrethroids would become important. Now that nearly all the relevant toxicological and residue studies have been completed the potential of pyrethroids for this market should be realised.

The good activity of pyrethroids against *Boophilus* is particularly surprising, since these compounds have the reputation in the agricultural field of being inactive or having a resurgent effect against phytophagous mites, as we have demonstrated. This is in fact a remarkable instance of selectivity: haematophagous acarids have a much greater susceptibility to pyrethroids than phytophagous acarids. Moreover, this also applies to predatory spiders of the rice leafhopper and other mites. Here again there is an apparent trophic effect on the activity of pyrethroids, which appears to be much more pronounced than with organophosphates and organochlorines. Likewise, activity against Sercoptideae (scab of the genus *Sarcoptes* and *Psoroptes)* has been observed to be excellent.

In the veterinary field favourable effects have also been observed with pyrethroids. For example, the repellency effects of deltamethrin on *Stomoxys calcitrans* have been clearly demonstrated to last for 20 days after dip or shower applications at doses of 25 p.p.m., according to Escuret & Scheid (1982). Similarly in the control of *Melophagus ovinus*, an ectoparasite of sheep, it has been found that unshorn sheep treated in a bath containing a 5 p.p.m. dose of deltamethrin are protected for many (23) weeks, although this dose does not retain its lethal effects over this period. This activity, which is quite remarkable, has been confirmed by Wellcome–Cooper research workers in "pour on" applications, that is, a localised deposition of 40 to 80 ml along the spine of the animal.,

On cattle in Canada, Shemanchuk (1981) has shown that permethrin is highly repellent to blackfly, preventing any take-up of blood for 11 days following treatment at 12 mg a.i./kg in ethanol or at a dose of 6 a.i./kg in the form of an emulsifiable concentrate. Likewise, cypermethrin also used in the same form at a dose of 2 mg/kg, induces a repellency effect lasting for more than 5 days after a generalised treatment to bullocks when applied by a sprayer

Table 6.46. Relative toxicities of some important pyrethroids and other insecticides to five species of insect by topical application.[a] Elliot et al. (1978)

Compound	Musca domestica[b]	Periplaneta americana[b]	Glossina austeni[c]	Boophilus microplus	Stomoxys calcitrans[c]
Bioresmethrin (standard)	100	100	100	100	100
LD50 of standard in ng/insect	5	2500	2·6	0·00014%[d]	2·2
Natural pyrethrins	2	(100)	(20)	170	—
Allethrin	3	—	—	2	—
Bioallethrin	6	150	—	12	—
S-Bioallethrin	10	—	—	—	—
Tetramethrin*	2	—	—	—	(14)
Resmethrin	42	—	79	120	79
Cismethrin	42	500	260	—	180
RU 11679 (experimental pyrethroid)	140	230	81	—	—
Kadethrin	34	80	—	520	—
Phenothrin	30	—	26	—	—
Permethrin	60	290	87	200	65
Cypermethrin	210	—	350	330	55
Deltamethrin	1500	3000	3300	240	300
Fenvalerate	38	200	31	—	54
DDT	12	(15)	3	3	—
Dieldrin	20	(100)	26	21	22
Carbaryl	—	—	<5	0·1	<5
Malathion	1	—	<2	—	—
Dimethoate	45	60	5·5	—	18

[a] Results in this table are intended for general comparison only: figures in parentheses are from very approximate comparative data, not necessarily directly against bioresmethrin. None should be used or quoted without this warning.
[b] From data obtained at Rothamsted by A. W. Farnham (Musca), and P. E. Burt (Periplaneta).
[c] From data supplied by C. J. Lloyd, Pest Infestation Control Laboratory, London Road, Slough, Berks.
[d] Mean lethal concentration by Shaw Immersion Technique. From data supplied by Matthewson, Wellcome Res. Lab.
*Produced also in the "forte" form, containing the IR, cis and IR, trans isomers only.

Table 6.47. Repellency, knockdown and mortality values for four pyrethroids against larvae of *R. evertsi*. Matthewson et al. (1981)

Chemical (10 mg/ml)	Repellency (%) after 10 min	Knockdown (%) after 1 h	Mortality (%) after 24 h
Cypermethrin[a]	14	100	15
Deltamethrin	16	100	45
Fenvalerate	20	81	29
Permethrin[b]	20	99	73

[a] *Cis:trans* isomer ratio, 40:60.
[b] *Cis:trans* isomer ratio, 25:75.

dispensing 2 litres of solution per animal. The animals are then brushed to make the product penetrate their hair. A repellency effect is detected when a low number of satiated blackfly females are found after one hour's contact with the product, in the absence of fly mortality. The observations of Shemanchuk agree with those of Horiba et al. (1977) and Burt et al. (1974) who demonstrated that the *cis* forms are about twice as active as the *trans* forms in terms of repellency against houseflies and cockroaches.

One of the main reasons for the expansion of pyrethroid use in the veterinary field, as was the case for agricultural use, is the optimisation of control through an augmentation of target pest kill with many beneficial secondary effects against other animal parasites. For example, protection against ticks provides additional protection at the same time against various species of flies for several days.

The application against ticks represents one of the most important uses of pyrethroids. It is surprising that such excellent results are obtained, considering the capability pyrethroids have of absorbing onto organic matter, their relative susceptibility to biodegradation and the fact that they retain activity in dips prepared with water very heavily loaded with organic water. Flumethrin, a new synthetic pyrethroid, has excellent activity from 30 mg/l upward against all stages of *Boophilus microplus* on cattle, when applied by spraying or in a dip. This was demonstrated under practical conditions in Australia by Hopkins & Woodley (1982). Against another species, *Rhipicephalus evertsi*, Matthewson et al. (1981) produced the three principal effects of pyrethroids, namely, repellency, knockdown and killing, with a single dose of each, which shows the excellent knockdown activity of the products tested, with the exception of fenvalerate, which is probably active at a higher dose (Table 6.47). In the future it should be possible to exploit the above effects to the full in tick control programmes.

Roberts et al. (1980) have shown that it is possible to control the lone star tick (*Amblyomma americanum* L.) by the direct spraying of fields which harbour them. Table 6.48 shows that relatively high doses, compared with the

J. J. Hervé

Table 6.48. Control of nymphs and adults of the lone star tick with various formulations of acaricides, 1977–78. Roberts et al. (1980)

	Average dose (kg a.i./ha)	Total no. of plots treated	Pretreatment mean (ticks/plot)	Mean percentage control (6-week period)
Permethrin	1·11	7	242	96 ± 3
	0·56	7	265	90 ± 5
	0·29	7	315	71 ± 13
Deltamethrin ULV	0·11	4	272	98 ± 3
	0·056	6	222	90 ± 6
	0·028	4	140	83 ± 5
	0·012	3	225	26 ± 5
Deltamethrin granules	0·11	3	280	91 ± 4
	0·056	4	325	60 ± 10
Deltamethrin WP	0·1	2	210	95 ± 3
	0·053	2	120	94 ± 5
Carbaryl ULV	2·22	3	376	76 ± 9
	1·03	3	221	45 ± 21
	60	3	236	40 ± 13
Chlorpyrifos granules 10·6% slow release	0·56	3	200	12 ± 17
Chlorpyrifos EC (standard)	0·57	10	242	79 ± 5

usual insecticidal doses of pyrethroid, can give good results lasting for 6 weeks, which is superior to the results obtained with standard products, like carbaryl or chlorpyrifos. Considerable improvement was obtained by ULV application which, by increasing the number of toxic microdroplets over the whole of the foliage and preventing run-off, optimises the general activity of a synthetic pyrethroid like deltamethrin against such pests.

Similar results had been obtained by Roberts & Zimmerman (1980) against chigger mites *(Eutrombicula alfreddugesi)* with deltamethrin by residual treatment on grassland at a dose of 0·12 kg a.i./ha, which is uneconomic when compared with the standard product, propoxur, used at a rate of 1·1 kg a.i./ha. In this trial, EC formulations gave better results than ULV applications. Finally, on the subject of miticides, good results have also been obtained with permethrin applied to the vent area as a 0·05% a.i. spray of the diluted EC or WP at 40 ml per bird, a 0·1% a.i. mist of the diluted EC at 20 ml per bird or as a dust at 4·5 g of 0·25 a.i. per bird for the control of the northern fowl mite on domestic fowl (Arthur & Antell, 1982).

Pyrethroids have always given excellent results against detriphagous lice of horn cattle *(Bovicola bovis)*, for example with 5 p.p.m. deltamethrin, as well as against sheep lice *(Damalinia ovis* and *Linognathus ovillus)*, for a period of at least 3 weeks. Pyrethroids are also effective against sheep strike *(Lucilia cuprina* and *Lucilia sericata)*, protecting against any re-establishment of larvae over several weeks; for example, 500g a.i./sheep of deltamethrin provided protection for 19 weeks against *Lucilia sericata* and for 12 weeks against *Lucilia cuprina.*

As in the agricultural field, it is interesting to note that *cis* forms of the photostable synthetic pyrethroids are often more effective than the *trans* forms, for kill but not for knockdown. Schnitzerling et al. (1982) found that in general, *cis* isomers are more effective than *trans* isomers and that cypermethrin isomers penetrated more slowly than permethrin isomers, especially in DDT-resistant larvae of *Boophilus microplus*. In this DDT-resistant strain, metabolic detoxification through esterases was, surprisingly, the most rapid form of detoxification, and *trans* isomers were consistently metabolised faster than *cis* isomers. Stubbs et al. (1982) noted the excellent effects of a new synthetic pyrethroid, cyhalothrin, against ticks. In Australia, treatment applied by dipping with a 0·007% dose provided protection for at least 7 days before infestations re-appeared, while buffalo fly infestations *(Haematobia irritans exigua)* did not re-appear until 28 days later. Bay et al. (1976) reported that permethrin was very effective against tabanids *(Tabanus subsimilis, T. salcifrons, T. proximus)* on horses and calves when applied as an emulsion spray containing 0·05 and 0·1% active ingredient at rates of 1 litre/animal. The results showed that 90% of the flies exposed to treated animals died 9 and 14 days later, respectively. Dust formulations containing 0·025% toxicant were less effective. In Australia, Nolan & Bird (1977) noted an interesting synergistic effect between permethrin, cypermethrin or deltamethrin and certain organophosphates such as ethion, chlorfenvinphos and bromophos ethyl. Organophosphate and synthetic pyrethroid combinations provided a 97·5% control of tick populations as compared with 65–80% for synthetic pyrethroids alone and 23–36% for organophosphate preparations alone – including organophosphate-resistant strains of *Boophilus*. Laboratory tests indicated that this effect was due to potentiation which ocurred when combinations of these two types of compound were used.

In Texas (USA), Schmidt et al. (1976) noted that when 30 to 50 mg of a 2·5% market formulation of deltamethrin was applied as a spot treatment to animals weighing at least 700 kg, a systemic effect occurred against *Stomoxys calcitrans* (stable flies).

At the end of their study on animal dipping against *Boophilus* in Argentina, Bulman et al. (1981) noted a phenomenon which had already been observed in agriculture: the surprising long-range effect of synthetic pyrethroids. It was found that the untreated control cattle, when left in contact with the deltamethrin-treated animals, later showed insignificant levels of infestation. Though this phenomenon requires further investigation, it was postulated that the transfer of the acaricide from the dipped cattle happened not only through contact between the animals, but also through particles suspended in the air.

One of the most surprising outlets for pyrethroids has been their use as an insect repellent in the form of ear tags, consisting of PVC impregnated with the product. Ear tags impregnated with the volatile organophosphate insecticide dichlorvos give good control for 1 month (Harvey & Brethour, 1970). However, ear tags impregnated with fenvalerate, which has insignificant volatility, nevertheless also gives 100% efficacy against cattle flies, and the control lasts for up to 20 weeks. These studies were continued by Williams & Westby (1980) with permethrin and deltamethrin against *Musca autumnalis* (face fly) and *Haematobia irritans* (horn fly). Table 6.49 shows that a 99% horn fly reduction was observed after 13 weeks, a period which is often sufficient to prevent this from being a nuisance to animals during hot weather. On the other hand, deltamethrin, when 1·5% of it was incorporated in ear tags, gave only a 47% face fly reduction over the whole of the 13 weeks covered.

To conclude on the subject of veterinary applications, it is quite reasonable to believe that pyrethroids, because of their acute toxicity, will take over a substantial part of the market, and that their secondary properties, such as repellency, assure them a promising future now that toxicological studies as well as studies on residues in meat and milk have been completed, and should enable new strategies for the control of livestock parasites to be developed.

Protection of stored foodstuffs

Numerous studies are currently in progress on the use of pyrethroids for the

Table 6.49. Face fly and horn fly counts on cattle at the Scholet-Purdue Agricultural Centre, 1979. Williams & Westby (1980)

	Face fly				Horn fly			
Weeks post treatment	1·5% deltamethrin PVC ear tags	10% permethrin PVC ear tags	5% permethrin PVC ear tags	Control	1·5% deltamethrin PVC ear tags	10% permethrin PVC ear tags	5% permethrin PVC ear tags	Control
0	2·8	2·6	2·1	4·2	23·5	15·2	14·1	21·7
1	5·4 (46) b	8·6 (13) a	9·4 (5) a	9·9 a	1·5 (98) c	3·0 (95) b	1·3 (98) c	63·7 a
2	6·0 (29) ab	4·5 (46) b	2·2 (74) c	8·4 a	1·5 (98) c	2·1 (97) c	4·3 (93) b	59·3 a
3	5·3 (58) c	8·9 (29) ab	7·5 (40) bc	12·5 a	2·7 (97) b	4·2 (95) b	5·6 (93) b	84·0 a
4	7·7 (0) a	7·0 (0) a	5·3 (19) a	6·5 a	3·1 (95) c	3·7 (95) bc	8·4 (88) b	68·0 a
5	4·3 (48) c	5·2 (37) bc	9·0 (0) a	8·3 a	2·6 (98) b	4·5 (96) b	4·2 (96) b	106·7 a
6	3·5 (65) b	7·6 (24) a	4·8 (52) b	10·0 a	3·1 (98) c	9·9 (92) b	5·0 (96) c	127·3 a
7	7·9 (48) b	11·8 (23) ab	11·0 (28) b	15·3 a	2·1 (99) b	4·2 (97) b	4·2 (97) b	164·7 a
8	8·4 (51) c	11·9 (31) b	9·8 (43) bc	17·2 a	3·7 (96) b	5·1 (94) b	3·7 (96) b	84·7 a
9	5·0 (72) c	5·1 (71) c	12·1 (32) b	17·7 a	2·7 (97) b	2·4 (98) b	3·0 (97) b	101·3 a
10	6·1 (59) b	6·6 (56) b	13·3 (11) a	14·9 a	2·3 (97) b	4·6 (95) bc	6·7 (92) b	85·3 a
11	4·0 (55) b	4·6 (48) b	12·4 (0) a	8·8 a	1·2 (99) c	2·6 (97) b	3·3 (96) b	92·0 a
12	6·5 (0) ab	7·2 (0) ab	10·7 (0) a	4·4 b	2·1 (97) c	1·4 (98) c	6·6 (91) b	74·0 a
13	2·2 (37) b	4·5 (0) a	—	3·5 ab	0·4 (99) b	1·3 (96) b	—	36·3 a
	5·6 (47) b	7·2 (32) ab	9·0 (15) ab	10·6 a	2·2 (98) c	3·8 (96) b	4·7 (95) b	88·3 a

[a] Numbers in parentheses give percentage reduction from control.
[b] Means within rows for each fly species followed by same letter are not significantly different at 5% level (Student-Neuman-Keuls Test).

protection of stored foodstuffs. However, the Codex Alimentarius and the FAO have so far laid down only a few provisional tolerance and directives, in, line with the progress of the relevant toxicological studies (Table 6.50). In addition it is becoming increasingly evident to food specialists that obtaining good yields to resolve the problems of famine is not sufficient, and foodstuffs must also be protected up to the time when they are used. It has been estimated that losses during storage amount to 30–50% and sometimes even more. This need has not gone totally unheeded: very often legislation requires that the foodstuffs concerned are treated at the processing or shipping stage but, for economic reasons, malathion is very frequently chosen. However, its half-life is relatively short, and all the shorter when grain contains moisture which promotes the hydrolysis of this compound, whereas a synthetic pyrethroid, like bioresmethrin, is not affected by this type of problem and provides complete protection so long as there is no exposure to light, as is the case in silos and ships' holds. It is for this reason that photostable synthetic pyrethroids are in fact opening a new era in the protection of food stocks, including that of produce on farms, like maize in cribs. A number of uses will now be summarised, but not exhaustively because the applications of pyrethroids seem to be endless, ranging from conventional treatment to the protection of fish during drying in Africa, and incorporation in paper or plastics in order to prevent insects from penetrating storage containers. Apart from the protection of foodstuffs, another related use should be mentioned, which is the protection of materials such as felled trees and timber. Finally, to illustrate the size of the problem, the total value of the pesticides market for the protection of foodstuffs – wood, textiles, and so on – has been estimated at 150 million US dollars.

Grain protection

This is the most important sector which includes the control of *Sitophilus granarius* (Granary weevil) for which synergists like piperonyl butoxide (PB) are used quite generally to reduce the rates of pyrethroid required appreciably and thereby make treatment more economic. In 1979 Coulon & Barres published a very detailed study of the relative efficacy of bioresmethrin and deltamethrin with and without PB. They showed that deltamethrin, in contrast to organophosphates and even bioresmethrin, was sufficient to prevent the appearance of a new generation of weevils when used at rates of 0·5 to 1 p.p.m. with PB as the synergist in a ratio of 1:4 or at rates of 0·25 and 0·5 p.p.m. with PB in a ratio of 1:10. This study was all the more interesting as it was carried out against *Sitophilus granarius* not only in its mobile stage but also when concealed in the grain. The same authors continued these studies in 1980 with various pyrethroids obtaining the results given in Figures 6.12 and 6.13. These clearly show the importance of deltamethrin in this application, because

Table 6.50. ADIs (acceptable daily intake) and MRLs (maximum residue level) for insecticides used for the protection of grain. 1982 Joint FAO/WHO meeting on pesticide residues

Insecticide	Evaluated by JMPR (not all evaluations are included for all compounds)	Maximum ADI (mg/kg) bodyweight	Maximum residue levels[a], mg/kg									
			Cereal grain	Milled cereal products	Bran (raw)	Bran (processed)	Wholemeal flour	Flour (white)	Wholemeal bread	White bread	Rice (milled)	Cooked cereal foods
Piperonyl butoxide	1973	0·03	20									0·05[b]
Bioresmethrin	1975–76	NE	5	5								
Bromophos	1972–75–77	0·04	10		20			2	2	0·5		
Carbaryl	1965–73–75–77	0·01	5		20		2	0·2	2			
Chlorpyrifos-methyl	1975–79	0·01	10		5			2	2	0·5		
Deltamethrin	1980–81–82	0·01	2					2	2	0·5		
Dichlorvos	1965–66–67–70	0·004	2	0·5			10	2				
Etrimfos	1980	0·003	10		20		5	3				
Fenitrothion	1969–74–76–77–79	0·005	10		20	2	2	0·2			1	
Fenvalerate	1979–81–82	0·06	5		5		2	2				
Malathion	1963–65–66–67	0·02	8		20		10	2				
Methacrifos	1980	0·0003	10		20		2	0·5				
Permethrin	1979–80–81–82	0·03	2		10		2					
Phenothrin	1978–80	0·2	5		15							
Pirimiphos-methyl	1974–76–77–79	0·01	10		20		5	2			1	
Pyrethrins	1965–66–70–74	0·04	3									

[a] Guideline levels for bioresmethrin and deltamethrin.
[b] At or about the limit of determination.
NE, not yet estimated (bioresmethrin only)

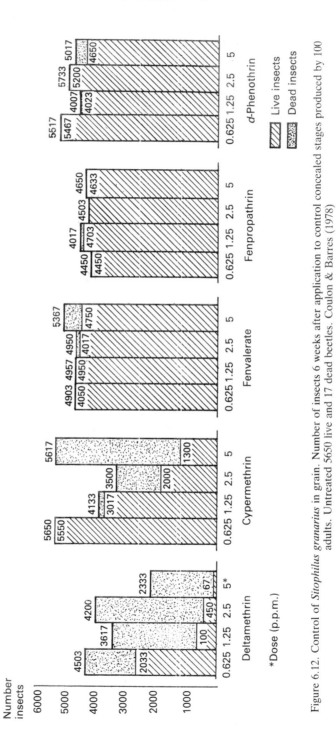

Figure 6.12. Control of *Sitophilus granarius* in grain. Number of insects 6 weeks after application to control concealed stages produced by 100 adults. Untreated 5650 live and 17 dead beetles. Coulon & Barres (1978)

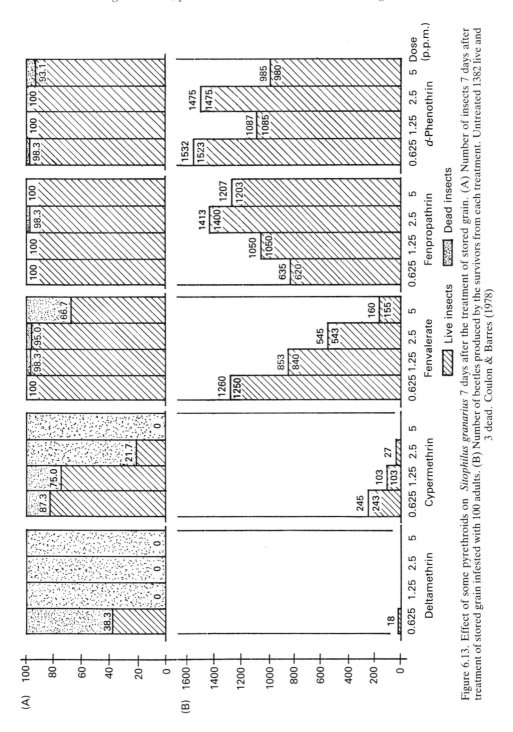

Figure 6.13. Effect of some pyrethroids on *Sitophilus granarius* 7 days after the treatment of stored grain. (A) Number of insects 7 days after treatment of stored grain infested with 100 adults. (B) Number of beetles produced by the survivors from each treatment. Untreated 1382 live and 3 dead. Coulon & Barres (1978)

of the very low doses at which it is used. Cypermethrin proved to be the most active compound after deltamethrin at moderate doses.

In the protection of wheat stocks, particularly in Australia (a great exporter of cereals), one is confronted with the problem of controlling *Sitophilus oryzae* (saw-toothed grain beetle), *Tribolium castaneum* (flour beetle) and *Rhizoperta dominica* (lesser grain borer). Bengston (1978) published a study on the activity of various pyrethroids against the Australian pest complex, the results obtained being shown in Table 6.51. These results show that the *Sitophilus* strain QSO, which is resistant to organophosphates and organochlorines, also resists pyrethroids, including deltamethrin, which is the most active compound against *T. castaneum* and *R. dominica*. On the other hand, in this trial, this compound was poorly effective against *S. oryzae*, and cypermethrin proved to be the most active. As before temperature plays an important role, as shown by Table 6.52, based on results obtained by P. R. Carle et al., (unpublished results).

A feature of the pyrethroids is their very great stability so that they retain their excellent activity throughout the storage period; this has been confirmed by determinations of residues. The quantity of product found after 9 months is the same as that initially deposited, for example Table 6.53, which shows the results of a trial carried out jointly by Wellcome–Cooper and Roussel–Uclaf. It was surprising to observe that the quantity initially deposited in the wheat bins was much lower than the dose of application.

In Brazil, coffee is stored in bags and because of this the treatment consists of spraying the filled bags and thermal-fogging the interior of the coffee bean stores. These two methods have the advantage of leaving practically no residues. However, when coffee beans are treated by dusting or spraying it is found that practically all the residues are destroyed by the roasting process; in the case of treatment with 1 p.p.m. deltamethrin, this amount is reduced to 0·01 p.p.m. It can therefore be considered that the treatment of coffee will not present a problem for pyrethroid use.

An interesting method has been described by Bitran et al. (1980) in Brazil for the protection of maize grain in cobs against *Sitophilus zeamais* and *Sitotroga cerealella* by a combination of fumigation with phosphine and dusting with deltamethrin. This method combines the persistence of a pyrethroid with a fumigant knockdown effect, particularly useful against the concealed instars of the insects such as *S. zeamais*.

Table 6.54 demonstrates that this complementary effect is absolutely necessary for treatment because the pyrethroid, which cannot penetrate bags, prevents further attack, and the fumigant has a curative effect on the grain inside the bag.

The work of Webley & Kilminster (1981) showed the potential for applying pyrethroids to braided polyethylene bags to protect stored grain. With malathion there is no residue left for analysis on the plastic 8 weeks after spraying, whereas with permethrin and deltamethrin, on average, 60 to 70%

Table 6.51. LC99·9 of various insecticides for adults beetles deposited on freshly treated wheat. (Bengston 1978)

Species	Strain	Bioresmethrin	d-Phenothrin	Fenvalerate	Permethrin	Cypermethrin	Deltamethrin
S. oryzae	LS 2	15·8	10·7	9·8	3	1·5	2·9
	QSO 56	10·4	8·3	7·1	2·8	2·3	6·6
	QSO 231	352	52·5	108·1	11·5	19·7	35·7
T. castaneum	QTC 39	7·5	17·8	18·3	17·7	1·6	0·3
	QTC 34	26·1	62·8	43·8	13·0	1·6	0·5
R. dominica	QRD 14	0·6	0·9	0·7	0·7	0·1	0·05
	QRD 2	—	0·8	0·5	0·8	0·4	—
	QRD 63	0·6	2·5	0·7	1·7	0·2	0·05

J. J. Hervé

Table 6.52. Lethal doses and thermodependence of deltamethrin and bioresmethrin on grain beetles. P. R. Carle (unpublished results)

Adult insects	Products		Value in p.p.m. after 7 days contact		
		LC	10 °C	20 °C	30 °C
	Deltamethrin	LC 50	3·9	3·1	1·8
		LC 95	5·6	4·9	4·5
T. confusum	Deltamethrin	LC 50	1·7	1·6	1·0
	+ PB (1/5)	LC 95	2·8	2·8	2·4
	Bioresmethrin	LC 50	4·5	4·5	4·7
	+ PB (1/10)	LC 95	8·2	6·7	8·9
	Deltamethrin	LC 50	0·63	0·34	0·25
		LC 95	1·35	0·55	0·53
T. castaneum	Deltamethrin	LC 50	0·20	0·17	0·11
	+ PB (1/5)	LC 95	0·50	0·40	0·25
	Bioresmethrin	LC 50	1·3	1·4	1·4
	+ PB (1/10)	LC 95	1·9	2·2	1·9
	Deltamethrin	LC 50	0·66	0·45	0·24
		LC 95	1·4	0·86	0·72
S. granarius	Deltamethrin	LC 50	0·25	0·18	0·11
	+ PB (1/5)	LC 95	0·49	0·32	0·22
	Bioresmethrin	LC 50	0·99	0·97	0·46
	+ PB (1/10)	LC 95	1·4	1·4	0·94
	Deltamethrin	LC 50	1·0	0·40	0·36
		LC 95	1·7	0·65	0·73
S. oryzae	Deltamethrin	LC 50	0·54	0·32	0·15
	+ PB (1/5)	LC 95	1·1	0·45	0·50
	Bioresmethrin	LC 50	0·99	0·69	0·58
	+ PB (1/10)	LC 95	2·50	1·20	1·10
	Deltamethrin	LC 50	—	—	0·013
		LC 95	0·054	0·045	0·050
R. dominica	Deltamethrin	LC 50	—	—	0·0073
	+ PB (1/5)	LC 95	0·0156	0·0156	0·032
	Bioresmethrin	LC 50	0.0,3	0·10	0·11
	+ PB (1/10)	LC 95	0·17	0·20	0·24

PB = piperonyl butoxide

of the product remains. These residues were still 100% effective against *S. oryzae* and *T. castaneum*, 12 weeks after treatment following an initial deposit of 40 mg/m^2 of permethrin and 6·5 mg/m^2 of deltamethrin, whereas the OP fenitrothion at 500 mg/m^2 and pirimiphos methyl, also at 500 mg/m^2, had a 100% activity for only 2 weeks and lost all their activity after 4 weeks. On the other hand when insects were placed in contact with the maize grain in the treated bags 100% mortality was obtained with the OPs and 0% with the SPs since the latter did not diffuse through the plastic and reach the grain. An

Table 6.53. Residue levels (ppm) on wheat according to storage time

Intended application dose (ppm)	Start (13.11.79)[a]	3 months (11.02.80)	6 months (12.05.80)	9 months (06.08.80)
1·0	0·44	0·50	0·48	0·41
2·0	0·80	1·46	0·96	1·06

[a] Sampling date.

OP + SP combination would undoubtedly enable the preventive effect to be combined with the curative effect and, in addition, might avoid or delay the onset of resistance problems.

When disinfecting stored foodstuffs, the difficulty of extrapolating from laboratory LC50 values must be borne in mind, because in this field an LC99·9 is an ideal rarely achieved outside the well-defined conditions of the laboratory. This degree of effect can be attained today with some synthetic pyrethroids at doses which, depending on the acceptable daily dose (ADD) laid down by the Codex Alimentarius, should ensure them a large share of this market in the future, despite their slightly higher treatment costs, particularly when persistence is taken into account. It remains to be seen whether or not food stocks will continue to be treated with less persistent products for reasons of immediate (but probably false) economy. It should be possible by combining pyrethroids under certain conditions with fumigant insecticides or organophosphates to prevent 30 to 50% of annual losses. It is important that users are fully conversant with the objectives of the treatment before a choice of products is made: these are to protect grain not for a week, but often for several months. It is within the scope of pyrethroids to be able to achieve this with a minimum toxicological risk both to the user and the consumer.

Many investigations are currently in progress on:

1. The control of *Dermestes maculatus* during the drying of fish; promising results have been obtained with permethrin and deltamethrin, but the studies on residues have not yet been completed.

2. On wool, with the object of obtaining permanent mothproofing.

3. The protection of felled trees and timber. This is already a commercial reality in many countries, particularly with permethrin and deltamethrin. For example, spraying 25 g a.i./100 l results in the deposition of about 7·5 mg a.i./m² of deltamethrin, which protects felled trees in a tropical environment for 4 to 6 weeks. Similarly timber is protected through dipping in a bath containing 0·005% of the same product, as against 0·5 to 1% of lindane.

Finally, very good results have been obtained by Rutherford et al. (1980)

Table 6.54. Results of trials at Sao Paulo on the protection of maize cobs. Bitran et al. (1981)

Insecticide used	Dose (p.p.m.)	Test after 4 months			Test after 8 months		
		Product loss (%)	Infest. *S. zeamais* (%)	Infest. *S. cerealella* (%)	Product loss (%)	Infest *S. zeamais* (%)	Infest. *S. cerealella* (%)
Malathion	8	4·4	16·7	8·5	14·3	46·8	15·2
Deltamethrin	1	3·0	12·6	3·7	5·2	17·2	11·6
Phosphine	1 g/m^3 [a]	1·3	4·5	1·6	12·6	39·2	13·7
Phosphine + Malathion	1 + 8	1·4	3·9	2·1	10·0	34·4	11·1
Phosphine + Deltamethrin	1 + 1	1·0	2·9	1·3	1·7	4·5	3·6
Control		5·7	22·6	11·9	17·6	54·0	21·6

[a] During 72 h of gas contact.

with deltamethrin, permethrin and cypermethrin against the termite *Reticuli-termes santonensis*. Deltamethrin at a dose of 0·01% is active under laboratory conditions for 72 weeks; to obtain similar results 0·1% of cypermethrin and 1% of permethrin are required.

Many applications still remain to be discovered or specified, but it can be justifiably predicted that pyrethroids, because of their outstanding properties, will in the next 5 years come to cause an upheaval comparable to or even greater than that which occurred with foliar insecticides, now that the toxicology and residue evaluations in this field are close to being completed for this group.

Great endemics and public health

Application of plant protection products in the great endemics sector is of fundamental importance for the control of many diseases at an acceptable level. In particular, it is certain that the restrictions on DDT (which are more or less obeyed in the relevant countries) in certain areas stem from resistance to this insecticide, resulting in the resurgence of diseases that were declining. It must nevertheless be pointed out that DDT, a product which is so much decried and considered to be ineffective against flies and mosquitoes, many of which have become resistant according to many authors, is still included in tenders of recommendation of the World Health Organisation (WHO), the tonnages involved being significant (40 to 50% of forecasts up to 1985) and constant for many years. Because of the problems associated with DDT, the advent of pyrethroids has raised new hope in this field, however great care must be taken to use these products rationally so as to delay the onset of resistance which may have been preselected by DDT, for as long as possible.

WHO has just published (1984) the fourth revision of its guide for the use of chemical products against insects important in public health. One of the objectives of this revision was to include the new compounds, among which are the pyrethroids, in the range of products available for controlling the great endemic vectors. This comprised an exhaustive updating of the information on these problems all over the world, and it is for that reason that we strongly recommend consulting this remarkable publication. Nevertheless we shall consider some of these problems and the advantages that pyrethroids can bring, especially in the control of great endemics.

Three examples would appear interesting to discuss:

1. Malaria and mosquitoes.
2. Chagas disease and triatomid bugs.
3. The fight against sleeping sickness in Africa through the control of tsetse flies.

Malaria

A very large number of studies are being carried out at present all over the world under the aegis of WHO with the object of testing pyrethroids and determining whether they can replace DDT and other organochlorine compounds, both for the control of adults inside and outside dwellings, and of larvae.

In the case of control inside dwellings, excellent results have been obtained on a large scale in the treatment of all the huts in a number of villages for two consecutive years, the interior wall surface of the huts being treated every six months. These tests, some of which were followed from a medical viewpoint, led to a reconsideration of the control methodology used. This is because, whether in Guatemala, Nigeria, Upper Volta or the Philippines, a reduction in the number of cases of malaria has regularly been observed without any relation to the percentage mortality of the mosquito population. In fact, the first test in Africa showed that mosquitoes, having fed, would not settle on walls treated with pyrethroids. There was, however, a great drop in the number of mosquitoes entering huts and therefore a drop in the number of bites recorded on humans every night. It is therefore clear once again, that because of secondary effects, like repellency and knockdown, a pyrethroid cannot be assessed in the same way as an organo-phosphorus or organochlorine compound. It is often with high doses that low mortality is observed because the repellency effect is such that mosquitoes will not settle on treated surfaces. In contrast, with lower doses, the legs of the insects come into contact with the toxic film, in which case a knockdown effect can be observed and the insect recovers and leaves the hut. Thus an insect entering a treated hut may become highly disturbed and this may explain the drop in the number of bites and degree of contamination recorded.

It would be interesting to determine whether the *Plasmodium* cycle is modified in insects exposed to a sublethal dose of pyrethroid. Such an effect could also explain the drop in the number of cases of malaria, even though no significant reduction in the number of people infected was observed.

Some pyrethroids currently under evaluation can provide protection for 6 to 12 months:

- deltamethrin at 0·025 and 0·05 g a.i./m^2
- cypermethrin at 0·5 g a.i./m^2
- permethrin at 0·5 g a.i./m^2

These hut treatments can be combined with direct application of insecticides to adults and also larval control measures which doubtless would enable greater efficiency to be achieved, particularly by choosing a different active ingredient in order to prevent the selection of resistant insects. Doses of 2·5 to 10 g a.i./ha for deltamethrin and 5 to 10 g a.i./ha for permethrin are recommended by WHO for larvicidal treatments, whereas in adulticide treatment excellent results are obtained with 0·5 to 1 g a.i./ha with deltamethrin and 5 g a.i./ha with permethrin (or bioresmethrin if a residual effect is not required). There does not seem to be much variation in the doses required for different species of mosquito. Final choice of product and dose should be based on the insect's behaviour and the characteristics of the chemical.

Chagas disease

This serious disease, which occurs especially in Tropical Africa, is due to *Trypanosoma cruzi*. The vector is a group of Reduvidae whose haematophagous members include *Triatoma infestans*, *Panstrongylus megistus* and *Rhodnius prolixus*, particularly abundant in the northern part of South America. The first studies carried out on triatomid bugs, in particular Busvine & Barnes (1947) with natural pyrethrins, had led the authors to observe the erratic movements of bugs in contact with these compounds. These studies were recently taken up again by Pinchin et al. (1980) who investigated the "flushing out" activity of many pyrethroids against *Panstrongylus megistus* (5th instar), a vector of Chagas disease. All these products were studied in 0·5% solutions in kerosene (Table 6.55). Results relate to a given dose and time of assessment; it will now be necessary to try to optimise the flushing-out effect for each compound concerned. This effect partly explains the difficulties encountered

Table 6.55. The flushing-out effects of pyrethroids on *Panstrongylus megistus* (5th instar). Pinchin et al. (1980)

Treatments (0·5% solutions in kerosene)	Percentage flushed-out after 5 min	Time to flush out 50% of insects (min)	Percentage mortality (7 days)
Kadethrin	73	1·0	80
d-Tetramethrin	67	1·2	35
Deltamethrin	57	1·7	100
Permethrin	52	4·7	100
Tetramethrin	52	4·7	5
Bioallethrin	51	4.5	15
Pyrethrum	49	5·8	35
Bioresmethrin	49	4·8	100
Phenothrin	46	7·2	60
Fenvalerate	29	10·0	70
Allethrin	27	—[b]	35
Deet (diethyltoluamide)	0	—[b]	0
Isobutyric acid	0	—[b]	0
Piperonyl butoxide[a]	0	—[b]	0

[a] 4% solution in kerosene.
[b] Flushing-out activity absent, or less than half of the bugs flushed out during the 15 min observation period.

during the first trials with a product like deltamethrin and the problems of monitoring morality, which is difficult to do because of the repellency effect. In fact, when treating huts with deltamethrin bugs could no longer be found, the time required to free a dwelling from triatomids being dependent on the dose:

- 7 days at 75 mg/m^2
- 15 days at 50 mg/m^2
- 30 days at 25 mg/m^2

But, what is quite remarkable is that 180 days after an application at 75 mg/m^2 no insects were observed in the huts, and it is possible that the bugs had returned to feeding on wild or domestic animals in the vicinity. This gives a real hope of eradicating this disease by a mechanism quite different from that involved in the mortality effect. In the meantime, provided an active deposit remains in the huts insects cannot be found. It even seems that a single treatment of a 50-cm high strip is sufficient to make all the bugs flee the huts. We believe there is scope for development in this field, as was the case for indoor mosquito control, and these phenomena should be brought to the attention of public health specialists concerned with the control of great endemics.

Sleeping sickness and tsetse fly control

Tsetse flies are haematophagous flies comprising many species and are vectors of *Trypanosoma gambiense,* the sleeping sickness pathogen of West Africa, but also that of many cattle trypanosomes, and are generally found near water courses and around villages. Another group of tsetse flies, called savannah tsetse flies, encountered in East Africa, are essentially vectors of *Trypanosoma rhodesiense* and some animal trypanosomes. Numerous eradication campaigns have had some success, but centres of infection still remain and may at any moment become active again. It is for that reason that studies have been carried out with pyrethroids. Excellent results have been obtained by aerial ULV treatment at night in Central Africa against adults. However it is when blue screens impregnated with pyrethroids are used that the best results were obtained for the protection of villages and this provided the possibility of recommencing livestock breeding. What seems curious is that again it is the "flushing-out" effect linked to these blue, flexible plastic panels, which eliminates the flies, which seem (just as in the case of bugs) to disappear from the village areas, since traditional trapping does not result in any catches. Pyrethroids can be used in many ways to control tsetse flies but, once again, it is the secondary effects, for example repellency, which distinguish these compounds from former products. This consideration of secondary effects must be extended to the evaluation of the impact of pyrethroids in the environment. This may be particularly relevant to the assessment of the effects on non-target aquatic fauna. For instance the investigation of Mulla et al.

(1982) at Riverside, shows that cypermethrin and fenvalerate are not very toxic to *Berosas metalliceps, Callibaetis* and *Erythemis simplicocollis* under natural conditions on ponds of about 200 m². Despite high intrinsic activity against these organisms it has been shown that recolonisation of the species which are affected takes place very rapidly. All this shows that laboratory studies alone are quite insufficient to investigate fully the multiple effects of pyrethroids and that it is absolutely essential, regardless of the results obtained (good or bad), to carry out trials on a sufficiently large scale under practical conditions, before drawing any final conclusions. Conventional methodologies are often unsuitable and it may be necessary to develop new control programmes which take the multiple effects of pyrethroids into account.

The longest history of pyrethroid use is in the field of public health which involves not only natural pyrethrum, but also non-photostable synthetic pyrethroids such as allethrin and its isomers bioallethrin and *S*-bioallethrin, or resmethrin and its isomer bioresmethrin. The contribution of photostable synthetic pyrethroids in this field is not as great as in the agricultural and veterinary fields for several reasons; for example most photostable synthetic pyrethroids cannot be used in the coil form because their vapour pressure is too low. Nevertheless, they can be quite effective against crawling insects such as cockroaches thanks to their flushing effect and their significant residual power, which comes from their photostability in treatments applied to various surface types. In cockroach control programmes the choice of formulation is very important due to the more or less absorbant surfaces to be treated, such as wood, stone, tile, cement, and so forth.

Another fact often pointed out is that the synergistic effect obtained in combining piperonyl butoxide with photostable synthetic pyrethroids is less than when combined with non-photostable synthetic pyrethroids, except for oxidase-type resistant strains. A synergistic knockdown effect does occur regularly, however, even if a killing effect does not.

Conclusion

This review of the many uses of pyrethroids is much more a statement of the philosophy of using pyrethroids on the basis of the knowledge acquired by the author than an exhaustive review of all the possible uses of every pyrethroid. I believe that, apart from a few exceptions, when systemic products or very high vapour activity are essential, pyrethroids can be universally effective when applied at the correct dose.

In order to decide on the correct dose and utilisation strategy for a pyrethroid, consideration should be given to the following points:

1.   Obviously the dose level and dosing regime must control the pest. However, where possible the dose level and application strategy should be designed to have a minimum impact on non-target organisms.

2.   Care should be taken to avoid dose levels that are liable to differentiate between susceptible and resistant insects. Such avoidance of high selection pressure can minimise the rate of development of resistance populations

3.   Because of their high intrinsic activity, lower rates and fewer applications, relative to previous generations of insecticides, can be used.

As well as economic benefits, this also results in lower residues of these insecticides on the harvested crops.

It is also important to note that because of their favourable insect/mammal toxicity ratio pyrethroids are very safe products for the user and the environment in the broad ecological sense of the word. In fact, within the range of between 0·5 and 150 g/ha, the extreme application doses of the principal pyrethroids commercially available today, the products are non-toxic to the user in the sense of acute toxicity. They are also non-toxic to birds, not transmittable in the food chain, non-soil-contaminating because of their biodegradability and non-hazardous to the water-table, because they are absorbed by organic matter and colloids in general.

Apart from a few exceptions, all the major light-stable pyrethroids on the market today have a similar spectrum of insecticidal activity, if they are used at their correct application dose against a given insect or insect complex. Moreover their level of activity is relatively constant with respect to one another, regardless of the climatic conditions encountered. It is only sometimes that insect pressure levels may cause these ratios to vary around the average values suggested by Wood-Mackenzie. Thus a result obtained using one of the major commercial products should be achievable using the other pyrethroids mentioned at the beginning of this chapter, using a suitably adjusted dose. An important difference, however, arises when good control relies upon the optimisation of secondary effects, which are often greatly enhanced by the stereochemistry of a pyrethroid as shown, for example, by Gibson et al. (1982) for the repellency and knockdown against aphids of the *cis* derivatives as compared with the *trans* derivatives.

Presently only one of the major products mentioned exhibits good miticidal activity and this is fenpropathrin, on condition that the correct dose is used, which is different from the insecticidal dose (e.g. registered for use against *Eudemis and Cochylis* at 70 g/ha, but at 200 g/ha against *Tetranychus* in France). It is encouraging to note that a number of novel pyrethroids have been shown to exhibit miticidal activity at much lower rates, representing a real breakthrough in relation to the spectrum of this group.

So what of the future for synthetic pyrethroids? To quote Wood-Mackenzie (1984b) "Further developments are being undertaken by many companies to overcome some of the deficiencies of the current products.... While the pyrethroids group may never encompass the diversity of structural types seen in the organophosphates, it is now apparent that there is greater scope for chemical, and hence pesticidal diversity in the photostable pyrethroids than was envisaged when NRDC 143 was discovered".

I believe that what really differentiates pyrethroids from other products are the additional possibilities provided by secondary effects, which are seen for a wide range of species. These effects are more often than not a real benefit, particularly when they permit a reduction in the number of applications required. This in turn reduces the risk from residues, already minimal because of the low doses generally used per unit area or volume. All these considerations result in pyrethroids being one of the most effective tools for intelligent pest management because to each dose there is a corresponding specific effect, which enables a logical crop protection strategy to be drawn up. Thus pyrethroids should be used in conjunction with resistant crop varieties, microbiological insecticides, better cultural systems and, perhaps when our understanding of the complex interactions involved has improved, with predators and parasites.

Finally, I would say that pyrethroids have in most situations replaced organochlorine compounds with all the advantages of these products but without any of their disadvantages; this is the greatest compliment one can pay to this family of compounds. It only remains to endeavour to use pyrethroids rationally so as to ensure that they will continue to be used for as long as possible. Thus I conclude with a final wish: may pyrethroids be the first example of concerted action with a view to drawing up a utilisation strategy for a new chemical family so that, at least, the development of resistance to an insecticide group can be successfully held in check.

References

Ahrens, E. H. & Cocke, J., 1979, Season long horn fly control with an insecticide impregnated ear tag. *J. Econ. Entomol.*, **72**, 215.

All, J. N., Ali M., Hornyak E. P. & Weaver, J. B., 1977, Joint action of two pyrethroids with methyl-parathion, methomyl and chlorpyrifos on *Heliothis zea* and *Heliothis virescens* in the laboratory and in cotton on sweetcorn. *J. Econ. Entomol.*, **70**, (6), 813–7.

Angelini A. & Couilloud R., 1976, Premiers résultats obtenus en Côte d'Ivoire avec les pyréthrinoïde dans la lutte contre les ravageurs du cotonnier. *Coton et Fibres Tropicales*, **31**, (3), 323–6.

Angelini, A., Trijau, J. P. & Vaissayre, M., 1982, Activité comparée de trois pyréthrinoïdes de première génération et d'un certain nombre de pyréthrinoïdes nouveaux contre les chenilles de la capsule du cotonnier. *Coton et Fibres Tropicales*, **37**, (4), 359–64.

Arthur, F. M. & Antell, R. C., 1982, Comparisons of permethrin formulations and application methods for northern fowl mite control on caged laying hens. *Poultry Sci.*, **61**, 879–84.

Atkins, E. L. & Kellum, D., 1978, Effects of pesticides on apiculture, Project 1499 Annual Report University of California Riverside California.

Basha A. A., Chelliah, S. & Gopalan, M., 1982, Effect of synthetic pyrethroids in the control of brijal fruit borer (*Leucinodes orbonalis* Guen). *Pesticides*, **16**, (9), 10–11.

Bastos, H. S., Rosa, N. O. V. & Nakano, O., 1979, Economic use of the synthetic pyrethroid decamethrin on cotton to control tobacco budworm, *Heliothis virescens*. *Solo*, **71**, (1), 7–12.

Bay D. E., Ronald N. C. & Harris, R. L., 1976, Evaluation of a synthetic pyrethroid for tabanid control on horses and cattle. *Southwest. Entomol.*, **1**, (4), 198–203.

Benedek, P., 1983, Toxicity of synthetic pyrethroid insecticides to honeybees. 10th International Congress of Plant Protection, November 20–25, *Plant Protection for Human Welfare*, (Croydon, UK: BCPC Publications), Vol. 2, p. 717.

Bengston, M., 1978, Potential of pyrethoids as grain protectants. CSIRO, Division of Entomology, Ed. by D. E. Evans, Australia Contributions to the *Symposium on the Protection of Grain Against Insect Damage During Storage*, Moscow 88–91.

Bitran, E. A., Campos, T. D, Oliveira, D. A. & Araujo J. B. M., 1981, Avaliacao experimental da acao do piretroide decamethrin no tratamento e conservacao de milho nas beneficiado em paiol. *Anais da Sociedade Entomologica do Brasil*, **10**, (1), 105–117.

Bocquet, J. C., L'hôtellier M., Fèvre M. & Baumeister, R., 1982, A five year study on the effect of deltamethrin on bees under natural conditions. *2nd Symposium on the Harmonisation of Methods for Testing the Toxicity of Pesticides to Bees*, International Commission for Bee Botany, Hohenheim, West Germany 21–23 September.

Bocquet, J. C., Roa, L. & Beaumeister, R., 1983, Jaunisse de la betterave, contribution à la mise au point d'une méthode d'études en plein champ de spécialités insecticides. Résultats obtenus avec l'association deltaméthrine + heptenophos, *Med. Fac. Lanbouww. Rijksuniv. Gent*, **48**, (2), 317–330.

Bourdouxhe, L. & Collingwood, E. F., 1982, Efficacité de trois pyréthrinoïdes photostables à l'égard des principaux hôtes d'insectes et des acariens nuisibles aux cultures maraîchères. *Agron. Tropicale*, **37**, (4), 379–88.

Brettel, J. H. & Chipfuda, D. E. M., 1980, Annual Report, Cotton Reseach Institute, Zimbabwe.

Bulhozer, F. & Mabrouk, A., 1982, Problems of controlling *Spodoptera littoralis* on cotton with the synthetic pyrethroids. *Z. Angewandte Entomol.*, **94**, (4), 359–62.

Bulman, G. M., Aguilar, M. & Diaz, C. R., 1981, The special acaricidal action of decamethrin. *Revista Medicina Veterinaria*, **62**, (2), 110–6.

Burris, G., Clower, D. F. & Rogers, R. L., 1981, Results on cotton against boll weevil. *Insecticide and Acaricide Tests*, **6**, 120.

Burt, P. E., Elliott, M., Farnham, A. W., Janes N. F., Needham, P. H. & Pullman, D. A., 1974, Geometrical and optical isomers of 2,2-dimethyl-(2,2-dichloro-vinyl)-cyclopropanecarbolic acid and insecticidal esters — i.e. 5-Benzyl-3-furylmethyl and 3-phenoxybenzyl alcohols. *Pestic. Sci.*, **5**, 791.

Busvine, J. R. & Barnes, S., 1947, Observations on mortality among insects exposed to dry insecticidal films. *Bull. Entomol. Res.*, **38**, 81–90.

Butler, G. D. & Las, A. S., 1983, Predaceous insects: effect of adding permethrin to the sticker used in Gossyplure applications. *J. Econ. Entomol.*, **76**, (6), 1448–51.

Butter, N. S., Singh, J. & Sukhiya, H. S., 1982, Use of synthetic pyrethroids against bollworms of arboreum cotton in the Punjab. *Indian J. Entomol.*, **44**, (2), 113–6.

Caswell, G. H. & Raheja, A. K., 1977, Preliminary trials using synthetic pyrethroids to control pests to cotton and cowpea. *Samaru Agricultural Newsletter*, **19**, (2), 46–9.

Cauquil, J. & Guillaumont, M., 1979, Etude de la fructification du cotonnier sous protection par deux pyréthrinoïdes, IRCT, Congrès sur la lutte contre les insectes en milieu tropical 13–16 Mars, Chambre de Commerce et d'Industrie de Marseille, C.R. des travaux, premiére partie, pp. 127–144.

Cobb, L. J. & Poe, S. L., 1981, Mean percentage of apples damaged by San José Scale. *Insecticide and Acaricide Tests*, **6**, 9.

Coulibaly, L. & Decazy, B., 1981, Action de deux pyréthrinoïdes de synthèse, deltaméthrine et cyperméthrine sur quelques ravageurs des cacaoyers en Côte d'Ivoire. In *Compte-rendu des Séances du Colloque International sur la Protection des Cultures Tropicales Concernant les Cacaoyers et les Caféliers*, Lyon, 8–10 July 1981.

Coulon J. & Barres, P., 1978, Résultats obtenus en laboratoire avec quelques pyréthrinoïdes nouveaux appliqués au charançon du blé. INRA Laboratoire de phytopharmacie, *Bulletin CILDA*, (9), Avril.

Cox, J., 1982, Les pyréthrinoïdes dans la prochaine décade. *Cahiers Roussel-Uclaf*, No. 31, 20–2.

Cranshaw, W. S. & Radcliffe, E. B., 1983, Control of caterpillars in peas, *Insecticide and Acaricide Tests*, **8**, 89.

Croft, B. A. & Hoyt, S. C., 1978, Considerations for the use of pyrethroid insecticides for deciduous fruit pest control in the USA. *Environ. Entomol.*, **7**, (5), 627–30.

Davis, J. W., Harding, J. A. & Wolfenbarger, D. A., 1975, Activity of a synthetic pyrethroid against cotton insects. *J. Econ. Entomol.*, **68**, (3), 373–4.

Davis, J. W., Wolfenbarger, D. A. & Harding, J. A., 1977, Activity of several synthetic pyrethoids against the boll weevil and *Heliothis* spp. Southwest. Entomol., **2**, (4), 164–9.

Delattre, R., 1978, Efficacité des pyréthrinoïdes en culture contonnière. *Phytiatrie Phytopharmacie*, **1**, 53–71.

El-Guindy, M. A., Madi, S. M. & Abdel-Sattah, M. M., 1981, Joint action of two pyrethroids with several insecticides on the Egyptian cotton leafworm (*Spodoptera littoralis*). *Int. Pest Control*, **23**, (4), 99–101.

Elliott, M., Janes, N. F. & Potter, C., 1978, The future of pyrethroids in insect control. *Ann. Rev. Entomol.*, **23**, 443–69.

El-Okda M. M. K., Youssef, K. E. H., Eleva, M. A. S. & El-Assar, M. R., 1979, Toxicity and joint action for four synthetic pyrethroids combined with certain insecticides against the cotton leaf-worm, Alexandra strain. *Proceedings of the Fourth Conf. of Pest Control*, September 30 – October 3.

Escuret, P. & Scheid, J. P., 1982, Control of arthropods in veterinary medicine. *Deltamethrin Monograph*, (Roussel-Uclaf) pp. 275–85.

Ford, M. G., Reay, R. C., Pert, D., Ellis, P. E., 1977, Toxicity of pyrethroids to larvae of the Egyptian cotton leafworm *Spodoptera littoralis*. I. Relative toxicities and knockdown activities of benzyl-3-furylmethyl cyclopropane-carboxylates. *Pestic. Sci.*, **8**, (3), 203–10.

Forsythe, H. Y., 1982, Effectiveness of some chemicals against apple maggot. *Insecticide and Acaricide Tests*, **7**, 12.

Gerhardt, P. D., 1982, Efficacy of some S.P. against *Bemisia tabacci* on lettuce. *Insecticide and Acaricide Tests*, **7**, 93.

Gerig, L., 1982, Appendix 12 "Field trials with Cymbush (cypermethrin) and Cybolt (flucythrinate) in Switzerland during May 1982", *Second Symposium on the Harmonisation of Methods for Testing the Toxicity of Pesticides to Bees*, International Commission for Bee Botany, Hohenheim, West Germany, 21–3 September.

Gibson, R. W., Rice, A. D. & Sawicki, R. M., 1982, Effects of the pyrethroid deltamethrin in the acquisition and inoculation of viruses by *Myzus persicae*. *Ann. Appl. Biol.*, **100**, 49–54.

Harris, C. R., Svec, H. J. & Chapman, R. A., 1978, Potential of pyrethroid insecticides for cutworm control. *J. Econ. Entomol.*, August, 692–6.

Harvey, T. L. & Brethour, J. R., 1970, Horn fly control with dichlorvos-impregnated strips. *J. Econ. Entomol.* **63**, 1688–9.

Hervé, J. J., 1982, Le mode d'action des pyréthrinoïdes et le problème de la résistance à ces composés. *Monographie Deltaméthrine* (Roussel-Uclaf) pp. 67–107.

Hervé, J. J., Lerumeur, C. & Gamal, A. M., 1983, Influence of various agronomic and climatic parameters on toxicity of some synthetic pyrethroids. *Cotton Belt Conference*, Atlanta, Georgia.

Hervé, J. J. & Delabarre, M., 1979, Possibility of a better cocoa crop in a farming area by using deltamethrin—*Congrès sur la lutte contre les insects en milieu tropical*, Chambre de Commerce et d'Industrie de Marseille, 13–16 Mars.

Hislop, R. G., Rieldl, H. & Joos, J. L., 1981, Control of the walnut husk fly with pyrethroids and bait. *California Agriculture*, **35**, (9/10), 23–5.

Ho, S. H., Lee, B. H. & See, D., 1983, Toxicity of deltamethrin and cypermethrin to the larvae of the diamond back moth, *Plutella xylostella* L. *Toxicol. Lett.*, **19**, 127–31.

Hofmaster, R. N. & Francis, J. A., 1979, Foliar sprays to control fall armyworm in sweet corn. *Insecticide and Acaricide Tests*, **4**, 107–8.

Hopkins, T. J. & Woodley, I. R., 1982, The efficacy of flumethrin (Bayticol) against susceptible and organophosphate resistant strains of the cattle tick, *Boophilus microplus* in Australia. VMR, Vet. Med. Rev. 130–9.

Horiba, M., Kobayashi A. & Murano, A., 1977, Gas-liquid chromatographic determination of a new pyrethroid permethrin (S-3151) and its optical isomers. *Agric. Biol. Chem.*, **41**, 581–86.

Hull, L. A., 1982. Apple, concentrate airblast insecticide test. *Insecticide and Acaricide Tests*, **7**, 15.

Ishaaya, I., Ascher, K. R. S. & Casida, J. E., 1983, Pyrethroid synergism by esterase inhibition in *Spodoptera littoralis* (Boisduval) larvae. *Crop protection*, **2**, (3), 335–43.

Ishaaya, I., Elsner, A., Ascher, K. R. S. & Casida J. E., 1983, Synthetic pyrethroids: toxicity and synergism on dietary exposure of *Tribolium castaneum* (Herbst) larvae. *Pestic. Sci.*, **14**, 367–72.

Ishaaya, I., Gurevitz, E. & Ascher, K. R. S., 1983, Synthetic pyrethroids and avermectin for controlling the grapevine pests *Lobesia botrana*, *Cryptoblabes gnidiella*, and *Drosophila melanogaster*. *Phytoparasitica*, **11**, 3–4, 161–8.

Jayaswal, A. P. & Saini, R. H., 1981, Effect of some synthetic pyrethroids on pink bollworm incidence and yield of cotton. *Pesticides*, **15**, (1), 33–5.

Keerthisinghe, C. I., 1982, Synthetic pyrethroids and cotton bollworm control in Sri Lanka. *Trop. Pest Manag.*, **28**, (1), 33–6.

Koziol, I. & Witkowski, J. F., 1982, Synergism studies with binary mixtures of permethrin plus methyl parathion, chlorpyrifos and malathion on European corn borer larvae. *J. Econ. Entomol.*, **75**, (1), 28–30.

Kuppuswany, S. & Balsubramanian, M., 1980, Efficacy of synthetic pyrethroids against brinjal fruit borer *Leucinodes orbonalis* Guen. *South Indian Horticulture*, **28**, (3), 91–3.

Leigh, T. F. & Wynholds, P. F., 1981, Topical toxicity of insecticides to Lygus bugs. *Insecticide and Acaricide Tests*, **6**, 125.

Lerumeur, C., 1973, Quelques lépidoptères déprédateurs du cotonnier du Tadla, au Maroc. Unpublished PhD Thesis, University of Paris–Orsay.

Lhoste, J. & Piedallu, C., 1977, Control of insects in cotton crops in Africa with some pyrethroids. *Pestic. Sci.*, **8**, (3) 254–7.

Linduska, J. J., 1982, Foliar sprays to control the Colorado potato beetle on tomatoes, *Insecticide and Acaricide Tests*, **7**, 120.

Linduska, J. J. & Bagley, P. C., 1981, Control of the cabbage looper and imported cabbage worm on collards. *Insecticide and Acaricide Tests*, **6**, 67.

McClanahan, R. J., 1981, Effectiveness of insecticides against the Mexican bean beetle. *J. Econ. Entomol.*, **74**, 163–4.

McCuaig, R. D., 1980, Synthetic pyrethroid insecticides: some studies with locusts. *Trop. Pest Manag.*, **26**, (4), 349–54.

McDonald, R. S., Surgeoner, G. A., Solomon, K. R. & Harris C. R., 1983, Effect of four spray regimes on the development of permethrin and dichlorvos resistance, in the laboratory by the house fly (Diptera: Muscidae). *J. Econ. Entomol.*, **76**, 417–22.

Masao, H., Kobayashi, A. & Murano A., 1977 *Agric. Biol. Chem.* **41**, 581.

Matthewson, M. D., Hughes, G. Macpherson, I. S. & Bernard, C. P., 1981, Screening techniques for the evaluation of chemicals with activity as tick repellents. *Pestic. Sci.* **12**, 455–62.

Morton, N., 1979, Synthetic pyrethroids on cotton: a spray application strategy. *Outlook Agric.*, **10**, (2), 71–9.

Morton, N., Smith, R. H., Vigil, O. & Van Den Mersch, C., 1981, Evaluation of the "toxic carpet" spray strategy with cypermethrin on cotton. *Proceedings of the British Crop Protection Conference (Pests and Diseases)*.

Mulla, M. S., Darwazeh, H. A. & Ede, L., 1982, Evaluation of new pyrethroids against immature mosquito and their effects on nontarget organisms. *Mosquito News*, **42**, (4), 583–90.

Nolan, J. & Bird, P. E., 1977, Co-toxicity of synthetic pyrethroids and organophosphorus compounds against the cattle tick (*Boophilus microplus*). *J. Austral. Entomol. Soc.*, **16**, (3) 252.

Ouattara, A. 1976. Resultats de quelques essais dans la lutte contre les sauteriaux. *Oclalav Report,* Senegal.

Paradis, R. O. & Simarg, L. G., 1980, Action de quatre insecticides pyréthrinoïdes et du Guthion sur le rendement des fraisiers lorsqu'employés contre l'anthonome et la punaise terne. *Pesticide Research Report*, Canada Commitee on Pesticides Use in Agriculture, Ottawa.

Pastre, P., 1984, Intérêt de la deltaméthrine dans la lutte contre *Dacus oleae*. *Journées Circum méditerranéennes*, Crête.

Pastre, P., & Hervé, J. J., 1983, Essais de lutte contre *Coccus* spp. en culture d'olivier. Rapport Expérimentation, Espagne (Roussel-Uclaf).

Pastre, P., Hervé, J. J., Roa, L. & Penchi, L., 1978, Intérêt de la deltaméthrine dans la lutte contre principaux ravageurs de la vigne. *Phytiatrie Phytopharmacie*, **27**, (1), 39–52.

Pettinga, J. J., 1982, Appendix 10 "Investigations about the size and the duration of toxicity to bees of the pesticide Ambush (Permethrin) Decis (Deltamethrin), Gusathion (Azinphos-methyl) and Pirimor (Pirimicarb). *Second Symposium on the Harmonisation of Methods for Testing the Toxicity of Pesticides to Bees*, International Commission for Bee Botany, Hohenheim, West Germany, 21–3 September.

Pfrimmer, T. R., 1979, Control on cotton with pyrethroids, carbamates, organophosphates and biological insecticides, *J. Econ. Entomol.*, **72**, (4), 593–8.

Piedallu, C. & Roa, L., 1982, Results obtained against *Empoasca sp.* on cotton. *Monographie deltamethrin* (Roussel-Uclaf), p. 177.

Pinchin, R., Oliveiro Filho, A. M. & Pereira, A. C. B., 1980, The flushing-out activity of pyrethrum and synthetic pyrethroids on *Panstrongylus megistus*, a vector of Chagas's disease. *Transactions of the Royal Society of Tropical Medicine and Hygiène*, **74**, pp. 801–3.

Pitts, D. L. & Pieters, E. P., 1981, Efficacy of chemicals for control of cotton bollworm and tobacco budworm. *Insecticide and Acaricide Tests*, **6**, 126.

Plapp, F. W., 1979, Synergism of pyrethroid insecticides by formamidines against *Heliothis* pests on cotton. *J. Econ. Entomol.*, **72**, 667–70.

Pless, C. D. & Shamiyed, N. B., 1981, Control of tobacco flea beetles on tobacco. *Insecticides and Acaricide Tests*, **6**, 152.

Pollehn, E., 1982, *Résultats d'Essai avec Deltamethrine Réalisés au Yemen.* (Yemeni German Plant Protection project, P.O. Box 26, Sana'a, Yemen).

Rice, A. D., Gibson, R. W. & Stribley, M. F., 1983, Effects of deltamethrin on walking flight and potato-virus Y transmission by pyrethroid/resistant *Myzus persicae.* *Ann. Appl. Biol.*, **102**, (2), 229–36.

Roberts, R. H. & Zimmerman, J. H., 1980, Chigger mites: efficacy of control with two pyrethroids. *J. Econ. Entomol.*, **73**, 811–2.

Roberts, R. H., Zimmerman, J. H. & Mount, G. A., 1980. Evaluation of potential acaricides as residues for the area control of the lone star tick. *J. Econ. Entomol.*, **73**, 506–9.

Ruscoe, C. N. E., 1977, The new NRDC pyrethroids as agricultural insecticides. *Pestic. Sci.*, **8**, 236–42.

Ruscoe, C. N. E., 1979, The impact of the photostable pyrethroids as agricultural insecticides. *Proceedings British Crop Protection Conference,* Vol. 3, pp. 803–14.

Rutherford, D., Reay, R. C., Ford, M. G., Furtado, S. E. M. & Jones, E. D. G., 1980, Biology and control of some wood-boring invertebrates. *British Wood Preserving Association,* Cambridge, June 24th-27th.

Saad, A. S. A., Elewa, M. A., Aly, N. M., Auda, M. & El-Sebae, A. H., 1981, Toxicological studies on the Egyptian cotton leafworm *S. littoralis* I. Potentialisation and urea derivative insecticides. *Medelingen van de Faculteit Landbouwwetenschappen,* Gent, **46**, (2), 559–71.

Saad, A. S. A., El-Sebae, A. H. & Sharaf, I. M. F, 1981, AC 222705, a broad spectrum pyrethroid insecticides: performance in Egypt. *Proceedings British Crop Protection Conference.* Vol. 2, pp. 381–8.

Sanford, K. H., 1980, Effect of chemicals on three pest. *Pesticide Research Report, Canada Committee on Pesticide Use in Agriculture,* Ottawa.

Schmidt, C. D., Matter, J. J., Meurer J. H., Reeves, R. E. & Stelley, B. K., 1976, Evaluation of a synthetic pyrethroid for control of stable flies and horn flies on cattle. *J. Econ. Entomol.*, **69**, (4), 484–6.

Schnitzerling, H. J. et al. (1982). Resistance of the buffalo fly, *Haematobia irritans exigua* (De Meijere) to two synthetic pyrethroids and DDT. *J. Austral. Entomol. Soc.*, **21**, (1), 77–80.

Shamiyeh, N. N. & Pless, C. D., 1981, Control of *E. hirtipennis* on burley tobacco. *Insecticide and Acaricide Tests*, **6**, 155.

Shemanchuk, J. A., 1981, Repellent action of permethrin, cypermethrin and resmethrin against black flies *(Simulium* spp.) attacking cattle. *Pestic. Sci.*, **12**, 412–6.

Slawinski, A., 1983, Zabiegi z zakresu o chrony tytoniu. *Wiadomosci Tytoniowe,* August.

Sparks, T. C., Shour, M. H. & Wellemeyer, E. G., 1982, Temperature toxicity relationships of pyrethroids on three lepidopterans. *J. Econ. Entomol.*, **75**, 643–6.

Stubbs, V. K., Wiltshire, C. & Weber, L. G., 1982, Cyhalothrin: a novel acaricidal and insecticidal synthetic pyrethroid for the control of the cattle tick (*Boophilus microplus*) and the buffalo fly (*Hematobia irritans exigua*). *Austral. Vet. J.*, **59**, (5), 152–5.

Sukhoruchenko, G. I., Smirnova, A. A., Kapitan, A. I. & Vikar, E. V., 1981, Pyrethroids in cotton. *Zashchita Rastenii*, **10**, 31–2.

Tan, K. H., 1982, Sub-lethal effects of cypermethrin and permethrin on cabbage white caterpillars *Pieris brassicae* L. *Proc. Int. Conf. Plant. Prot. in Tropics*, pp. 383–390.

Van Steenvyk, R. A. & Hayashi, J. F., 1981, Lepidopterous larval control on celery. *Insecticide and Acaricide Tests*, **7**, 80.

Vanwetswinkel, G. & Seutin, E., 1978, Etude de quelques pyréthrinoïdes photo-stables en arboriculture fruitière. *Phytiatrie Phytopharmacie*, **27**, 15–26.

Vyman, J. A., Longridge, J. L., Chapman, R. K. & Talakoc, L., 1981, *Vegetable Crops Entomology Field Research*. College of Agricultural and Life Sciences Project Report (University of Wisconsin, Madison).

Webley, D. J. & Kilminster, K. M., 1981, The persistence of activity of insecticide sprays deposits on woven polypropylene fabric. *Pestic. Sci.*, **12**, 74–8.

Whitney, W. W. & Wettshein, K., 1979, AC 222705 a new pyrethroid insecticide performance against crop pests. *Proceeding British Crop Protection Conference*, Vol. **2**, pp. 387–94.

WHO, 1984, *Guidelines to the Use of the WHO Recommended Classification of Pesticides by Hazard* (Geneva: World Health Organisation), VBC 84.2.

Wilkinson, W., 1983, Permethrine, diméthoate et pyrimicarbe: effet des traitements insecticides de printemps sur les arthropodes des écosystèmes des céréales, *Journées d'études et d'informations ACTA,* Faune et Flore auxiliaires en Agriculture, 4 et 5 Mai.

Williams, R. E. & Westby, E. J., 1980, Evaluation of pyrethroids impregnated in cattle ear tags for control of face flies and horn flies. *J. Econ. Entomol.*, **73**, (6), 791–2.

Witkowski, J. F., 1979, Field corn, European corn borer control. *Insecticide and Acaricide Tests*, **4**, 135–6.

Wood-Mackenzie, 1983, *Agrochemical Monitor* **27**, 16 March.

Wood-Mackenzie, 1984a, *Agrochemical Monitor* **33**, 21 March.

Wood-Mackenzie, 1984b, *Agrochemicals Products Section*, **39**, February.

Wolfenbarger, D. A. & Harding, J. A., 1982, Effects of pyrethroid insecticides on certain insects associated with cotton. *Southwest. Entomol.*, **7**, (4), 202–11.

Zidan, Z. H., Sobeiha, A. M. K., Mahmoud, F. & Tantawy S., 1981, Response of susceptible and resistant strains of the cotton leafworm larvae, *S. littoralis* to certain synthetic pyrethroids, OP, carbamate insecticides and their mixtures. *Research Bulletin, Faculty of Agriculture*, **1669**, p. 16.

Index

427